MINNESOTA STUDIES IN THE PHILOSOPHY OF SCIENCE

MINNESOTA STUDIES IN THE PHILOSOPHY OF SCIENCE

Editorial Board
John Beatty (Ecology, Evolution, and Behavioral Biology,
   University of Minnesota)
Ronald N. Giere (Philosophy, University of Minnesota)
Geoffrey Hellman (Philosophy, University of Minnesota)
Helen Longino (Women's Studies and Philosophy,
   University of Minnesota)
C. Wade Savage (Philosophy, University of Minnesota)

Also in this series:

Quantum Measurement: Beyond Paradox
Richard A. Healey and Geoffrey Hellman, Editors
Volume XVII

Origins of Logical Empiricism
Ronald N. Giere and Alan W. Richardson, Editors
Volume XVI

Cognitive Models of Science
Ronald N. Giere, Editor
Volume XV

Minnesota Studies in the
PHILOSOPHY OF SCIENCE

C. KENNETH WATERS, GENERAL EDITOR

HERBERT FEIGL, FOUNDING EDITOR

VOLUME XVIII
Logical Empiricism in North America

GARY L. HARDCASTLE
AND
ALAN W. RICHARDSON, EDITORS

University of Minnesota Press
Minneapolis
London

"Hempel and the Vienna Circle," by Michael Friedman, was previously published in *Science, Explanation, and Rationality,* edited by James Fetzer (Oxford: Oxford University Press, 2000). Copyright 2000 by Oxford University Press, Inc. Reprinted by permission of Oxford University Press, Inc.

Copyright 2003 by the Regents of the University of Minnesota
"The Linguistic Doctrine and Conventionality: The Main Argument in 'Carnap and Logical Truth'" copyright 2003 by Richard Creath

All rights reserved. No part of this publication may be reproduced, stored in a retrieval system, or transmitted, in any form or by any means, electronic, mechanical, photocopying, recording, or otherwise, without the prior written permission of the publisher.

Published by the University of Minnesota Press
111 Third Avenue South, Suite 290
Minneapolis, MN 55401-2520
http://www.upress.umn.edu

Library of Congress Cataloging-in-Publication Data

Logical empiricism in North America / Gary L. Hardcastle and Alan W. Richardson, editors.
    p. cm. — (Minnesota studies in the philosophy of science ; v. 18)
    Includes bibliographical references and index.
    ISBN 0-8166-4221-4
    1. Logical positivism.   I. Hardcastle, Gary L.   II. Richardson, Alan W.   III. Series.
    Q175.M64 .L64 vol. 18
    [B824.6]
    501 s—dc21
    [146/.42/097]
                                                  2003010597

Printed in the United States of America on acid-free paper

The University of Minnesota is an
equal-opportunity educator and employer.

12 11 10 09 08 07 06 05 04 03   10 9 8 7 6 5 4 3 2 1

Contents

Introduction: Logical Empiricism in North America   vii
   Alan W. Richardson and Gary L. Hardcastle

1. Logical Empiricism, American Pragmatism, and the Fate of Scientific Philosophy in North America   1
   Alan W. Richardson

2. Two Left Turns Make a Right: On the Curious Political Career of North American Philosophy of Science at Midcentury   25
   Don Howard

3. Hempel and the Vienna Circle   94
   Michael Friedman

4. On Herbert Feigl   115
   Rudolf Haller

5. Edgar Zilsel in America   129
   Diederick Raven

6. Philipp Frank's History of the Vienna Circle: A Programmatic Retrospective   149
   Thomas E. Uebel

7. Debabelizing Science: The Harvard Science of Science Discussion Group, 1940–41   170
   Gary L. Hardcastle

8. Disunity in the *International Encyclopedia of Unified Science*   197
   George Reisch

9. Transfer and Transformation of Logical Empiricism:
   Quantitative and Qualitative Aspects   216
   Friedrich K. Stadler

10. The Linguistic Doctrine and Conventionality:
    The Main Argument in "Carnap and Logical Truth"   234
    Richard Creath

11. Languages and Calculi   257
    Thomas Ricketts

Contributors   281

Index   285

Alan W. Richardson and Gary L. Hardcastle

Introduction
Logical Empiricism in North America

Since the 1980s, the philosophy of science has taken a historical turn. We do not refer to the attention philosophers of science have paid to rich historical accounts of scientific episodes, a turn often taken to have been motivated by Thomas Kuhn's *Structure of Scientific Revolutions* ([1962] 1996) and to have importantly transformed philosophy of science. We refer, rather, to a more recent but equally significant development, in which philosophers of science have begun to recover the problems, solutions, and motivations of earlier projects in the *philosophy of science,* paying attention especially to how the historical figures engaged in these projects understood them.[1] Crucially, this work aims not to *disconnect* such historical projects from contemporary issues in philosophy of science but to *reconnect* contemporary philosophy of science with its history in a new way. Adapting what is perhaps the most famous sentence in the philosophy of science of the second half of the twentieth century, we can assert that the history of the philosophy of science is coming to be viewed as more than a repository for anecdote or chronology, and can, if we allow it, produce a decisive transformation in the *philosophy of science* we now possess.

This volume, *Logical Empiricism in North America (LENA),* is a contribution to this historical turn in philosophy of science. It contains essays that take up, in one way or another, the historical, sociological, and philosophical questions surrounding the particular intellectual movement of logical empiricism, both its emigration from Europe to North America in the 1930s and 1940s and its development in North America through the 1940s and 1950s. Although conceived as a companion to an earlier volume in the series Minnesota Studies in the Philosophy of Science, *Origins of Logical Empiricism* (*OLE,* 1996), *LENA* can be read independently of *OLE*. In this introduction, we explore, explain, and promote the historical turn in philosophy of science that is represented and reflected in this volume. We also relate the volume's contents to various larger issues in history of philosophy of science, issues that promise to transform philosophy of science itself.

The Historical Turn and Reformation

What has motivated this historical turn in the philosophy of science? Why have philosophers of science begun examining the history of philosophy of science in the way they have? Professional stagnation comes to mind as a possible answer. As Don Howard notes in chapter 2, the philosophy of science is not a leading (or even, perhaps, a growing) field in today's academic world. Within the general learned culture, philosophy of science is not even currently the most widely respected arena of reflection on science. Other branches of science studies—sociology of science, social history of science, and cultural studies of science, for example—are more widely read and debated among those interested in the study of science as a human practice. Perhaps, then, it is from the perspective of such doldrums that some philosophers of science are looking outward for new topics, methods, tools, and skills and looking, therefore, to historical figures in philosophy who concerned themselves with science. Considerable attention *is* being paid by philosophers of science, after all, to acknowledged historical figures (such as Hermann von Helmholtz, Charles S. Peirce, René Descartes, Henri Poincaré, Rudolf Carnap, Karl Popper, Immanuel Kant, and Hans Reichenbach), and lesser-known figures (such as Johann Friedrich Fries, Alois Riehl, and Hermann Cohen) are getting a first look. Perhaps this is all an attempt by philosophers of science to reinvigorate their discipline.

Attention to historical figures in philosophy of science is not novel, of course. For example, a typical training regime in the discipline includes exposure to the canonical issues—confirmation, explanation, the nature of theories, the empirical meaning of theoretical claims, the ontological status of theoretical entities, intertheoretical reduction—and this is often combined with some attention to what historically important philosophers said about those topics. Occasionally (perhaps in Peirce or William Whewell) unanticipated resources might be found for thinking through these issues; more often, one is invited to use an important historical figure (such as Carnap or Karl Pearson) as a whipping boy or dialectical opponent. In either case, however, the student in philosophy comes to the historical texts confident of already knowing the canonical philosophical issues surrounding science.

But when the canonical issues seem either misconceived or simply exhausted—when philosophy of science seems intellectually or professionally stagnant—the historical figures are read differently. When philosophers are fundamentally unsure of the philosophical project that ought to be associated with scientific explanation, for example, they are inclined to read, say, Emile Meyerson's *Identity and Reality* ([1908] 1962) not to find out what his account of scientific explanation was but to help with a rich set of concerns concerning the philosophical project of understanding scientific explanation. They will ask: Why did Meyerson have an account of scientific explanation at all? What resources did he employ in giving one? What relation did that account have to Meyerson's other concerns in philosophy of science? What scientific theories did he use as his explanatory exemplars or marshal as resources for his own work? To take another example, one can read the Marburg neo-Kantians not simply to find out what they thought were the foundations of exact science but to find out what they thought was the philosophical import of the task of giving the foundations of exact science. In these cases, the historical figures no longer simply provide views on the canonical topics or texts to think our way through and beyond. They provide philosophical projects to think *with*. The more stagnant the contemporary philosophical situation, the more interest we would expect in the historical figures, since these now appear as exemplars of fresh philosophical projects that we might *in some way* be able to take up and extend.

As a result of their contextualism and historicism, moreover, such historical approaches do not (and indeed, cannot) devolve into crude "back to *X*" movements, for any historical *X*. What such accounts show, indeed, is that we *cannot* go back to Kant, Helmholtz, Carnap, or Popper. Our philosophical and scientific world is not theirs. Nevertheless, the deepest issues in the philosophy of science are sufficiently open that we *can* still learn important lessons from these figures, especially regarding what it is to articulate a new philosophical project concerning science. There is an important difference between going "back to Kant" and going forward by keeping Kant firmly in mind.

No doubt a good deal of the work in this volume looks to historical figures for just these reasons (see, for example, the chapters by Howard, Thomas Uebel, and Alan Richardson). But this is not the only motivation for the historical turn in the philosophy of science. A philosopher confidently ensconced in one or another ongoing living enterprise in the philosophy of science (even one that appears entirely ahistorical) still needs to connect his or her enterprise with philosophical projects of the past, and that requires work in the history of philosophy of science. As Alasdaire MacIntyre (1984) and others (see, for example, Wilson 1992) have argued, philosophy in general is deeply historical, even when it expresses itself in a completely antihistorical fashion; there is simply no way to claim that one's interests are philosophical without finding some tradition of philosophy into which they fit. Thus W. V. O. Quine, although famous for erecting a distinction between philosophers and historians of philosophy, always in his own accounts embeds his philosophical projects in a well-worked-out story of "the empiricist tradition" (see, for example, Quine 1969, 1981, 1995). Even extreme philosophical revolutionaries have to find a way to tie their work to *some* philosophical tradition, on pain of being seen simply as having changed the topic or as having missed the point. A physicist, a counselor, a thief, or a gardener cannot simply *declare* him- or herself a philosopher. It is little wonder that some of the most effective revolutionaries in philosophy attempt less to argue against previous ways of doing philosophy than to "overcome," "deflate," or "turn away from" them. Here, at least, traditions of philosophy are not revealed as simply mistaken so much as interestingly and importantly misconceived and thus useful, at least as signs of roads no longer to be taken.

Whether philosophy of science is currently in crisis or not, then, philosophers of science can find ample justification for the historical turn that has in fact emerged in the philosophy of science. And although the scope of philosophy of science extends far beyond logical empiricism, it is no surprise that logical empiricism has been of particular interest to contemporary philosophers of science: It is, after all, not just a major part of the intellectual puzzle of the twentieth century but, for many philosophers of science, the core of our philosophical heritage. And with two decades of serious work in the history of logical empiricism behind us and with an active and well-established center for this work in the Vienna Circle Institute at the University of Vienna, a number of philosophical, historical, and historiographic issues are emerging. In the following section we will describe three such issues that, in one way or another, run through all the pieces in this volume. But first, we will quickly summarize the volume's eleven chapters.

The volume's first two chapters, Richardson's "Logical Empiricism, American Pragmatism, and the Fate of Scientific Philosophy in North America" and Howard's "Two Left Turns Make a Right: On the Curious Political Career of North American Philosophy of Science at Midcentury," address the history of logical empiricism in general ways and in terms of general themes. Although Richardson and Howard each argue for specific and provocative theses, their chapters also serve to introduce those new to the historical work surrounding logical empiricism to the set of figures, movements, and research problems currently on the table. Richardson, for example, raises the question of logical empiricism's relation to North American pragmatism. Simple characterizations of this question invite overly simple solutions: Logical empiricism *replaced* pragmatism, we might be inclined to say, and it did so because it solved a greater range of philosophical problems, because it was truer, or perhaps just because it was (at the time) more promising. In place of a simple formulation of the question, Richardson argues instead that logical empiricism and pragmatism were of a piece, that piece being scientific philosophy. Notably, in the course of his argument, Richardson brings to the fore Charles Morris, a figure many contemporary philosophers of science may view as only marginal to the logical empiricist program. Such a recovery of figures who are marginal by our present lights is indeed a theme of recent work in the history of philosophy of science, and one much in evidence in *LENA*.

Howard's extensive analysis of the complex philosophical and historical relationship among philosophy of science, politics, and political life introduces readers to a different but equally significant set of issues in the history of logical empiricism. Noting that "there was rather more politics in prewar philosophy of science than our contemporary image of the discipline usually acknowledges," Howard asks how it is that the philosophy of science became politically *disengaged* in the course of its professionalization (a disengagement Howard himself characterizes as "tragic") and how and why the political engagement of our predecessors was obscured in early histories of logical empiricism. Against the background of political histories of both the Vienna Circle and the journal *Philosophy of Science,* Howard identifies the lack of a "successor paradigm" to logical empiricism and, ultimately, the "loss of the sense of a cultural, social, and political mission" that philosophy of science ought to have as the chief causes of the discipline's political disengagement. Reengagement, Howard suggests, might take place via a reconsideration of "the naturalism of Neurath and Dewey."

Richardson's and Howard's respective essays set the stage for the four chapters that follow, each of which focuses on a figure significant to logical empiricism. Philosophers of science trained since the 1970s will readily and rightly associate the name of Carl G. Hempel with the movement, and they are furthermore likely to characterize his intellectual development over several decades as proceeding *from* logical empiricism and *toward* a view sympathetic to Kuhn's, a trajectory that culminated in his emphasis on "provisoes" in science (Hempel 1988). In a demonstration of how work in the history of logical empiricism can lead to revisions in its "received history," Michael Friedman argues in chapter 3, "Hempel and the Vienna Circle," that Hempel's later pragmatic and naturalistic views in fact had their roots in Hempel's earliest thinking, specifically, in his sympathy with Otto Neurath's position in the Vienna Circle's protocol-sentence debate of the 1930s. In a similar vein, Rudolf Haller's "On Herbert Feigl," chapter 4, reminds us that Herbert Feigl, a young member of the Vienna Circle and among the first of its members to emigrate permanently to the United States (Moritz Schlick had visited Stanford in 1929, a year before Feigl moved to Cambridge, Massachusetts), defended philosophical views more often associated with a later time, including a view of theories that emphasizes the extent to which they are "free constructions" and a conviction that probability plays a central role in science.

If the stories of Hempel and Feigl are stories of professional success (to which we could add the stories of Carnap, Reichenbach, and several other émigré logical empiricists), Diederick Raven's contribution to *LENA*, chapter 5, "Edgar Zilsel in America," reminds us that not all such emigrations were successful. After recounting Zilsel's life in Europe and the United States (culminating in his self-inflicted death in 1944), Raven attends to the specific and complicated matter of what went wrong for Zilsel and to the more general matter of what his trajectory can tell us about the philosophical and historical dimensions of logical empiricism in North America. In all, these three chapters remind us that the success of the logical empiricists in North America (as well as significant aspects of their philosophical views) was a contingent matter.

The twin issues of contingency and success recur in Thomas Uebel's chapter 6, "Philipp Frank's History of the Vienna Circle: A Programmatic Perspective." Via a comparison of two instances in which Frank told the story of the Vienna Circle and logical empiricism (first in his 1941 *Between Physics and Philosophy* and eight years later in his *Modern Science and Its Philosophy*), Uebel argues that Frank strove, without success, to carve a place for the social-historical concerns that were championed by Neurath but that were not well represented after Neurath's 1945 death. In the process, we are led not just to Frank's role in logical empiricism in North America but to Neurath's as well.

The history of logical empiricism in North America is a history not just of individuals but of cooperative ventures. *LENA*'s next two chapters take up separate cooperative efforts of some significance to logical empiricism. In chapter 7, "Debabelizing Science: The Harvard Science of Science Discussion Group, 1940–41," Gary Hardcastle recounts the workings of the short-lived Harvard Science of Science Discussion Group (SSDG) and argues that the SSDG reflected a commitment to a particular notion of scientific unity, one best associated with Neurath. Although the group lasted just one academic year, the threads it shared with Frank's later Inter-Science Discussion Group and ultimately with the Institute for the Unity of Science suggest that an important aspect of Neurath's thinking did, in fact, make it to North America. In chapter 8, "Disunity in the *International Encyclopedia of Unified Sciences*," George Reisch gives a detailed history of logical empiricism's most prominent cooperative effort, the *International Encyclopedia of Unified Science (IEUS)*. He documents the somewhat ironic disunity between the *IEUS*'s editors, Carnap, Neurath, and Morris, and employs this tension (among others) to explain why the *IEUS* never realized Neurath's extensive plans for it. In Reisch's hands, this example serves to introduce a dispute between what Reisch calls "large-large" and "small-large" (or as Reisch points out, "Neurathian") explanations in the history of philosophical movements. For Reisch, the story of the *IEUS* is small-large; it is the story of specific people and the decisions they made.

The logical empiricists were, of course, not the only intellectuals forced to flee Europe in the 1930s. Friedrich Stadler's chapter 9, "Transfer and Transformation of Logical Empiricism: Quantitative and Qualitative Aspects," applies the framework of emigration studies to understand logical empiricism. After establishing that the emigration of the Vienna Circle was "manifold and conflicting, involving both success and failure, acculturation and disintegration, diffusion and isolation," Stadler explores a variety of perspectives that might be brought to bear on the emigration and suggests that these perspectives themselves are continuous with philosophy of science.

For many, the most challenging philosophical issue raised by logical empiricism is analyticity. Indeed, if one were asked to locate a point around which logical empiricism has seemed to turn, historically and philosophically, one would be well-advised to select Carnap and Quine's 1950s debate over this very topic. *LENA*'s final two chapters, Richard Creath's "The Linguistic Doctrine and Conventionality: The Main Argument in 'Carnap and Logical Truth'" (chapter 10), and Thomas Ricketts's "Languages and Calculi" (chapter 11), take up the Quine-Carnap debate. Each underscores analyticity's central role in logical empiricism while suggesting that the issue was not, *pace* popular opinion, fruitfully engaged by Quine and Carnap. Creath, for example, explores an early argument of Quine's against the "linguistic doctrine of logical truth," one that has received less attention than the others in Quine's well-known "Two Dogmas of Empiricism" (1951) but that forms, Creath argues, the "basis of much of Quine's subsequent writing." By thinking carefully about Quine's reliance on the claim that logic is "obvious," Creath stakes out two deeply different epistemic perspectives associated with Quine and with Carnap, and he argues that Quine does not argue for his picture so much as presume it. In "Languages and Calculi" Ricketts engages this matter from Carnap's perspective rather than Quine's. After tracing Carnap's account of analyticity as it applies especially to mathematics, Ricketts locates a deep contrast between Quine and Carnap over the relationship of logic to languages, artificial or natural.

LENA thus presents a broad array of work in the history of logical empiricism. Taken altogether, this work raises a number of philosophical and methodological issues, to which we now turn.

Boundary Work, Philosophical Schools, and the Social History of Philosophy

The work represented in *LENA* (indeed, all work in the history of logical empiricism) apparently presupposes that there is something named by the term 'logical empiricism'. This immediately raises questions about the characteristics, family resemblances, boundaries, and so on of this philosophical school. The canonical logical empiricists—the members of the Vienna Circle, Reichenbach, Hempel—did not, however, all agree on very many things, and the search for defining features or family resemblances has generally been fruitless. Even the movement's name is vexed: Reichenbach wielded 'logical empiricism' in the 1930s in opposition to the 'logical positivism' (which was Feigl's term; see chapter 4) of the Vienna Circle, while Schlick came to prefer to call his philosophy 'consistent empiricism', and Neurath occasionally flirted with 'scientific rationalism'. Moreover, the boundaries of the movement were porous and contested even at the time: Persons now not canonically understood to be logical empiricists, such as Morris, took themselves, and were taken by various others, to be inside the movement (for more on Morris, including a few caveats about his place in the movement, see chapter 8). Others, such as C. I. Lewis, officially distanced themselves from logical empiricism but were nevertheless widely regarded as promoting notions that were certainly within the family of views appropriately denominated as logical empiricist (on Lewis, see chapters 1 and 4; and see chapter 7 for yet other figures at the margins of logical empiricism). There are problems, then, when it comes to fixing the subject matter of the history of logical empiricism in North America. Philosophical, sociological, and historical issues are bound together.

By way of increasing the difficulties surrounding these issues, consider the question of the relations of philosophical schools to each other in the 1930s. Investigating the historical relations of logical empiricism and American pragmatism, for example, requires some sense of who the pragmatists in the 1930s were and what pragmatism in the 1930s was. But, as Richardson documents in chapter 1, these are hard questions. Morris had the appropriate pedigree and claimed to be a pragmatist; Lewis, also, had both pragmatist pedigree and self-identification. Ernest Nagel is *now* understood as a sort of pragmatist, but in the 1930s his views were more closely related to the philosophical position of Morris R. Cohen—a sort of naturalistic realism that found fault with John Dewey's pragmatist positions (see Cohen and Nagel 1934). If we *assume* that pragmatism and positivism were opposed, we can write the history of American philosophy in the 1930s and 1940s as pragmatism's betrayal by Morris, Nagel, and other young American philosophers (see Giere 1996). Richardson here rejects that assumption, though, and attempts to write the history as it was understood by both the logical empiricists and the pragmatists while it was happening. He finds a higher level, the level of "scientific philosophy," in which the coming together of the projects makes sense. Howard, in chapter 2, reminds us of some of the costs to both projects resulting from their combination and calls our attention to some remaining political tensions between the projects. Reisch, in chapter 8, meanwhile reminds us of some of Morris's second thoughts about logical empiricism.

These issues are going to sort themselves out only as philosophers become more adept at social and institutional history. There may not be—in fact, we conjecture that there *is* not—any deep philosophical or conceptual coherence to movements as broad as American pragmatism or logical empiricism. By way of illustration, consider the name of the philosophical program the readers of (and contributors to) this volume are most likely to describe themselves as engaged in; that is, consider 'analytic philosophy'. The term is philosophically opaque. 'Analytic philosophy' denotes a social structure, a group held together not by any substantial philosophical commitments—there is no one metaphysical, epistemological, ethical, or even metaphilosophical project in analytic philosophy—but by an amorphous group of issues, texts, and canonical historical interpretations of great figures in the history of philosophy and, similarly, by a rather murky list of issues, texts, and figures it *excludes*.[2] Most of this volume's readers would find it philosophically comforting if we could locate a deep coherence to logical empiricism, pragmatism, or (especially) analytic philosophy, but it simply does not seem to be in the cards. As a way of going forward, philosophically, though, we can recognize that this desire for coherence is itself a dimension of analytic philosophy, and we can seek to understand where it came from and why it is ours.

In this regard, the best current tool for understanding 'analytic philosophy' must surely be sociology of knowledge, especially the notion of boundary work (Gieryn 1999). From this perspective, 'analytic philosophy' as a term is used principally for boundary work, and it acquires its meaning in that use. Consider analytic philosophers of mind. They use the term '*analytic* philosophy of mind' in order to distance themselves from "Continental" concerns with mind or with subjectivity, and in so doing they claim for themselves virtues such as intellectual rigor, attention to logical argument, connection to the current sciences of mind, and hardheaded empiricism in order to promote their projects at the expense of phenomenology, psychoanalysis, and other expressions of Continental tender-mindedness. They will, however, use 'analytic *philosophy* of mind' in a different way, namely, to set themselves off from *scientists* of mind, and here they will make much of their place in the tradition of philosophical issues regarding mind (the mind-body problem, the subjective nature of conscious states), their connection to historical figures in philosophy, and, finally, their own distinctive interest in and interpretation of the empirical results of the neurosciences. Arguments aimed at policing these different boundaries cannot be easily combined into one global account of the project of "analytic philosophy of mind," for reasons Thomas Gieryn (1999) notes and that any rhetorician would point out: The arguments on behalf of "analytic *philosophy* of mind" move the project in the direction of philosophy generally and thus are not easily employed against "Continental *philosophy* of mind." Conversely, arguments for "*analytic* philosophy of mind" move the project in the direction of scientific accounts of mind and cannot be easily employed against neuroscience. In general, one's reasons for doing X rather than Y are typically different from one's reason for doing X rather than Z. This fact becomes problematic only in contexts where there is a drive to give a complete list of "the reasons for doing X" and present those reasons as the defining features of X.[3]

To return to logical empiricism and American pragmatism, we offer the following suggestion: We are not going to understand logical empiricism, pragmatism, or their relation until we ask questions characteristic of social history. We need to know why terms like 'pragmatism' or 'logical empiricism' arose in American philosophy, what those terms were introduced to do, and how they came to be banners under which various philosophers gathered. We need to know what sorts of techniques, knowledge, and skills these terms were taken as characteristic of, how they were taught or transferred, how the various contrasts and commonalities changed over time, why they sometimes go out of philosophical discourse, and, finally, why they reemerge in new contexts. We have to seek answers to these questions by looking in detail into the places, times, and contexts in which these things happened, and happened as actions of human beings. The story of analytic philosophy in North America is the story of the appearance, disappearance, or reappearance of crucial philosophical terms, among them 'pragmatism', 'naturalism', 'metaphysics', and 'the a priori'.[4] Some of this work is represented in this volume, but there is much more to do. In general, Lorraine Daston's (1994) and Arnold Davidson's (2001) "historical epistemology" could find no more fertile field than that offered by recent history of analytic philosophy and philosophy of science.[5]

Philosophical Schools and Their Margins

This volume takes a broader view of logical empiricism than is usual. It contains discussions of, for example, Morris, Frank, Zilsel, and Harvard's SSDG. It also treats of canonical figures such as Feigl, Hempel, and Carnap. Behind this is an interesting tension. The volume seeks to open up new understandings of logical empiricism by attending to the work of logical empiricists who are not now considered to be central figures while at the same time accounting for the historical fact that in logical empiricism some figures and not others came to be canonical. Why, for example, did the more technical work of Carnap, Reichenbach, and Hempel come to capture the attention of the philosophical community and come to be taken as the core of logical empiricist philosophy of science while, for example, Frank's more historical and cultural understanding of philosophy of science came to be more marginalized within it?[6] This volume contains some partial—and partially competing—answers to this question in particular. For Howard, the construction of the logical empiricist canon depends on professionalization and specialization. Reisch suggests, on the other hand, that Frank simply did not have the local resources to become Neurath's successor at the helm of the Unity of Science movement. Uebel argues that there was a rather important subgroup within the left wing of the Vienna Circle, one that was rooted in the "first Vienna Circle" of the pre–World War I era (of which Frank was a member), and that Frank's efforts to restore this group's project in America in the 1950s failed.

The availability of several answers to the question of Frank's apparent marginalization in the United States in the 1950s suggests the not very interesting fact that the scholarly community does not yet know why it happened. There are, however, more interesting issues. Howard, Uebel, and Reisch all seek to broaden our understanding of the range of positions taken within logical empiricism, but they all suggest that that wider range of positions cannot be combined into a coherent program the way the logical empiricists had hoped. Yes, these accounts claim, Neurath and Frank were logical empiricists, and so were Carnap and Reichenbach, but the moral to be drawn is that *the movement was deeply divided and not philosophically coherent from the start*. In retrospect, philosophy of science proceeded under the influence of the Carnap-Reichenbach wing of logical empiricism. Howard, Uebel, and Reisch speculate about a short-lived philosophy of science that attended (or attends) to the Frank-Neurath wing, and they wonder why its life was not longer.

There are other ways to tell the story. One could argue, as Peter Galison (1998) has, that Frank was caught up after World War II in a new notion of scientific unity and was in fact promoting, under the rubric "unity of science," a *different* project from any the logical empiricists had promoted before the war. Or one might attempt to argue that Neurath and Frank, to the end of their lives, took their main philosophical mission to be the promotion of the "scientific world conception," whereas within the American context of the 1940s and 1950s, Carnap, Reichenbach, Hempel, and Feigl all came to do work inside that conception, no longer expending much effort arguing for it. On such an account, Frank finds a place within a certain division of labor in logical empiricism, a division that explains why he did not do much of the detailed work in philosophy of science that the students of Carnap, Reichenbach, and Hempel would find important or useful. Frank's philosophical task, on this account, concerned *arguing for* something that the mainstream of philosophers of science came simply to take for granted.

The point is not simply to proliferate narratives but to point to a crucial aspect of historical work in philosophy of science. Which narrative, if any, comes to be accepted regarding the marginalization of Frank and Neurath in philosophy of science will inform future understanding of who Frank and Neurath were as philosophers. On some narratives, they provide the road not taken (at least not taken yet) in philosophy of science. On other narratives, they are an older generation whose philosophical work paved the way for Reichenbach and Carnap and who were naturally superseded by the more mature and technical philosophy of science they made possible. On yet other narratives, Frank and Neurath were forward-looking and creative thinkers who constantly remade the Unity of Science movement. The narratives constructed will set the conditions under which Frank's and Neurath's work will be approached and used in the coming years. We write them back into the history of philosophy of science by writing stories of how they were written out of it. And depending on how we do that, they look like very different philosophers.

Attending to such figures as Frank, Morris, and Zilsel raises further questions surrounding the facts our historical accounts are expected to explain. Is Morris a key figure in the demise of American pragmatism at the hands of logical empiricism? Are Zilsel and Frank figures whose life histories indicate that not all European logical empiricism was moved successfully to North America? Was logical empiricism already insulated from science in the 1930s, or was it deeply engaged with science right through the 1950s?

Here, again, Frank is an interesting figure to think with. Arguably, Frank was no marginal figure in the American academic scene after World War II. Frank organized and participated in various activities with important long-term effects in philosophy of science, including laying the foundation for the Boston Colloquium for Philosophy of Science and the series Boston Studies in Philosophy of Science (see chapter 8). Frank organized and participated in many influential conferences, including the Conferences on Science, Religion, and Philosophy held in the 1940s and 1950s, several conferences on science education, his own conference on the validation of scientific theories, and the 1955 Conference on History, Philosophy, and Sociology of Science sponsored jointly by the American Philosophical Society (APS) and the National Science Foundation (NSF).[7] Only one other philosopher of science in America appears to have been present in such venues in anything like this frequency: Ernest Nagel.

But how are we to construe these facts? Are these the activities of a man who had become alienated from professionalized logical empiricist philosophy of science and who, in light of these activities, was rendered marginal in philosophy of science? Or are these the actions of a leading logical empiricist, actions that give the lie to any claim that logical empiricism was disconnected from history and sociology of science and from the larger social and cultural contexts of science and its philosophy?

These questions are important because, among other things, Frank, his courses in philosophy of science (which he taught in the Physics Department at Harvard), and his books were well known to James B. Conant and, at least through Conant, to the young Thomas Kuhn. He was also well known to George Sarton, a colleague of Percy Bridgman, and a key mentor of Gerald Holton, Marx Wartofsky, and Robert S. Cohen. Frank was, that is to say, the member of the Vienna Circle who worked most closely with physicists, historians of science, and sociologists of science in the American context. It is enormously important to figure out, therefore, whether his relations with those groups were part of a *turning from* logical empiricism or part of a *commitment to* logical empiricism. The issue is central to understanding the place of logical empiricism not merely in the history of philosophy of science and not merely in the history of philosophy but in the history of our culture's twentieth-century attempts to understand science as a human activity.

How do we decide these issues? With painstaking and subtle historical work, of course. We must, we propose, determine whether Frank understood himself to be acting as a logical empiricist in undertaking his work and whether he was understood by his colleagues and readers—both those who described themselves as logical empiricists and those who did not—as a logical empiricist. On the first issue, there is good reason to believe that through the end of his life, Frank understood himself to be a logical empiricist. As chapter 6 argues, Frank's revisionary histories were in support of an alternative vision of logical empiricism, not in support of overthrowing, transcending, or renouncing logical empiricism. As late as *Relativity:*

A Richer Truth (1950), Frank seems both to be presenting logical empiricist doctrines and presenting them *as* logical empiricist doctrines. The book is one of those in which strong connections are made between logical empiricism, operationalism, and pragmatism, and Frank mentions all three movements approvingly and by name. Indeed, all are named in the titles of chapters of the book: pragmatism in chapter 5, operationalism in chapter 6, and logical empiricism in chapter 7.

The evidence is, admittedly, more ambiguous in Frank's 1957 textbook, *Philosophy of Science: The Link between Science and Philosophy*. Much of what Frank says in the book is unobjectionable from the point of view of Carnapian or Reichenbachian logical empiricism, and Frank cites Carnap, Reichenbach, and Richard von Mises with greater frequency than he cites any other twentieth-century philosophers. But here Frank rarely, if ever, uses the term 'logical empiricism', and his historical concerns seem more aligned with Conant's Harvard Case Histories in Experimental Science, the key texts in the Harvard General Science Education program as taught by Conant, Kuhn, Leonard Nash, and others, than with the technical concerns of Carnap or Reichenbach in the 1950s. Moreover, Frank hinted at dissatisfaction with recent logical empiricist work in the book's preface:

> Presentations in [philosophy of science] have very often started from a concept of science that is half vulgar and half mystical. Other presentations have linked science with a philosophy that has actually been a mere system of logical symbols without contact with the historical systems of philosophy. But these very philosophies have served as support for ways of life and, specifically, for religious and political creeds. (1957, iv)

The import of this statement becomes clear only at the end of the book, when Frank talks of "extrascientific" reasons for the acceptance of "scientific theories of high generality" (342–60). Frank there employs underdetermination to argue that the historical philosophies have provided "extrascientific" reasons for the acceptance of high-level theories, such as theories of causality. Thus, the historical philosophies have to be at least a topic for philosophy of science, which Frank conceived of as a part of a general science of science that includes a "sociology of science" or a consideration of the "humanistic background of science" (Frank 1957, 359).[8]

Regarding the second issue—the reception of Frank's work—we note that a preliminary reading of some reviews of Frank's books indicates that he was understood to be a logical empiricist, even a vulgar logical empiricist, but, more interestingly, that logical empiricism during its alleged era of philosophical preeminence received plenty of bad reviews in important journals. Frank was routinely condescended to by his reviewers, and his *Relativity: A Richer Truth* was especially pilloried. A young Stephen

Toulmin thought the book naive, writing that "its sub-title would perhaps best be: Logical Empiricism told to Children" (1951, 181), and A. P. Ushenko used his review to condemn the whole of logical empiricism:

> I am urging students to read the book on account of Part One. . . . because its simplicity and clarity of presentation exposes the inadequacy and confusion of the author's philosophical affiliations where the more technical writings are protected by a camouflage of symbolic notation or pedantic belaboring of detail. (1951, 587)

These reviews suggest a corrective to any view that logical empiricism dominated American philosophy in the 1940s and 1950s. Ushenko's review even announces a "decline in [the United States]" of logical empiricism (587). A preliminary reading of some of the reviews of Frank's books suggests that logical empiricism was not only not accepted but was, at least at times, deeply resisted and even resented by philosophers working in North America in the 1940s and 1950s.

Narrative Structure and Historical Explanation in Philosophy

The historical work in logical empiricism has also raised large historiographic issues, two of which we will mention here. The first has already been exemplified in some of the concerns raised; the second is raised explicitly by Reisch in chapter 8 in his remarks on Peter Galison's 1996 "Constructing Modernism."

Consider the terms in which historical accounts in philosophy, especially philosophy of science, are often given. The terms in which the career of logical empiricism is treated seem taken from an inquest: "When did it die and how? Was there foul play?" Richardson took the death of logical empiricism to be "a social fact of analytic philosophy" and cited as reasons for this view the further fact that "few, if any, consider themselves to be continuing the project" of logical empiricism (1996, 13 n. 4). But such comments do not suffice to explain why the use of "death" is helpful in historical investigation of discontinued social practices or intellectual movements. We think of logical empiricism as having a life cycle—it is born, it flourishes, and it dies—but that may be the *wrong* way to think, especially if it is embedded in a Hobbesian discourse of the war of all against all, where a philosophical project flourishes at the expense of other projects and dies at the hands of younger, more vital projects.

A variation on the death theme is told in terms of failure rather than death. In this variant the question becomes, Why did logical empiricism fail? George Reisch in chapter 8 speaks in terms of the failure of the *International Encyclopedia of Unified Science,* and he further articulates what he means by saying that the failure of the project is captured by "the facts that the project never recaptured the success it enjoyed before the war and that only . . . twenty monographs appeared" even though the original plan expected hundreds of monographs. Failure to achieve one's goals *is* failure of a sort; we are rightly sick and tired of hearing sports commentators tell us that the Canadian competitor, who had set his goal to come in first in the slalom, had great success despite coming in forty-seventh. It is not at all clear, however, that the *Encyclopedia* was or is *less* influential or important for comprising only twenty monographs rather than Neurath's originally envisioned hundreds. A solid and lean *Encyclopedia* is much more likely to have actually been read than would have been the bloated monster of Neurath's dreams. As anyone who has written a Ph.D. thesis can attest, sometimes one avoids failure and achieves a modicum of success precisely by setting aside one's early ambitions.[9]

Leaving aside the particular case of logical empiricism, the more general question is whether we really want to have a history of philosophy in which the historically given philosophical projects or schools either continue to inform (and are known to continue to inform) current philosophical practice or, if not, have died or failed. It is, after all, an unusual narrative structure for fields of human endeavor. Neither accumulationist nor Kuhnian stories of the history of science need say that classical electromagnetism importantly *failed.* Theories do not need to fail before they are improved, no more than do computers or refrigerators.[10] Histories of artistic, literary, culinary, or any other type of achievement need not speak of death and failure in order to recount the movement from one school or style to the next. Moreover, philosophers' unreflective willingness to speak of their own history in terms of death and failure gives other disciplines both motive and opportunity to find philosophy a very strange and irrelevant discipline, a demoralized discipline giving off the scent of decay. Historians of philosophy come to look like spectators caught up in the grandeur of a historical procession of death and failure; philosophers come to look like cheerful or dutiful marchers *in* that procession. "Join us," we seem to say to our students, "so that someday soon your work, too, can be seen as dead and failed."[11]

Indeed, the pathos of this heroic philosophical failure through thousands of years is unbearable. A history of philosophy that explains why Plato failed, Aristotle failed, Aquinas failed, Descartes failed, Spinoza, Leibniz, Locke, Hume, Kant, Hegel, all failed, all failures, all dead, dead, dead—such a history is the intellectual equivalent of the weary academic job seeker reciting the history of her years on the market or of the dinner companion whose conversation consists of an exhaustive and exhausting, demoralized and demoralizing, account of why his romantic relationships have all failed. Enough; no more of these images of failure and death and of philosophers' pathetic struggles against certain failure and death. We do not admire the Romans who mocked Christ on the cross, and we ought not happily take up their role with respect to the philosophers, however ungodly they may be.

The final issue we want to scout concerns the levels of historical explanation. Reisch, for example, contrasts his "small-large" account of the troubled career of the *IEUS* with a "large-large" account he sees in Galison's "Constructing Modernism." That is, Reisch argues that the appropriate level at which to explain what happened to the *Encyclopedia* is a fine-grained one, describing in detail the interactions among the editors, between the editors and the authors, between the editors and the University of Chicago Press, and between the editors and their readers, whereas he sees Galison moving in the rarified world of idea complexes that have trouble moving from Europe to America, *zeitgeistige* changes in the cultural landscape, and so on. Reisch attributes Galison's search for expansive explanations of the decline of logical empiricism to a quasi-Cartesian principle of causation: Large effects must have large causes. Reisch himself, conversely, offers a narrative in the manner of the butterfly effect: For want of an editorial team, one might suggest, a philosophical program was lost. Interestingly, in this, Reisch appears as the hardheaded positivist, linking the particular events that constitute the world together into chains of causation. Galison appears as the slippery idealist, trying to pull the wool over our eyes with soothing words and grand schemes.

It is certainly true that particularism and materialism of a sort are in vogue these days in history of science and intellectual history. "Ideas" are now seen as ghostly things to be replaced by practices, texts, and so forth. Grand narratives have given way to minutely detailed descriptions of particular people in particular places doing particular things with particular instruments. But one sort of grand *meta*narrative persists, a *metahistorical* metanarrative that pours value over the move from grand narratives (or the history of disembodied ideas) to localism, particularism, contextualism, and so on by arguing that this move is a methodological or philosophical triumph. Historians of science write fascinating histories, but they may be more fascinating as the particular histories they are than as arguments by example for a certain account of causation or an "ontology."

In fact, history of philosophy of science needs both Reischian explanations and Galisonian explanations. After all, what is at stake is historical understanding, not necessarily causation or "images of the world" or anything deeply metaphysical, and from this point of view the explanations are more complementary than competing. Consider this volume. It has appeared in the world in its current form due to the individual actions of individual people; it has a history. That history includes a detailed Reischian account of why some people did and others did not submit papers, why the editors took certain actions and not others, and so on. It could suggest different actions that would have led to different volumes, and it would indicate critical points in the volume's construction. The story of *LENA* is the story of those actions. But, of course, those actions do not occur in a cultural void. Larger "standing" conditions would have to be cited in rendering the actions of the actors intelligible: for example, professional considerations such as the relative value of having a paper in an edited volume rather than a journal or having an edited volume on one's curriculum vitae. The rise of historical interest in logical empiricism since the 1980s would also have to be cited as a precondition of the possibility of this volume (as, indeed, evidenced by the first paragraph of this introduction). Such larger features of the situation are, in this case, conditions of intelligible action and are the sorts of features that Galison has sought to elucidate regarding logical empiricism in its places and times.

Both sorts of history are crucial to breaking the hold of a sort of textual hegemony in history of logical empiricism. Philosophers typically think they can "read the texts" and "figure out the arguments" and, on this basis, construct the rational history of philosophy. Logical empiricism, on such views, declined because it was refuted by Quine or Kuhn or because it was wrong about the nature of scientific theories, confirmation, realism, naturalism, or whatever. Reisch reminds us of the extent to which the individual actions and commitments of individual people *actually* form the history of logical empiricism in America. The *Encyclopedia* was not refuted; it was abandoned. Or, if you prefer, abandonment of projects and alteration of commitments while citing arguments as the reasons for such abandonment is the empirical face of refutation: If no one were to change their philosophical commitments or projects after "Two Dogmas of Empiricism," it would not have "refuted" anything. Galison, on the other hand, reminds us that patterns of such actions and commitments—and also the texts themselves—make sense only when placed in larger contexts. Indeed, only by combining the cultural and the individual can the poignancy of the actions taken or not taken become manifest.

The Role for History

Earlier, we adapted to our own purposes the most famous sentence in the philosophy of science in the second half of the twentieth century, the first sentence of Thomas Kuhn's *Structure of Scientific Revolutions*: "History," Kuhn wrote, "if viewed as a repository for more than anecdote or chronology, could produce a decisive transformation in the image of science by which we are now possessed" ([1962] 1996, 1). Kuhn obviously took himself here to be arguing not against science but against an *image* of science. Moving to philosophy of science, it seems quite clear that there is no *image* of philosophy of science that currently possesses philosophers. What philosophy of science is, what its topics, methods, and epistemic status are, are matters of dispute and dissent. If there is a role for history of philosophy of science, it is not a role that mirrors Kuhn's "role for history" of science in our understanding of science.

We suggested above that the history of philosophy of science has a deeper role: It can transform philosophy of science itself. Philosophers of science might, while lacking an image of their own practices, nevertheless be fairly considered to possess a philosophy of science: a set of unarticulated working methods, canonical topics, intellectual values, and ways of arguing. It is perhaps characteristic of today's analytic philosophy more generally that it consists of highly specialized sets of practices, argumentative procedures, and textual traditions without a compelling account of the meaningfulness of those practices, procedures, or traditions.[12] History of philosophy of science can transform contemporary philosophy of science by tying past practice in philosophy of science to the actions of individual philosophers and scientists set in their cultural contexts and by exploring the relationships between the culturally available understandings of science and of philosophy throughout our history. History of philosophy of science does not only suggest that this is the way for contemporary philosophy of science to acquire a focused image of itself; it also suggests that the very process of acquiring that image of our own activities will alter those activities. It will do this by making the cultural task of philosophy of science an explicit topic of consideration.

Those in the postphilosophical camp have argued that there are no roles or tasks for philosophy these days. Often this view is presented as the conclusion of a historical argument that claims to lay bare the nature of the philosophical project and indicate its poverty, impossibility, or irrelevance. Interestingly, those historical stories have been some of the worst caricatures of the historical projects in philosophy that anyone has ever written. But despite this reliance on bad history, the postphilosophers *might* be right. After all, it is implausible that every human culture has had a role for philosophers within it. Those of us engaged in serious history of philosophy of science look to the historical figures not to discover and take up a cultural mission more pertinent to times past but to forge new cultural missions and philosophical projects. We must be prepared to fail in these endeavors, but we must be, therefore, all the more committed to succeeding in them.

In 1946, Neurath was writing an essay that was left unfinished by his sudden death. It concludes with this uncharacteristically simple sentence: "The difficulty is that, whatever changes we accept, it is a kind of venture" ([1946] 1983, 246). The changes we create are even more difficult than those we simply accept, and *our* philosophical venture may yet come to nothing. But if it does, we will let down more than just ourselves. Historical and philosophical awareness induces a knowledge of our obligations to the past and fills us with our sense of obligation to the future. The interest we current philosophers have in logical empiricism can perhaps be explained by another image from Neurath: Wanderers lost in a forest do not purposefully return to their previous places, but they do well to remind themselves of the advice they have received in the past regarding the best way out of the forest. This is no time to return to logical empiricism, but it may well be time to return to the social spirit of philosophy of science of the 1930s. Perhaps that is our best philosophical venture in a world of anxious social and technological Maybes.

Notes

1. Early exemplars of the new historical interest in philosophy of science are Michael Friedman's review of the collected papers of Moritz Schlick (Friedman 1983a) and the first chapter of his book on space-time theories (Friedman 1983b). In both these places he began to tell a very different story of the relation of logical empiricism to relativistic physics than had become by then canonical. In the Austrian context, 1982, the centenary year of Schlick and Otto Neurath, brought new historical interest in philosophy of science (see Haller 1982). In addition to this literature, the historical turn has institutional and organizational superstructure. In 1990, the History of Philosophy of Science (HOPOS) Working Group was founded; in 1991, the Vienna Circle Institute was founded at the University of Vienna. These two entities came together for the HOPOS 2000 conference hosted by the Vienna Circle Institute, a conference that matched the number of presentations offered at the 2000 Philosophy of Science Association meetings in Vancouver.

2. Indeed, the *exclusionary* practices of analytic philosophy may come first. Hilary Putnam comments on this in an interview: "What happened to me, as to many other young American philosophers, was that in graduate school one learned what *not* to like and what not to consider philosophy" (Borradori 1994, 57). John McCumber (2001) comments on the larger social conditions attending such exclusionary practices as well as on their detrimental effects.

3. And here the move to "family resemblances" helps not a whit. Boundary concepts are typically terms of contrast rather than resemblance—which means that they combine in a way that either rules out nothing or rules out everything. That is, either analytic philosophy of mind is not phenomenology, not neuroscience, and so on, or else thought about one way, it sort of looks something like phenomenology, but thought about another way, it is sort of like neuroscience, and so on.

4. This is, in fact, something of a metaphilosophical exemplar of the noble history

of words that McCumber offers as a way for philosophy to climb out of "the ditch" (2001, 132–56).

5. An example of why issues such as those considered in this volume demand institutional and sociological perspectives, and yet of how halfhearted the reliance on a robust version of sociology of philosophy can be, is provided by Robert Butts. Butts's question concerns "the reception of German scientific philosophy"—the reception, that is, of logical empiricist accounts of science—in the United States. His answers divide into the internal and the external: His internal answer is that logical empiricist philosophy of science has "no proper alternative" (Butts 2000, 200), and his external answer is that "by the time that logical empiricism moved to the United States, the soil had been already well-prepared" (Butts 2000, 201) by American philosophical movements that shared an interest in science. As it happens, what matters to Butts is to argue for the truth of his internal reason for the positive reception of logical empiricism: Logical empiricism has, he argues, no proper alternatives. Butts's essay thus trades upon a fatal ambiguity in the notion of "proper alternative": His own argument is that logical empiricist accounts of theories are true and thus lack properly true alternatives, but this notion of "proper alternative" cannot play any explanatory role in the history of philosophy. After all, if logical empiricists were right about theories, then there never has been a proper alternative in this sense to their views. But that putative fact does not at all explain why these views only became canonical in the 1930s rather than the 1830s or the 1630s. If "proper alternative" is to play any explanatory role in the history of logical empiricism, it must mean something like "alternative that could at the time be adopted without losing one's standing as a philosopher." But then the question of the reception of logical empiricism becomes the question of how logical empiricism came to have no proper alternative, and Butts's account is seen as containing a canonical *petitio principii* of history of philosophy: He takes his more sociologically robust version of the problem to be a cause to be cited in the problem's solution.

6. After Neurath, Frank is perhaps the best (but not the only) example of a major yet marginalized figure in the received history of logical empiricism. Morris is a close second, and perhaps Zilsel comes in third. For this reason, and to demonstrate the fruitfulness of considering his work in the context of the history of philosophy of science, we draw attention to Frank in this introduction.

7. Frank's *Relativity: A Richer Truth* (1950) is derived from his talks at the Conferences on Science, Philosophy, and Religion in the 1940s, and he continued attending the conferences and publishing in their proceedings into the 1950s (Frank 1953, 1954). The conference on validation of theories in science had a proceedings volume (Frank 1956). His remarks at the APS/NSF conference are in Frank 1955. His views on science teaching can be found, for example, in Frank 1947.

8. Interestingly, the very slender Frank papers in the Harvard University Archives (UAP 4406.xx) consist largely of several hundred heavily edited typescript pages of an unpublished book manuscript by Frank entitled "The Humanistic Background of Science."

9. It could be argued that the size constraints of the individual monographs of the *Encyclopedia* greatly contributed to the success of its most famous one, Thomas Kuhn's *Structure of Scientific Revolutions*. If one keeps track of all Kuhn's claims in the text regarding things he wished to include but for which was not given the space, one sees a very different book emerge. Indeed, the book Kuhn *wanted* to write looks more like a prize-winning "thick description" of the 1990s than the most significant and widely read book in history and philosophy of science in the 1960s.

10. Popper's insistence that science progresses through repeated failure under severe testing is a philosophers' history par excellence. It overturns the value judgments implicit in canonical history of philosophy and the history of science by making failure the ultimate

epistemic virtue. Significantly, Popper's "enemies of the open society" are also supremely confident philosophers who do not think failure of prediction is the hallmark of epistemic success. Friends of the open society have a distinctly Protestant air about them. Falsificationism inscribes a narrative that involves endless yearning after the truth but certain knowledge that one has fallen short of it.

11. For a recent example that shows how natural this way of framing an historical narrative of philosophy is among analytic philosophers, see Stroll 2000; for a review of Stroll's book that takes its narrative structure as topic, see Richardson 2001.

12. McCumber, interestingly, begins his account of the current "dysfunction" of American philosophy by noting the "ongoing and *general* absence of reflection on the discipline" in American philosophy (2001, 8), as emblematized by the change in format of the American Philosophical Association presidential addresses from occasions to reflect on philosophy to occasions to engage in philosophy.

References

Borradori, G. 1994. *The American Philosopher: Conversations with Quine, Davidson, Putnam, Nozick, Danto, Rorty, Cavell, MacIntyre, and Kuhn.* Chicago: University of Chicago Press.

Butts, R. 2000. "The Reception of German Scientific Philosophy in North America: 1930–1962." In *Witches, Scientists, Philosophers: Essays and Lectures,* ed. G. Solomon, 193–204. Dordrecht: Kluwer.

Cohen, M. R. 1940. "Some Difficulties in John Dewey's Anthropocentric Naturalism." *Philosophical Review* 49: 196–228.

Cohen, M. R., and E. Nagel. 1934. *An Introduction to Logic and Scientific Method.* New York: Harcourt, Brace, and Company.

Conant, J. B., ed. 1957. *Harvard Case Histories in Experimental Science.* Vols. 1 and 2. Cambridge, Mass.: Harvard University Press.

Daston, L. 1994. "Historical Epistemology." In *Questions of Evidence,* ed. J. Chandler, A. I. Davidson, and H. Harootunian, 282–89. Chicago: University of Chicago Press.

Davidson, A. 2001. *The Emergence of Sexuality: Historical Epistemology and the Formation of Concepts.* Cambridge, Mass.: Harvard University Press.

Frank, P. 1947. "The Place of Philosophy of Science in the Curriculum of the Physics Student." *American Journal of Physics* 15: 202–18.

———. 1950. *Relativity: A Richer Truth.* Boston: Beacon Press.

———. 1953. "The Role of Authority in the Interpretation of Science." In *Freedom and Authority in Our Time: Twelfth Symposium of the Conference on Science, Philosophy, and Religion in Their Relation to the Democratic Way of Life,* ed. L. Bryson et al., 361–63. New York: Harper.

———. 1954. "Non-scientific Symbols in Science." In *Symbols and Values: An Initial Study. Thirteenth Symposium of the Conference on Science, Philosophy, and Religion,* ed. L. Bryson et al., 341–48. New York: Harper.

———. 1955. "Conference on the History, Philosophy, and Sociology of Science: Summarizing Remarks." *Proceedings of the American Philosophical Society* 99: 350–51.

———. 1956. *The Validation of Scientific Theories.* Boston: Beacon Press.

———. 1957. *Philosophy of Science: The Link between Science and Philosophy.* Englewood Cliffs, N.J.: Prentice-Hall.

Friedman, M. 1983a. "Critical Notice: Moritz Schlick, Philosophical Papers." *Philosophy of Science* 50: 498–514.

———. 1983b. *Foundations of Space-Time Theories*. Princeton, N.J.: Princeton University Press.
Galison, P. 1996. "Constructing Modernism: The Cultural Location of Aufbau." In *Origins of Logical Empiricism,* ed. R. N. Giere and A. Richardson, 17–44. Minneapolis: University of Minnesota Press.
———. 1998. "The Americanization of Unity." *Daedalus* 127: 45–71.
Giere, R. 1996. "From *Wissenschaftliche Philosophie* to Philosophy of Science." In *Origins of Logical Empiricism,* ed. R. N. Giere and A. Richardson, 335–54. Minneapolis: University of Minnesota Press.
Gieryn, T. 1999. *Cultural Boundaries of Science*. Chicago: University of Chicago Press.
Haller, R. 1982. "Neurath und Schlick: Ein Symposion." *Grazer Philosophische Studien* 16/17.
Harvard University Archives. Frank Papers. UAP 4406.xx. Cambridge, Mass.
Hempel, C. G. 1988. "Provisoes: A Problem Concerning the Inferential Function of Scientific Theories." *Erkenntnis* 28: 147–64.
Kuhn, T. [1962] 1996. *The Structure of Scientific Revolutions,* 3rd ed. Chicago: University of Chicago Press.
MacIntyre, A. 1984. "The Relationship of Philosophy to Its Past." In *Philosophy in History: Essays on the Historiography of Philosophy,* ed. R. Rorty, J. Schneewind, and Q. Skinner, 31–48. Cambridge: Cambridge University Press.
McCumber, J. 2001. *Time in the Ditch: American Philosophy and the McCarthy Era*. Evanston, Ill.: Northwestern University Press.
Meyerson, E. [1908] 1962. *Identity and Reality,* trans. K. Loewenberg. New York: Dover Publications.
Neurath, O. [1946] 1983. "Prediction and Induction." In *Otto Neurath: Philosophical Papers, 1913–1946,* ed. R. S. Cohen and M. Neurath, 243–46. Dordrecht: Reidel.
Quine, W. V. O. 1969. "Epistemology Naturalized." In *Ontological Relativity and Other Essays,* 69–90. New York: Columbia University Press.
———. 1981. *Theories and Things*. Cambridge, Mass.: Harvard University Press.
———. 1995. *From Stimulus to Science*. Cambridge, Mass.: Harvard University Press.
Richardson, A. 1996. "Introduction: Origins of Logical Empiricism." In *Origins of Logical Empiricism,* ed. R. N. Giere and A. Richardson, 1–13. Minneapolis: University of Minnesota Press.
———. 2001. Review of *Twentieth-Century Analytic Philosophy,* by Avrum Stroll. *Philosophy in Review* 21: 156–58.
Stroll, A. 2000. *Twentieth-Century Analytic Philosophy*. New York: Columbia University Press.
Toulmin, S. 1951. Review of *Relativity: A Richer Truth,* by P. Frank. *Philosophical Quarterly* 1: 180–81.
Ushenko, A. P. 1951. Review of *Relativity: A Richer Truth,* by P. Frank. *Philosophy and Phenomenological Research* 11: 587–90.
Wilson, M. 1992. "History of Philosophy in Philosophy Today; and the Case of the Sensible Qualities." *Philosophical Review* 101: 191–243.

Alan W. Richardson

1
Logical Empiricism, American Pragmatism, and the Fate of Scientific Philosophy in North America

The history of logical empiricism in America appears to be quite out of the ordinary. An understanding has grown up around that history. It goes something like this:[1] A small group of exiled, technically minded philosophers somehow was able radically to change the philosophical scene in America within a single generation. Logical empiricism became the dominant philosophical project in the United States within ten years or so of the first arrival of self-described logical empiricists in roughly 1930. The issues and methods of logical empiricism set the agenda of analytic philosophy until roughly 1960, when the first serious rivals began to appear in the form of the naturalism of W. V. O. Quine and the historicism of Thomas Kuhn. Despite having lost its dominance, however, logical empiricism still strongly influences the concerns of those working in technical areas of philosophy. A central puzzle is suggested by this story: Logical empiricism seems to have achieved its dominance by supplanting a home-grown philosophical tradition, American pragmatism, that in the 1930s had long-standing and deep roots and a strong public leader in John Dewey. This tradition differed radically from logical empiricism on the central issues of philosophy and on the appropriate methods for tackling those issues. It seems particularly surprising that American philosophy lost its American nature at just the time that America took center stage in the world's political and intellectual scene.

This chapter takes issue with this standard account and the puzzle it raises. The success of logical empiricism in America is indubitable, and I will have something to say about how it was achieved. My account will suggest, however, that it is misguided to embed this success in a story in which logical empiricism struggled with and (temporarily) vanquished a diametrically opposed project of American pragmatism. I will stress, rather, certain convergences of thought between logical empiricism and American pragmatism that led many in the 1930s to see them as kindred rather than opposing projects. The locus of the puzzle will then shift to why we have subsequently come to see a sharp divide of interests and methods between logical empiricism and American pragmatism.[2]

My story will suggest one way in which the ongoing reappraisal of logical empiricism can provide insight into the development of central themes in American philosophy in the period of the diaspora, World War II, and into the 1950s.[3] I emphasize an aspect of logical empiricism that has become salient in recent scholarship: the self-conscious framing of logical empiricism as *scientific* philosophy.[4] This is a key to understanding the career of logical empiricism in North America because, I shall argue, in the 1930s logical empiricism was received by important pragmatists as scientific philosophy and, *in that most central regard,* as a kindred project in philosophy. This sense of kinship was due to some central shared concerns, especially with language and meaning, and their relations to confirmation and the a priori, as well as a general agreement on the importance of a scientific philosophy for social and political life. In large measure, the era of success for logical empiricism coincided with the era in which its scientific credentials were acknowledged and promoted, its decline with a change in the understanding of the point and status of logical empiricism—a rhetorical shift most noticeable and important in the work of W. V. O. Quine in the 1950s and 1960s.[5] Quine taught us to see logical empiricism as the last best hope for a traditional foundationalist epistemology—one that must be overcome by a new *scientific naturalism*. In inviting us to question the scientific status of logical empiricism, Quine also presented a neutralized version of scientific philosophy; the pragmatism espoused in "Two Dogmas of Empiricism" (Quine [1951] 1980) shares none of Dewey's, Charles Morris's, or Rudolf Carnap's sense of the importance for society of achieving a scientific philosophy and a scientific habit of mind.[6]

The Reception of Logical Empiricism: Clarifying the Explicandum

In his postscript to *Origins of Logical Empiricism* (Giere and Richardson 1996), Ronald Giere (1996) sought to raise certain historical issues that scholars should think about as they attend to the North American context of mature logical empiricism. In section five of that essay, Giere asked about the relation between the rise of logical empiricism and the contemporaneous decline of American pragmatism. The crux of the matter for Giere is, perhaps, to be found in this passage of the postscript:

> Part of our question, then, is: How did a naturalistic pragmatism incorporating an empirical theory of inquiry get replaced by a philosophy that regarded induction as a formal relationship between evidence and hypothesis? In the 1950s, many philosophers of science would have given such a question very short shrift. Many would simply have asserted that pragmatism was mistaken and logical empiricism correct. . . . But such a response will not suffice in the 1990s. Ever since Quine advocated naturalizing epistemology, philosophical sentiment has been moving back in Dewey's direction. So the question now is why philosophers then *believed* that pragmatism was so obviously wrong and logical empiricism so obviously right. (347)

Among Giere's own attempts at an "internal" answer to this question and plea for future research are conjectures about the influence of Alfred Tarski's semantic account of truth as an alternative to a pragmatic account, and Hans Reichenbach's accusation in his essay "Dewey's Theory of Science" (1939) that Dewey's pragmatism conflated the contexts of discovery and justification. Such answers, in a sense, merely postpone the question. Now we must ask, for example, whether Reichenbach was right that pragmatist naturalism did not or could not allow such a distinction, and why philosophers at the time would adopt Reichenbach's distinction. (Giere himself seems to think that it is not a well-founded distinction.) In any case, the general tenor of Giere's concerns is clear: Quinean naturalism in epistemology and philosophy of science is in many ways more Deweyan than Reichenbachian. Why, then, did American philosophers steeped in Dewey in 1940 allow Dewey's naturalism to fall to logical empiricist formalism, only to be revived a quarter-century or so later by Quine's naturalistic conclusion to his rejection of the analytic/synthetic distinction?[7]

Giere's question seems to be the right one (but see below). Certainly, Dewey's pragmatism was a going concern in 1940, and his theory of inquiry in, for example, his *Logic: A Theory of Inquiry* (1938) was naturalistic, biologistic, and social. Thus, by certain (albeit highly contested) contemporary standards in science studies, Dewey circa 1938 looks terrifically sexy and seems a potential source of genuine insight into today's problems. By 1950, however, work such as Carl Hempel's "Studies in the Logic of Confirmation" ([1945] 1965a), Hempel and Paul Oppenheim's "Studies in the Logic of Explanation" ([1948] 1965), and Carnap's (1950) and Reichenbach's (1949) technical work in induction, probability, and confirmation served as the exemplars of philosophical work in scientific methodology. This work is formalistic in the sense that Giere's suggests: It offers a vision of philosophy of science as the logic of scientific and metascientific language and proffers methods consistent with this vision. If a formalist, nonnaturalist conception of philosophy is a mistake, then this is a misstep in the history of philosophy. We can, however, investigate the situation without ourselves taking sides on naturalism. How the change in view came to pass is interesting in its own right.

Giere's question is, however, not *exactly* the most pertinent one to ask. His discussion at times takes on a melodramatic air: We see a titanic struggle for philosophical dominance between two opposing camps. The struggle ends, or perhaps merely continues, in act 3, which we might entitle, to borrow a phrase, "The Naturalists Return."[8] It is, however, unclear to what extent American pragmatism was a *dominant* project in American philosophy in 1930, when logical empiricism first started to be taken seriously in America, or in 1940, when Reichenbach, Carnap, Hempel, Herbert Feigl, Philipp Frank, Gustav Bergmann, Kurt Gödel, and others were safely on American soil. The so-called new realism, for example, was newer than pragmatism, and in some locales (Ann Arbor, Michigan, for example), it was taken to be more vital.[9] More importantly, by 1930, having jointly beaten back absolute idealism, at least some of the erstwhile new realists and pragmatists thought it had become rather pointless to pledge allegiance to such programmatic camps. What reigned in American philosophy by 1930 was, on this view, a new eclecticism. One of the principal architects of new realism, R. B. Perry, wrote:

> A contemplative observer of the times would have great difficulty in describing its characteristic philosophical activity in terms of the doctrinal cleavages that were so well marked at the opening of this century. Its most conspicuous feature is, I think, an avoidance of dualisms and disjunctions with which the influence of Descartes is associated. This attitude is due in part to recent changes in science, in part to a revival of interest in ancient and medieval philosophy, and in part to a growing sense of the inadequacy of any of the sharply antithetical alternatives that divided the thought of the last century. (1930, 200)[10]

Nor is it clear to what the extent logical empiricism was dominant in philosophy of science even in the 1940s and 1950s. The answer depends, in part, on what counts as logical empiricism. Recent scholarship has stressed the variety of views put forward by self-proclaimed logical empiricists. Certainly, if the project of unified science as put forward in the *International Encyclopedia of Unified Science,* for example, is taken to be the crucial project of logical empiricism (Otto Neurath would have thought so), then the project was already in decline in the war years and by 1950 was a mere shadow of its early ambitions.[11] Logical empiricism may not have become the received view in philosophy of science until the 1960s, when it became "the received view in philosophy of science" (Suppe 1974, 1)—a value-laden term that insists that a critical thinker find reason to reject whatever is so labeled.[12]

Giere, though, as we have already noted, had in mind logical empiricism's dominance not as regards some particular topic in philosophy of science, however large scale, but in the way philosophy of science is done. His concerns were, after all, primarily methodological, and he raised what is without question an important puzzle: What are we to make of the relations of American pragmatism and logical empiricism, given the vexed relations between the naturalism of pragmatism and the formalism of logical empiricism? I want, here, to focus on this key methodological issue, naturalism versus formalism, and three principal actors, Carnap, Quine, and Morris. I think we should not follow Giere's own suggestions that the principal point of dispute was Reichenbach's justification/discovery distinction or Tarski's account of truth. Rather, insofar as we view this dispute as revolving around naturalism, we should fasten on a topic different from but related to both of Giere's conjectures. The issue, I believe, most pertinent to the history of logical empiricism in America is the analytic/synthetic distinction. Quine's rejection of this distinction was, he claimed, the reason he embraced naturalism. Moreover, it is clear that Carnap's vision of philosophy, which was the guiding vision for working logical empiricists, required this distinction.

I find of particular interest two aspects of the reception of Carnap's work by Morris in the 1930s and its similarities to and differences from the situation of the later Carnap/Quine dispute. First, Morris presented a version of pragmatism that accepts a robust role for the a priori and, with it, both an analytic/synthetic distinction and an appreciation for the intent of Carnapian philosophy. Second, Morris and Carnap agreed that Morris's pragmatism and Carnap's logical empiricism were versions of an importantly new project—scientific philosophy.

This suggests an important lesson for our understanding of the reception of logical empiricism in American philosophy: It might be *our* understanding of the point of logical empiricism that leads us to see it as *fundamentally* opposed to naturalism. Logical empiricism, at least in the form associated with Carnap and Reichenbach, is opposed to naturalism, but naturalism is not its most fundamental opponent. That opponent is a vision of philosophy as something other than a scientific discipline. Perhaps by uncritically accepting not so much Quine's arguments against the analytic/synthetic distinction (many of us have been asked to assess those arguments critically in the course of our philosophical training) as his account of the philosophical point of the distinction in the work of Carnap, we are led to be puzzled by a fraternal linking of hands between Carnap and Morris in the 1930s.

There is, however, a way of reading the mutual citations and influences of the logical empiricists and the pragmatists in the 1930s that pointed to a true commonality of project—a joint insistence on scientific method in philosophy, an insistence that led directly to the rejection of traditional metaphysics and traditional epistemology. Many of the differences in approach that characterize the Carnap/Quine dispute in the 1950s can be found *in ovo* in remarks by Carnap and Morris in the 1930s. For Carnap and Morris, however, these were differences of detail. They had a bigger issue, the scientific status of philosophy itself, on which they agreed and of which they most wanted to convince the philosophical community. Morris, in the 1930s, granted the scientific status of Carnap's philosophy; Quine, in the 1950s, framed his rejection of Carnap's philosophy by pointing to alleged foundationalist motives that revealed that philosophy to be unscientific.

I want to make this sketch plausible by first briefly reviewing some central characteristics of the Carnap/Quine debate and then seeing how the shape of that debate was prefigured by, but not carried out in, the work of Carnap and Morris in the 1930s. I will attempt to diagnose why the debate did not happen in the 1930s. The story may help us come to some deeper understanding of the relationships between American pragmatism in the 1930s and both logical empiricism and Quine's naturalism. This deeper understanding will, I hope, stimulate further thinking about two issues. First, what are the most pertinent questions that historians of philosophy could be asking about the development of philosophy of science in the twentieth century? How, for example, should we come to understand the rise of logical empiricism and its subsequent decline? More particularly, do the standard terms—"struggle," "dominance," "failure," "death"—express the most useful structure within which to tell the story of logical empiricism?[13] Second, what is the relation of contemporary naturalism to other twentieth-century philosophical projects that presented themselves as scientific? More particularly, why is naturalism now taken in many portions of the philosophical community to be the one candidate for a genuinely scientific philosophy? This sentiment seems well out of proportion to the arguments offered, and the successes achieved, by contemporary naturalism.

Two Issues Dividing Carnap and Quine

In order to focus attention on some issues as they do and do not arise between Carnap and Morris in the 1930s, I want to review a couple of the central issues in the analytic/synthetic debate of the 1950s. I make no claim that these are the only issues under debate, though they are among the most important ones.[14] The two issues are first, Quine's assertion that Carnap had to have a criterion of analyticity if he was to avoid "creative reconstruction," ultimately amounting to no more than "make-believe" (Quine 1969, 75), in epistemological matters; and second, the relation for Quine between the rejection of analyticity and the adoption of naturalism.

On the first issue, Quine demanded a behavioral criterion of analyticity from Carnap. Absent such a criterion, there is no determinate way to ascribe analyticity to the sentences of a language in use and, thus, no cash value from an epistemological point of view to the analytic/synthetic distinction. No such criterion being forthcoming, Quine argued that the analytic/synthetic distinction is a philosophical artifact demanded only because of unexamined presuppositions about meaningfulness. The distinction plays no role in a descriptively adequate account of the reasons why people come to accept what they accept.

On the second point, Quine almost invariably presents his naturalism as the only place the empiricist has left to go once the analytic/synthetic distinction has been dropped. If there are no analytic sentences, then there is no plausible remaining candidate for an a priori element in knowledge. If this is the case, then all knowledge-producing enterprises, including mathematics and philosophy, are on an equal footing with empirical science. Having no place else to stand, if the epistemological question—How do we come to accept what we accept?—comes up, we should feel free to use whatever empirical resources might help us to an answer. Naturalism, in the sense of adopting an empirical methodology and using empirical results to answer philosophical questions, follows from a rejection of an a priori element in scientific knowledge.

Rather than pursuing Quine's understanding of the situation here in greater depth, let us remind ourselves that Carnap did have responses to Quine's criticisms. On the first issue, Carnap sensed philosophical sleight of hand. Quine was issuing a very curious challenge, demanding a pragmatic analysis for a semantic notion. Semantic notions are semantic precisely because they are not pragmatic (nor are they syntactic), and thus Carnap saw no reason to meet this challenge. Meaning, for Carnap, was neither in the head nor in the behavior; it was in the structure of the language as revealed in the metalanguage in a Tarskian way. Semantic concepts were not explications of antecedently given pragmatic concepts. Semantic notions, especially as applied to formal languages specially formulated for certain purposes, could, nonetheless, do real work in the logic of science—rendering precise certain confused epistemological questions involving confirmation, for example—even absent any correlation between those notions and antecedent human behavior.[15]

On the second point, from a Carnapian perspective, one can begin seriously to wonder whether Quine's objections to the analytic/synthetic distinction do not presuppose rather than lead to his wholly a posteriorist naturalism. After all, the circularity concerns re analyticity, synonymy, convention, semantic rules, and so on raised early in "Two Dogmas" could also be read as objections of the form: According to Carnap, semantic concepts are semantic and not pragmatic. Once one has any of them, Quine claimed, one can have all of them. He placed severe constraints, however, on when one can claim to have made appropriate sense of any of them. Perhaps Quine was demanding from the start that all linguistic notions be given behavioral criteria simply because there is nothing other than behavior that could ground the notion of language or linguistic rule. This seems to be a robust presupposition about language that already evinces a commitment to a very strong version of naturalism.

Quine did not raise such objections to Carnap's views on analyticity haphazardly. Quine's objections stem from an understanding of the philosophical burdens that Carnap, in his view, placed upon the notion of analyticity. Quine saw Carnap's commitment to analyticity as a commitment to a certain sort of answer to an antecedent epistemological question: What accounts for the certainty of mathematics? For Quine, then, Carnap had foundationalist epistemological ambitions for his account of analyticity. Analyticity yields a priority, and a priority is needed for certainty. Moreover, Quine saw those epistemological ambitions connecting to semantic questions through Carnap's (and his own) commitment to verificationism. His diagnosis of the philosophical point of analyticity, therefore, was that, for Carnap, the analytic truths of a language are those that are true by virtue of meaning and, thus, verified come what may. (It is in this sense that Carnap, according to Quine, saw mathematical and logical truths as certain: They are immune to empirical disconfirmation.) At this point, Quine attempted to trump Carnap's philosophical maneuvers through his own argument from holism: If any relatively high-level theoretical claim can be "verified come what may" (with suitable changes elsewhere), then Carnap's account of analyticity does not fulfil its epistemological function; it does not divide mathematics and logic from empirical science. If it does not do that, there is surely no other need for it. Therefore, epistemology may set it aside.[16]

For Carnap, however, the order of priority is reversed. The point of Carnap's philosophy is that no question on any topic is possible antecedently to the adoption of a linguistic system. This is true of questions of verification and confirmation as much as of any other kind of question. Thus, there cannot be an antecedent epistemological perspective within which to ask for or give an account of analyticity. Rather, all epistemological questions are questions within a language, which must be specifiable in advance and for which, therefore, an analytic/synthetic distinction must be in hand. This was, indeed, the order exhibited in his work: His work on confirmation always presumed that confirmation is relative to an antecedently given linguistic framework. None of his technical accounts of analyticity, therefore, could or did rely on confirmation theory. Thus, Carnap rejected Quine's claim that his (Carnap's) technical accounts of analyticity are explications of the informal notion of sentences verified come what may.[17]

The framework relativity of epistemological questions indicates the sense in which Carnap, far from thinking that he must answer a particularly difficult epistemological question about logic, thought that no epistemological questions about logic could be raised. A logical framework is in place before any questions of confirmation on the basis of experience even make sense. This does not show foundationalist desires to find a firm epistemological grounding for science in a realm outside of science. It shows, rather, a desire to explicate and employ procedures of formalization in philosophy, so that formalization can clarify philosophical questions just as it had clarified scientific ones. Carnap, as much as Quine, viewed his own philosophy as importing science into philosophical method. Carnap introduced formal methods; Quine, empirical methods.

The crucial issues, then, are these: Is it possible to make sense of analyticity without connecting analyticity of sentences to some aspect of human behavior? Is it possible for an a priori philosophy also to be a scientific philosophy? Carnap answered both questions affirmatively. Philosophy is the mathematics of language, sharing in the framework constitutive nature of mathematics and providing the conceptual means to explicate precisely the obscure and confused terms of epistemological discussion. Quine answered both negatively. The failure to provide behavioral criteria for analyticity reveals the a priori to be a sham. No science is a priori. Any philosophy claiming a priori status reveals thereby its extrascientific ambitions while giving up its claim to our allegiance.

By the early 1950s, then, Quine succeeded in putting two aspects of Carnap's philosophy into question: its purpose and its adequacy, given that purpose. If I am right, then whatever the merits of Quine's arguments against Carnap's notion of analyticity, given the purposes that Quine saw it serving, we should not take for granted that Quine fully accurately enunciated Carnap's own purposes. Indeed, from a Carnapian point of view, Quine can be seen as having helped himself to an unproblematic and informal epistemological perspective, whereas Carnap was attempting to raise questions about the availability of just such a perspective. Indeed, throughout his career, Carnap argued that the seeming availability of such informal philosophical perspectives had led philosophers into pseudo problems and endless controversies of just the sort that he was seeking to avoid. Without taking sides in the issue, our attention is called to the subtle relations that various notions must play in the philosophies of both Carnap and Quine. The semantic, pragmatic, logical, epistemological, scientific, "natural," formal, and metaphysical are at stake all at once.

Charles Morris, Semiotic Pragmatist and Scientific Empiricist

Before taking up these issues again in the context of Morris's reception of Carnap's philosophy, some brief remarks on Morris are in order. Morris is today not nearly as well known in the philosophical world as are Quine and Carnap. Morris was born in 1902, making him eleven years younger than Carnap and six years older than Quine. He had been an engineering student as an undergraduate and then did his graduate work in philosophy at the University of Chicago under George Herbert Mead and the Chicago school of pragmatism in the 1920s. After spending some years at Rice University in the early 1930s, he returned to the University of Chicago in the mid-1930s, when several positions opened up in the Philosophy Department. (Several members of the department had resigned to protest the hiring of Mortimer Adler into the department, without the department's consent, by the university's president, Robert M. Hutchins.) At Chicago, Morris became an early American spokesman for scientific philosophy, logical empiricism, and unified science. He was instrumental in helping several of the logical empiricists get jobs in the United States, most notably Carnap, who became his colleague at the University of Chicago in 1936. Morris became a coeditor of Neurath's *International Encyclopedia of Unified Science* in the mid-1930s—as did Carnap—and had the unfortunate experience of being the principal liaison between Neurath and University of Chicago Press until Neurath's death in December 1945.[18]

Philosophically, Morris is principally remembered for the tripartite distinction of the theory of signs—semiotics—into syntax, semantics, and pragmatics.[19] The introduction of pragmatics was, in part, a response to Carnap's philosophy, which in the mid-1930s was a semantics and syntax of scientific language. Pragmatics concerned the relations of signs to their users. (Thus, questions of the relative priority of pragmatics and semantics were, as we have seen, the issues at stake in the question of a behavioral criterion of analyticity.) Morris introduced the term "pragmatics," which was meant to pay homage to the role of American pragmatism in making salient the problems studied in pragmatics. For example, Morris explained the importance of investigating the relations of signs to their users through citation of Mead's understanding of meaning as involving expectable reactions to signs. For Mead, the deliberation about whether to shout "Fire!" in a burning movie theater—how should one alert the audience without panicking them?—revealed the sense in which the reactions of one's audience play into the precise nature of one's speech acts. Thus, pragmatic concerns exhibit and elucidate very subtle distinctions of meaning. A general account of the meaning of signs, according to Morris, therefore, could not abstract away from the speakers and audiences of acts of sign-giving.[20]

Carnap and Morris had, as these brief remarks indicate, similar interests. Both were principally interested in providing a general theory of language as the perspective from which philosophy was done. Both had, as a result, a linguistic understanding of the unity of science—unity of science involved unity of the language of science. Both argued, moreover, that this linguistic understanding of philosophy could lead to the dissolution of seemingly insurmountable philosophical boundaries. Carnap continually attempted to use his metalogical perspective to push for "tolerance" of various philosophical projects. His tolerance stopped just this side of metaphysical projects precisely because metaphysical projects were recognizable as metaphysical because of the impossibility of finding or formulating a formal language that captured what was allegedly being said. Morris was very explicit in his attempt to comprehend what was right in the historical versions of empiricism, rationalism, and pragmatism while also rejecting metaphysics, in his vision of "scientific empiricism."[21]

Finally, both Carnap and Morris stressed a role for "attitude" or "stance" in their philosophies. For Carnap, "the scientific world conception" was an attitude that one adopted for pragmatic reasons in one's philosophical work. It was not a general truth about how things are or about how philosophy must be done.[22] Similarly, Morris also stressed a "scientific habit of mind" as one of the principal goals of a project of unified science and a general scientific education. (This may be the most Deweyan element of his thinking.) They also shared, with Dewey and Neurath, a progressive political agenda. They saw a political role for a scientific philosophy that helped with the internationalist and progressive project of unifying scientific knowledge in support of social needs. Indeed, Morris and Carnap both felt that their scientific attitude was inseparable from a politically responsible philosophy, since, in the words of Morris, "this [scientific] habit of mind is the best guaranty of an objective consideration of the multiplicity of factors which enter into the complex problems of contemporary man" ([1938b] 1955b, 74).[23]

Morris and the Early Reception of Carnap's Logical Empiricism

This invites us back into the question of the early reception of Carnap's work in the writing of Morris. Morris did not diagnose Carnap's philosophy in the way that Quine did. Moreover, although Morris's views on these matters are, for those of us who have lived through the Carnap/Quine debate, fraught with tensions and perhaps ultimately untenable, it is surely important that Morris thought that his pragmatism, although naturalist, contradicted neither the letter of Carnapian philosophy nor the scientific spirit in which it was offered.

The most important rhetorical and systematic similarity between Morris and Carnap in the 1930s was their mutual insistence on the scientific status of philosophy. Morris thought that his vision of scientific philosophy was more broad-minded than Carnap's, but I have found no place within his early writings where Morris ever attributed the sort of foundationalist motives that are the hallmark Quine's rejection of Carnap's philosophy as unscientific. Indeed, Morris was at pains to present Carnap's philosophy as one of a limited number of options available once one has *dropped* any philosophical perspective outside of science. Thus, in a 1935 paper in *Philosophy of Science,* tellingly entitled "Philosophy of Science and Science of Philosophy," Morris wrote,

> We begin then by rejecting any conception of philosophy which regards philosophy as proceeding by methods other than those of science or as obtaining an order of certainty different from that obtained by science. This is essentially the same thing as to deny the existence of a priori synthetic judgments and any philosophy that rests upon the affirmation of such judgments. (1935, 271–72)

Having framed his discussion in this way, Morris presented "philosophy as the logic of science," that is, Carnap's vision of philosophy, as the first option consistent with this presupposition (272–75).

This was, as mentioned above, also Carnap's view, as is clear in the final pages of *The Logical Syntax of Language* (Carnap [1934b] 1937, 331–33), which Morris had read in the German original by the time he wrote his paper. Thus, if we were to think of naturalism as a methodological thesis—the insistence that philosophy can employ no methods other than those of science—then both Carnap and Morris would say that Carnap's philosophy is naturalist in that sense. This is because the philosophical tools that Carnap employed were ones that he claimed to find at work whenever scientists pause to clarify their concepts. Philosophy is simply the domain that clarifies this process of clarification, gives a general framework for understanding how logic and mathematics play this role, and brings such clarification to heretofore unclear metascientific concepts such as confirmation and explanation. Clearly, if there is an a priori element in science, then an a priori philosophy is not unscientific (or nonnaturalist in the sense above) provided its a priori is the very one found in the sciences. This is precisely the situation as Carnap saw it.[24]

Moreover, Morris agreed with Carnap that there is an a priori element in the sciences. Indeed, although his account of this element was informal and imprecise by Carnapian standards, it was remarkably more in tune with Carnap's understanding of analyticity than what one finds in Quine circa 1951. Thus, in a 1934 paper in *Erkenntnis* (one of the earliest papers in that journal to be printed in English), Morris presented the view as follows:

> What is here suggested can be generalized in the concept of the variable a priori. There is at any moment, for thinking beings, an a priori in the sense of a set of meanings in terms of which empirical data are approached, and logical analysis may be regarded as following the structural lines of the a priori which support inference. This a priori, however, undergoes change through contact with the new data which are encountered through its use, and through changes in human interests and purposes. With every such change the a priori is altered, and new content for logical analysis is provided. Acting on the new set of meanings brings new data and new purposes, which in turn affect the content and structure of the a priori. And so the spiral continues. (1934, 10)

One thing that Carnap would insist on more sharply here is that in genuine cases of change of language, human interests and purposes act independently of evidence or data. This is what makes the choice pragmatic rather than theoretical. Conversely, Quine would insist that no sharp distinction could be drawn. Thus, Morris pointed the way into one of the chief issues under contention in the debates about analyticity and the a priori in his very way of explicating those notions.

Regardless of our sense of the delicacy of Morris's own view here, there does seem to be an element of thought in pragmatism circa 1930 that came very close to Carnap's understanding of the analytic sentences of a language as providing the conditions of meaningfulness and rational inference for that language, where such conditions vary across languages. Lewis's investigations of different logical systems led him to a very similar conclusion in his "pragmatic a priori." He wrote, for example, in 1930 that he "began to see that the principles of logic will answer to criteria of the general sort which may be termed pragmatic, and that where empirical verification is not in point, and logical 'necessity' itself is not sufficient, no other kind of criterion can in any sense be final" (42).[25] That is, no questions of confirmation are relevant to questions of logical truth, nor do any discussions of strict necessity provide a way to decide which logical system is "correct."

Indeed, there is an important sense in which Dewey's own logical theory agreed with this. In setting logical principles down as operational postulates of inquiry, Dewey argued that inquiry is constituted as inquiry only by following those logical principles. Such principles were not, however, discovered as permanent a priori constraints on inquiry. He rejected any possibility of an independent and antecedent intuitive or transcendental search of the mind's conditions of thought. A logical law, wrote Dewey,

> is a stipulation. If you are going to inquire in a way which meets the requirements of inquiry, you must proceed in a way that observes this rule. . . . A postulate is neither arbitrary nor externally a priori. It is not the former because it issues from the relation of means to the end to be reached. It is not the latter, because it is not imposed upon inquiry from without, but is an acknowledgment of that to which the undertaking of inquiry commits us. It is empirically and temporally a priori. (1938, 17)

This is an a priori found in its constitutive role in a framework of inquiry. It is changeable over time and for reasons that have to do with unwieldiness as regards data or with a change of the desired ends of inquiry. In this very general sense, it agrees with the methodological point of Carnap's analytic sentences. Perhaps Carnap, Morris, Lewis, and Dewey all fall prey to Quine's arguments on this point, and perhaps not. From a historical point of view, however, what must be borne in mind is that a pragmatist of the 1930s would have found nothing either philosophically disturbing or methodologically suspect in Carnap's variable a priori as exhibited in the analytic sentences of different logical languages.

There is another way in which Morris clearly attempted to agree with Carnap and, by extension, disagree with Quine. Morris agreed with Carnap that semantic notions such as analyticity cannot be given pragmatic analyses. This is because, for Morris, the very notion of the pragmatic presupposes both syntax and semantics already in hand. Thus, one recognizes certain particles of speech as requiring pragmatic analysis because one recognizes that those particles do not contribute to the truth conditions of the sentences in which they appear. An example in English is "well" as used in sentences such as "Well, there are philosophers for whom a primitive notion of the normative does all the work." The initial word plays no role in the truth conditions of the sentence. It merely indicates the conversational place or point of the sentence. In a certain intonation, the initial "well" indicates an ironic exasperation, while with another intonation it could indicate puzzlement (as when this sentence is uttered as a response rather than as an answer to an unclear question about normativity). What is important is that the pragmatic account of "well" begins with the more or less informal sense that the presence of the word (as a particle, not as an adjective or adverb) does not change truth conditions or meaning. Thus, the semantic notion of truth conditions must be in hand for us even to recognize what falls within the purview of pragmatics.

Here again, though, on examination, Morris's position will strike us as delicate. In the very same section of his *Encyclopedia* article "Foundations of the Theory of Signs" in which he made the preceding point, he wrote,

> Any rule when actually in use operates as a type of behavior, and in this sense there is a pragmatical component in all rules. But in some languages there are sign vehicles governed by rules over and above any syntactical and semantical rules that may govern those sign vehicles, and such rules are pragmatical rules. ([1938a] 1955a, 113)

That is, the very notion of linguistic rule is a pragmatic one (in the sense that explicating the notion requires ineliminable reference to language users), but only some linguistic rules are pragmatic (in the sense that certain signs contribute to the meaning of an utterance but not to its truth conditions). A similar tension is found in his endorsement of the variable a priori. The very section of the *Erkenntnis* essay in which he introduced the notion of the variable a priori, he also argued that metalogic is both the study of the a priori so conceived and itself an empirical science (Morris 1934, 12). In cases such as these, Morris seems to have wanted to maintain doctrines closer to those of Carnap than to those of Quine, but he explained his views in ways that give more aid and comfort to Quine's rejection of those notions than to Carnap's acceptance of them.

Given this, the stage was already set in 1935 for a dispute between Carnap and Morris about the appropriate methods for philosophy. It did not happen. My suggestion is that it did not happen precisely because of the common presupposition they shared: that both were engaged in an attempt to provide a scientific philosophy within which there was no room for traditional philosophical preoccupations such as metaphysics and epistemology. Each, in his own way, wanted to show that all legitimate questions of philosophy were questions of language and that such a perspective did not allow for some traditional philosophical worries precisely because a linguistic philosophy does not allow certain concerns to be expressed. This way of connecting certain fundamental commitments of logical empiricism and pragmatism had already been expressed before either Morris or Carnap came on the scene. For example, in 1920 Dewey, though not making all philosophy linguistic, argued for "the release of philosophy from sterile metaphysics and sterile epistemology" ([1928] 1948, 126). By 1935, even if from a rather different point of view, Carnap was also advocating an end to metaphysics and an end to epistemology.[26]

Such calls for the end of traditional philosophy are also the connecting point between logical empiricism and American pragmatism according to Reichenbach in *Experience and Prediction* (1938). Reichenbach saw the rejection of sensation as "observable facts"—that is, of experience as a veil of ideas—as the point in the 1930s where the physicalism of the logical empiricists could fruitfully come together with Dewey's arguments against the spectator view of knowledge of traditional epistemology (Reichenbach 1938, 163).

For Morris and Carnap, pragmatism and logical empiricism shared a certain fundamental understanding of what was at stake in a scientific philosophy. A scientific philosophy would show its scientific credentials by providing a way out of a bad game—the game of traditional metaphysics and epistemology. Thus, the general theory of signs, or a general semantics in Carnap's sense, would provide the conceptual means to find a new technical role for philosophy and a diagnosis of the ills of metaphysics and epistemology. Scientific philosophy was, for both these thinkers, a radical solution to a radical problem in traditional philosophy. It was also a solution that finally connected philosophy with the ongoing struggles of society. These were the deepest convictions about the intellectual and social value of scientific philosophy that bound Morris (and Dewey) together with Neurath and Carnap.[27]

On the other hand, as we have seen, Quine's naturalism is a scientific philosophy that shows its credentials precisely by allowing us to raise the question of epistemology in a new setting. Moreover, as Quine famously asserted, the way into naturalism via the rejection of the analytic/synthetic distinction brings "a blurring of the supposed boundary between speculative metaphysics and natural science" ([1951] 1980, 20), since Carnap's principled separation of the two is no longer available. In contrast to the scientific philosophy of Carnap and Morris, Quine's naturalism is intellectually conservative. It opens up a way back into metaphysics and epistemology and changes the revolutionary, forward-looking rhetoric of both logical empiricism and American pragmatism into a story of continuity going back all the way to Locke and Hume.[28]

Pragmatism: Carnap, Quine, and the Tradition

Quine claimed that the rejection of the analytic/synthetic distinction induced "a shift toward pragmatism" ([1951] 1980, 20). The reason for this is given in these words from the very end of "Two Dogmas of Empiricism":

> Carnap, Lewis, and others take a pragmatic stand on the question of choosing between language forms, scientific frameworks; but their pragmatism leaves off at the imagined boundary between the analytic and the synthetic. In repudiating such a boundary I espouse a more thorough pragmatism. Each man is given a scientific heritage plus a continuing barrage of sensory stimulation; and the considerations which guide him in warping his scientific heritage to fit his continuing sensory promptings are, where rational, pragmatic. ([1951] 1980, 46)

It is instructive to think about what 'pragmatic' means for Quine. A clue comes from his immediately preceding paragraph, where he speaks of our "vaguely pragmatic inclination to adjust one strand in the fabric of science rather than another" (46). Presumably, this inclination is pragmatic (however vaguely so) because there is no compelling logical reason to adjust one rather than another, given the argument from holism of theory testing. Such decisions, on Quine's view, are pragmatic because they express the sense in which rules of verification do not wholly determine the choice of what to believe. One takes the rules of verification as far as they can go, and then the pragmatic kicks in.

This is importantly different from the sense of pragmatic we have seen in common among Lewis, Dewey, and Carnap. For Lewis, decisions about logical laws are pragmatic because "empirical verification is not in point" (Lewis 1930), that is, because questions of empirical verification simply do not arise in the case of logic. For Dewey, verification or confirmation is a move in the context of inquiry and not, therefore, a relevant factor in setting the conditions of that same inquiry. Carnap rejected the idea that analytic sentences are "verified come what may," insisting instead that a linguistic framework and thus analytic sentences had to be in place before any clear sense could be made of questions of verification. None of them saw the pragmatic as filling a gap between the rules of confirmation and the dynamics of belief. Pragmatic concerns are, rather, those that explain belief when confirmational concerns do not arise.

Quine's thorough pragmatism is an insistence that logical principles do not determine the reallocation of belief among sentences in the light of recalcitrant experience. Such a thorough pragmatism seems also a thin pragmatism. Pragmatism in Dewey's sense had much more to do with an insistence that an adequate philosophy both understand and provide means for human agency. It is much more about what to do than what to believe. (Believing is only one species of doing.) It is hard to imagine Dewey being much impressed by a pragmatic naturalism that is all about the decisions taken about how to allocate confirmation values to sentences, given a barrage of sensory stimulations. Indeed, here we see a problem that Dewey would find with Quine's new setting for epistemology: "A barrage of sensory stimulations" sounds very much like a physicalized version of the passive experience that Dewey thought was incoherent.

On this score, Carnap's limited pragmatism might do better. The adoption of a linguistic framework is an undertaking for certain human purposes, amounting to a commitment to the use of an instrument in understanding the world. Carnap and Morris shared the view that they were providing conceptual technologies of new rigor in support of a project of unifying the sciences. Unified sciences of this sort would provide the way into genuine empirical research into the scientific means toward one or another of the humanly possible futures. For Carnap in particular, as Richard Creath (1991) has stressed, the metalogical perspective of philosophy provides an unlimited battery of technical tools available for shaping and understanding the world. Adoption of these tools must be motivated by the recommendation of certain purposes and values but cannot be seen as accountable to some transcendent realm of facts or values that force the choice. This is the sense in which the analytic/synthetic distinction induces the practical/theoretical distinction for Carnap. This, in turn, connects it to the agentive concerns of pragmatism and to the realm of practical reason.[29]

We began by taking up Giere's question of how a naturalist pragmatism could have been overtaken by logical empiricist formalism. If what I have claimed is correct, the focus of that question should be changed. We should ask why *we* have come to see naturalism as the most important aspect of pragmatism. Pragmatism circa 1930 might, rather, be seen primarily as a naturalist version of a particular subspecies of scientific philosophy. The point of "scientific" in this branch of scientific philosophy is to provide the place from which to reject traditional metaphysics and epistemology. They are rejected as disciplines because they claim to deal with extrascientific realms that are investigated by special philosophical methods. This rejection of a claim to special, expert knowledge in philosophy goes together with a desire to provide conceptual technologies that aid in the scientific development of solutions to social problems. An advocate of a pragmatism so conceived may well find a greater kinship with Carnap's formalist philosophy of science than with Quine's naturalized epistemology. Morris and Carnap shared a technological vision of philosophy that suggested the way this philosophy is offered as a benefit to humanity: Philosophy builds conceptual bridges for unified science.

None of this serves as a direct response to Quine's worries about analytic sentences and a priori knowledge. It suggests, however, that a changing vision of science and its social importance altered the face of scientific philosophy in America from the 1930s to the 1960s. In many ways, analytic philosophy has succeeded in organizing itself in accordance with an informal model of the scientific ethos. Analytic philosophers are narrow specialists engaged in fundamental research. They are just as divorced from social concerns as any such group of specialists.[30] The philosophical and social reasons that motivated Carnap and Neurath to adopt scientific philosophy, and which they shared with at least some of the American pragmatists, have, however, almost wholly disappeared.

Notes

Earlier versions of this essay were given at a Philosophy Department Colloquium at Stanford University and at the Logical Empiricism in North America Conference, hosted by the History of Science Department, Harvard University. I would like to thank my hosts and audiences, especially Lanier Anderson, Richard Creath, Michael Friedman, Peter Galison, Ronald Giere, Peter Godfrey-Smith, Warren Goldfarb, Gary Hardcastle, Don Howard, George Reisch, Thomas Ricketts, Ken Taylor, and Thomas Uebel. Thanks go to Gary Hardcastle also for his close reading of an earlier draft.

1. The following story is short and sketchy, meant more as a bit of tacit "knowledge" among contemporary analytic philosophers than as a summary of anyone's considered opinion. As we shall see, however, it is not wholly removed from remarks to be found explicitly in Giere 1996.

2. One aspect of the standard story that I will pause to mention only now is this: Consciously or not, that story suggests that something un-American happened in philosophy in America after World War II. Kuklick 1985 takes up the question of whether American philosophy ever was, in any significant sense, American. My story is made more plausible when we remember the engagement that philosophers such as Dewey, Clarence I. Lewis, and Josiah Royce had with German philosophy.

3. The central essays by the most significant scholar in the movement to reassess logical empiricism are in Friedman 1999; see also the summary remarks in Richardson 1996b. Some of the primary works in the reappraisal are Cartwright et al. 1996, Cotta 1991, the essays in Giere and Richardson 1996, Richardson 1998, and Uebel 1992, 1996a, as well as in the encyclopedic work, Stadler 1997.

4. For logical empiricism as scientific philosophy, see Galison 1990, 1996; Uebel 1996a, 1996b; and Richardson 1997a, 2002, forthcoming.

5. Most notably, in Quine ([1951] 1980, 1969).

6. When called upon to comment on whether philosophy had lost contact with the people, Quine (1981) in essence responded that it never did have contact with the people and that there was no reason to start now. It is hard to imagine Dewey or Carnap, who objected to the lack of usefulness of traditional philosophy, being happy with that answer. Incidentally, this 1981 essay is one of the few places where Quine employs the term

"scientific philosophy." He does so in a way that covers almost all modern philosophy. Thus, the term does not have the significance for Quine that it does for Carnap and Morris. If Richardson (1997a) is right, Quine is here closer to Martin Heidegger than to Carnap.

7. It is hard not to read Giere as suggesting that the American pragmatists lost nerve somehow, allowing misdirected research, which led, in turn, to wasted youths among philosophers of his generation. Giere and the gang were, when young, rebels *with* a (principle of the common) cause.

8. The phrase is, of course, borrowed from the title of Kitcher 1991. Act 4 seems to be starting; see Stroud 1996 and Friedman 1997.

9. The new realism was announced in Holt et al. 1910, 1912.

10. This was the main theme of Perry 1928.

11. As George Reisch (1995) has uncovered, Frank's Institute for the Unity of Science, founded only in 1949, was given a terminal three-year grant in 1952 by the Rockefeller Foundation, largely because the Unity of Science movement was seen to be filled with old-timers who neither could produce much interesting creative work themselves nor recruit younger scholars to the cause!

12. The term "received view" seems to have become prominent in discussions of logical empiricism with the publication of Suppe 1974. Suppe 1974 is a report of a conference held in 1969. Suppe himself cites Hilary Putnam as the source of the term.

13. Logical empiricism seems not only to have died but to have been killed! The main suspects are Quine, Kuhn, and Ludwig Wittgenstein. Evidence is marshaled for and against each. For examples of the ease with which such metaphors frame the discourse about logical empiricism, see Reisch 1991 and Richardson 1996b. The presumption seems to be that in the life of the mind there are no deaths by natural causes. Thanks to Judy Segal for helping me appreciate the importance of examining the terms in which historical problems and narratives are cast.

14. A fuller treatment of the analytic/synthetic debate from the point of view sketched here is offered in Richardson 1997b.

15. This argumentative line is most clear in the first few paragraphs of Carnap 1956, app. D.

16. This way of framing the argument against Carnap is most explicit in Quine 1963.

17. Carnap rejects Quine's claim explicitly in Carnap 1963.

18. Morris found mediating between Neurath and the University of Chicago Press to be very taxing, given Neurath's ambitions and the press's constraints. My sketch here relies on the detailed work of George Reisch (1994, 1995).

19. His scheme is outlined, for example, in his own *Encyclopedia* article, Morris ([1938a] 1955a.

20. The fire example was given in Morris 1935, 10. These remarks point to a crucial difference between Carnap and Morris. For Carnap, meaning was a semantic notion; for Morris, a semiotic one. That is, for Morris, meaning had syntactic, semantic, and pragmatic moments and, thus, could not take a univocal place within any one branch of semiotics.

21. Morris's vision of "scientific empiricism" as the synthetic *Aufhebung* of empiricism, rationalism, and pragmatism can be found, for example, in Morris ([1938b] 1955b, 63–71.

22. This Carnapian view expresses his distinction between internal and external questions. Carnap 1956, app. A provides the most famous expression of this distinction. The distinction, and its sociopolitical overtones, are already found in Carnap 1934a.

23. Informal remarks on the importance of the scientific attitude can be found in Morris ([1938] 1955b) and Carnap 1928, introduction.

24. That is, a commitment to methodological naturalism is simply a commitment to scientific philosophy. I find "scientific philosophy" a more useful term than "naturalism," and not merely because it was the term actually used by both Morris and Carnap: The contemporary discussion of naturalism frequently conflates two issues, one methodological (philosophy must use the methods of science) and one metaphysical (philosophy may cite only "natural" or causal processes in its explanations). Having made this conflation, the contemporary discussion makes it hard to see Carnap's philosophy as scientific. This is unfortunate, since Carnap insisted that a commitment to a method does not commit one to an ontology. As I use the terms, then, Carnap was a methodological naturalist, but he rejected the whole question of metaphysical naturalism as meaningless. Edmund Husserl, to take another salient case, was a methodological naturalist but a metaphysical antinaturalist. Sharply distinguishing the questions of the "natural" from those of the "scientific" is all to the good. For more on the ubiquity of the term "scientific philosophy" in the period from 1860 to 1940, see Richardson 1997b; the place of scientific philosophy in American philosophy is most fully discussed in Wilson 1990.

25. Lewis's account of the pragmatic a priori is given most fully in Lewis 1929, chaps. 8, 9.

26. Carnap's rejection of metaphysics is well known; his rejection of epistemology less so. For Carnap's views on the matter, see Carnap ([1934b] 1937, sec. 72; 1936). For a diagnosis of what the rejection of epistemology amounts to for Carnap, see Richardson 1996a; 1998, chap. 9.

27. The radical nature of the agenda of scientific philosophy for the left wing of the Vienna Circle is explored in Friedman 1996 and in the literature cited in note 3 above.

28. Quine's recitations of the history of "the empiricist tradition" are legion. Important examples are found in Quine 1969; 1995, chaps. 1, 2.

29. Nor, for Carnap, does knowledge ultimately rest on a passively received barrage of stimulations. The foundation of knowledge is found in the protocol languages, which are the protocol languages due to the decision to use them as protocol languages.

30. Such groups may be not at all divorced from social concerns. That is, the informal vision of science informing the self-conception of analytic philosophy may be radically mistaken about science.

References

Adams, G. P., and W. P. Montague, eds. 1930. *Contemporary American Philosophy*. 2 vols. New York: Russell and Russell.

Carnap, R. 1928. *Der logische Aufbau der Welt*. Berlin: Weltkreis.

———. 1934a. "Theoretische Fragen und Praktische Entscheidungen." *Natur und Geist* 2: 257–60.

———. 1936. "Von Erkenntnistheorie zur Wissenschaftslogik." In *Actes du Congrès Internationale de Philosophie Scientifique, Sorbonne, Paris, 1935*. Vol. 1, *Philosophie Scientifique et empirisme logique*, 36–41. Paris: Hermann and Cie.

———. [1934b] 1937. *The Logical Syntax of Language*. Trans. A. Smeaton. London: Routledge and Kegan Paul.

———. 1950. *The Logical Foundations of Probability*. Chicago: University of Chicago Press.

———. 1956. *Meaning and Necessity*. 2nd ed. Chicago: University of Chicago Press.

———. 1963. "W. V. Quine on Logical Truth." In *The Philosophy of Rudolph Carnap*, ed. P. A. Schilpp, 915–22. La Salle, Ill.: Open Court.

Cartwright, N., J. Cat, L. Fleck, and T. E. Uebel. 1996. *Otto Neurath: Philosophy between Science and Politics*. Cambridge: Cambridge University Press.
Coffa, J. A. 1991. *The Semantic Tradition from Kant to Carnap: To the Vienna Station*. Cambridge: Cambridge University Press.
Creath, R. 1991. "Every Dogma Has Its Day." *Erkenntnis* 35: 347–89.
Dewey, J. 1938. *Logic: A Theory of Inquiry*. New York: Holt.
———. [1920] 1948. *Reconstruction in Philosophy*. Enlarged ed. Boston: Beacon.
Friedman, M. 1996. "Overcoming Metaphysics: Carnap and Heidegger." In *Origins of Logical Empiricism*, ed. R. N. Giere and A. Richardson, 45–79. Minneapolis: University of Minnesota Press.
———. 1997. "Philosophical Naturalism." *Proceedings and Addresses of the American Philosophical Association* 71: 7–21.
———. 1999. *Reconsidering Logical Positivism*. Cambridge: Cambridge University Press.
Galison, P. 1990. "*Aufbau/Bauhaus*: Logical Positivism and Architectural Modernism." *Critical Inquiry* 16: 709–52.
———. 1996. "Constructing Modernism: The Cultural Location of *Aufbau*." In *Origins of Logical Empiricism*, ed. R. N. Giere and A. Richardson, 17–44. Minneapolis: University of Minnesota Press.
Giere, R. N. 1996. "From *Wissenschaftliche Philosophie* to Philosophy of Science." In *Origins of Logical Empiricism*, ed. R. N. Giere and A. Richardson, 335–54. Minneapolis: University of Minnesota Press.
Giere, R. N., and A. Richardson, eds. 1996. *Origins of Logical Empiricism*. Minnesota Studies in the Philosophy of Science, vol. 16. Minneapolis: University of Minnesota Press.
Hempel, C. G. [1945] 1965a. "Studies in the Logic of Confirmation." In C. G. Hempel, *Aspects of Scientific Explanation and Other Essays in the Philosophy of Science*, 3–51. New York: Macmillan.
———. 1965b. *Aspects of Scientific Explanation and Other Essays in the Philosophy of Science*. New York: Macmillan.
Hempel, C. G., and P. Oppenheim. [1948] 1965. "Studies in the Logic of Explanation." In C. G. Hempel, *Aspects of Scientific Explanation and Other Essays in the Philosophy of Science*, 245–95. New York: Macmillan.
Holt, E. B., W. T. Marvin, W. P. Montague, R. B. Perry, W. B. Pitkin, and E. G. Spaulding. 1910. "The Program and Platform of Six Realists." *Journal of Philosophy* 7: 393–401.
———. 1912. *The New Realism*. New York: Macmillan.
Kitcher, P. 1991. "The Naturalists Return." *Philosophical Review* 100: 53–114.
Kuklick, B. 1985. "Does American Philosophy Rest on a Mistake?" In *American Philosophy*, ed. M. G. Singer, 177–89. Cambridge: Cambridge University Press.
Lewis, C. I. 1929. *Mind and the World Order*. New York: Scribner.
———. 1930. "Logic and Pragmatism." In *Contemporary American Philosophy*, ed. G. P. Adams and W. P. Montague, vol. 2: 31–51. New York: Russell and Russell.
Morris, C. W. 1934. "The Relation of the Formal and Empirical Sciences within Scientific Empiricism." *Erkenntnis* 5: 6–14.
———. 1935. "Philosophy of Science and Science of Philosophy." *Philosophy of Science* 2: 271–86.
———. [1938a] 1955a. "Foundations of the Theory of Signs." In *Encyclopedia of Unified Science*, by O. Neurath, R. Carnap, and C. W. Morris, vol. 1: 77–137. Chicago: University of Chicago Press.
———. [1938b] 1955b. "Scientific Empiricism." In *Encyclopedia of Unified Science*, by O. Neurath, R. Carnap, and C. W. Morris, vol. 1: 63–75. Chicago: University of Chicago Press.

Neurath, O., R. Carnap, and C. W. Morris. 1955. *International Encyclopedia of Unified Science*. Vol. 1. Chicago: University of Chicago Press.
Perry, R. B. 1928. "Peace without Victory in Philosophy." *Journal of Philosophical Studies* 3: 300–312.
———. 1930. "Realism in Retrospect." In *Contemporary American Philosophy*, ed. G. P. Adams and W. P. Montague, vol. 2: 188–209. New York: Russell and Russell.
Quine, W. V. O. 1963. "Carnap and Logical Truth." In *The Philosophy of Rudolph Carnap*, ed. P. A. Schilpp, 385–406. La Salle, Ill.: Open Court.
———. 1969. "Epistemology Naturalized." In *Ontological Relativity and Other Essays*, 69–90. New York: Columbia University Press.
———. [1951] 1980. "Two Dogmas of Empiricism." In *From a Logical Point of View*, 2nd ed., 20–46. Cambridge, Mass.: Harvard University Press.
———. 1981. "Has Philosophy Lost Contact with People?" In *Theories and Things*, 190–93. Cambridge, Mass.: Harvard University Press.
———. 1995. *From Stimulus to Science*. Cambridge, Mass.: Harvard University Press.
Reichenbach, H. 1938. *Experience and Prediction*. Chicago: University of Chicago Press.
———. 1939. "Dewey's Theory of Science." In *The Philosophy of John Dewey*, ed. P. A. Schilpp, 157–92. Evanston, Ill.: Northwestern University Press.
———. 1949. *The Theory of Probability*. Berkeley and Los Angeles: University of California Press.
Reisch, G. 1991. "Did Kuhn Kill Logical Empiricism?" *Philosophy of Science* 58: 264–77.
———. 1994. "Planning Science: Otto Neurath and the *International Encyclopedia of Unified Science*." *British Journal of the History of Science* 27: 153–75.
———. 1995. "A History of *The International Encyclopedia of Unified Science*." Ph.D. diss. University of Chicago.
Richardson, A. 1996a. "From Epistemology to the Logic of Science: Carnap's Philosophy of Empirical Knowledge in the 1930s." In *Origins of Logical Empiricism*, ed. R. N. Giere and A. Richardson, 309–32. Minneapolis: University of Minnesota Press.
———. 1996b. "Introduction: Origins of Logical Empiricism." In *Origins of Logical Empiricism*, ed. R. N. Giere and A. Richardson, 1–13. Minneapolis: University of Minnesota Press.
———. 1997a. "Toward a History of Scientific Philosophy." *Perspectives on Science* 5: 418–51.
———. 1997b. "Two Dogmas about Logical Empiricism: Carnap and Quine on Logic, Epistemology, and Empiricism." *Philosophical Topics* 25: 145–68.
———. 1998. *Carnap's Construction of the World*. Cambridge: Cambridge University Press.
———. 2002. "Philosophy as Science: The Modernist Agenda of Philosophy of Science, 1900–1950." In *In the Scope of Logic, Methodology, and Philosophy of Science*, ed. P. Gärdenfors, J. Wolenski, and K. Kijania-Placek, vol. 2, 621–39. Dordrecht: Kluwer Academic Publishers.
———. Forthcoming. "The Scientific World Conception: Logical Empiricism, 1914–1945." In *The Cambridge History of Philosophy, 1870–1945*, ed. T. Baldwin. Cambridge: Cambridge University Press.
Schilpp, P. A. 1939. *The Philosophy of John Dewey*. Evanston, Ill.: Northwestern University Press.
———. 1963. *The Philosophy of Rudolf Carnap*. The Library of Living Philosophers, vol. 11. La Salle, Ill.: Open Court.
Stadler, F. 1997. *Studien zum Wiener Kreis: Ursprung, Entwicklung, und Wirkung des Logischen Empirismus im Kontext*. Frankfurt: Suhrkamp.

Stroud, B. 1996. "The Charm of Naturalism." *Proceedings and Addresses of the American Philosophical Association* 70: 43–55.

Suppe, F. 1974. *The Structure of Scientific Theories*. Urbana: University of Illinois Press.

Uebel, T. 1992. *Overcoming Logical Positivism from Within*. Amsterdam: Rodopi.

———. 1996a. "Anti-foundationalism and the Vienna Circle's Revolution in Philosophy." *British Journal for the Philosophy of Science* 47: 415–40.

———. 1996b. "The Enlightenment Ambition of Epistemic Utopianism: Otto Neurath's Theory of Science in Historical Perspective." In *Origins of Logical Empiricism,* ed. R. N. Giere and A. Richardson, 91–112. Minneapolis: University of Minnesota Press.

Wilson, D. J. 1990. *Science, Community, and the Transformation of American Philosophy, 1860–1930*. Chicago: University of Chicago Press.

Don Howard

2
Two Left Turns Make a Right: On the Curious Political Career of North American Philosophy of Science at Midcentury

When the philosophy of science as we know it today was first established, chiefly in German-speaking Europe in the 1920s and 1930s, many, if not most of its founders were motivated in large part by explicit social and political concerns, the dominant political orientation of those founders lying along a rather narrow spectrum of opinion, somewhere between Enlightenment liberalism and Marxist socialism. The companion movement in North America, at the center of which was the image of science crafted by John Dewey, was characterized by a liberal, social democratic political orientation that tended in a direction similar to that of its Viennese contemporaries, even if it found expression in a political vocabulary more attuned to North American habits of thought. For many of these thinkers, the association between one's politics and one's interest in the philosophy of science was not accidental. On the contrary, a specific image of science was often promoted in the service of political ends, and those political ends in turn frequently determined, to some extent, the image of science that one promoted.

Let me say as clearly as possible right here at the start that I am not asserting that the *content* of logical empiricist or pragmatist philosophy of science was straightforwardly determined by political ideology. To claim that would be crudely to oversimplify a much more interestingly complicated relationship. On the other hand, to deny categorically any influence of ideology on content would be to beg one of the historical questions at issue here. I shall not take up in this chapter the larger question of whether or not the content of one's philosophy of science should be understood as a social construction.[1] But I do want to assert that there was rather more politics in prewar philosophy of science than our contemporary image of the discipline usually acknowledges. More detail is given below, but for now just remember, for example, that while seemingly nonpolitical philosophical constructions such as Martin Heidegger's "Das Nichts nichtet" ["The nothing nothings"] represented one species of metaphysics targeted for elimination by means of logical analysis and the verifiability criterion of meaningfulness (Carnap 1932a, 229–32), metaphysical abstractions, such as *Volksgeist,* that served reactionary political interests were equally or even more a target for socialist logical empiricists like Otto Neurath, who grew up in a world where the association of Ernst Mach and Karl Marx was a commonplace of Austro-Marxist political ideology (O. Neurath 1930b, 120–21).

Social and political concerns continued to find explicit expression in the writings of émigré logical empiricists and homegrown pragmatist philosophers of science in North America from the time of the mid-1930s European emigration, through World War II, and well into the 1950s. The dominant political orientation began to shift more in the direction of liberalism during World War II and the early years of the Cold War, with many philosophers of science now championing an empiricist conception of science as perhaps the most important weapon in the battle to defend freedom and liberal democracy against perceived political foes of many different stripes, including fascism, communism, and even neo-Thomism. But while the political center of gravity among philosophers of science shifted a bit to the right, there was as yet no significant lessening of the conviction that the philosophy of science could and should serve political ends.

By the early 1960s, however, the situation had changed dramatically. Within a period of little more than ten years, starting in the early 1950s and with 1959 being, as we shall see, a watershed year, the expression of social and political concerns rapidly disappeared from the philosophy of science literature. Understanding how and why this change came about, how and why the explicit social and political concerns of the founder generation were read out of the philosophy of science during the process of its institutionalization in North America in the 1950s, is the aim of this chapter.

There is both irony and tragedy in this shift to an apolitical philosophy of science. The irony lies in the fact that the explicit championing of an empiricist philosophy of science in the service of liberal democracy played a major role in making possible the institutionalization of the philosophy of science in North America in the 1950s. The creation of academic departments in the philosophy of science, the establishment of privately endowed centers for the philosophy of science, the organization of a section for the history, philosophy, and sociology of science within the National Science Foundation—all of these causes were advanced by the philosophy of sciences being enlisted in the ideological battles of the day.

The tragedy lies in the fact that when the discipline of the philosophy of science disengaged itself from politics, there appeared a left-liberal political vacuum that was to be filled by other philosophical movements, much to the detriment of the left-liberal political cause, in my opinion.

By the late 1960s, any sense of the social and political relevance of the philosophy of science had largely disappeared, however much individual philosophers of science might have continued to see themselves as politically engaged individuals. In the 1950s, the locus of political interest and activity in philosophy was already beginning to shift in a Continental direction, Sartrean existential phenomenology being a major lure. Then, especially after 1968, the banner was taken up by social and political philosophers newly energized by the post-1968 revival of critical theory and Marxist humanism, and politically active, engagé students followed this trend. Herbert Marcuse and Jürgen Habermas, not Mach and Neurath, were now seen as the thinkers who held the key to making a better world.[2]

Today, debates over the place of science in society constitute one of the main battlegrounds in a highly politicized culture war. To me, it is astounding and dismaying that almost nowhere in this culture war does one hear the loud, clear voice of a credible, left-liberal, empiricist, naturalist philosophy of science, all the more so when one recalls the social and political concerns that motivated Neurath and Dewey in the 1930s or even Philipp Frank and Sidney Hook in the 1950s. The lack of an articulate, politically engaged empiricist philosophy of science in our day makes it possible for Paul Gross and Norman Levitt, authors of the polemical assault on critical science studies, *Higher Superstition* (1994), to collect together, under an invented category that they dub the "academic left," a diverse group of science critics—ranging from feminist philosophers of science, such as Sandra Harding, to postmodern theorists, such as Jacques Derrida—and then simplemindedly to deride the whole as representing "a rejection of the strongest heritage of the Enlightenment," claiming that among its members "irrationality is courted and proclaimed with pride" (Gross and Levitt 1994, 3). There is much to dispute in the analysis offered by Gross and Levitt, starting with their unsympathetic caricatures of what is in many cases very serious, critical scholarship on science. But for now I would only observe that they would have had a much harder time faulting an "academic left" that included Neurath, Dewey, and Frank for "irrationality" and a betrayal of the Enlightenment heritage. Such thinkers could hardly be styled "science bashers," and the continuing representation of their views in the community of left-wing critics of science would have the dual salubrious effects of making criticism of such new perspectives as feminist philosophy of science and social constructivism more difficult, thanks to the clearer articulation of their intellectual links to science-friendly naturalisms of the Neurathian and Deweyan varieties, and of making more sophisticated the debates growing out of the critical analyses put forward by these new theorists. I would love to see Barry Barnes and David Bloor, Simon Schaffer and Steven Shapin, Steve Woolgar and

Bruno Latour arguing not only with Larry Laudan but also with Neurath and Dewey.[3]

My major aim in this chapter is thus, again, to understand the shift to a socially disengaged, depoliticized philosophy of science in the 1950s. This change will be seen to have involved many factors, both internal philosophical ones and external institutional and political ones, ranging from the reconceptualization of "scientific philosophy" in a North American setting and the demise of naturalism and pragmatism after the death of Dewey in 1952 to the political tensions of the McCarthy period and the new patterns of funding for the philosophy of science in the GI Bill era of expansion in higher education and the post-*Sputnik* era of research funding. After sorting through these many aspects of the depoliticization of the philosophy of science, I want to ask whether some kind of repoliticization is either desirable or achievable.

First, however, I want simply to recall the extent to which a sense of social, cultural, and political mission infused the philosophy of science through the middle of the 1950s, for the purpose of better appreciating how different a field it became in later years.

Political and Social Issues in Pre-Emigration Logical Empiricism and Scientific Philosophy

The full story of the transition from the more socially and politically engaged prewar philosophy of science to the more disengaged philosophy of science of the later 1950s and beyond requires attention both to the logical empiricism and scientific philosophy that emerged chiefly in German-speaking Europe in the 1920s and 1930s and to the pragmatist tradition in the philosophy of science that developed in North America during those same decades. The history of pragmatism from a general political and cultural point of view has received a good deal of attention from scholars,[4] but the history of pragmatist philosophy of science has yet to be written.[5] By contrast, there is already a considerable literature on the history of logical empiricism and scientific philosophy,[6] and of late quite a lot of scholarly attention has been devoted to the social-political context in which they developed,[7] giving us a fairly clear picture of the political engagements and ambitions of the major figures associated with those movements in the decades of the 1920s and 1930s.

There is a long history of association between left-liberal political leanings and the empiricist and neo-Kantian philosophical sympathies that stand behind the rise of logical empiricism and scientific philosophy in German-speaking Europe in the late nineteenth and early twentieth centuries. In Vienna, in particular, this association was very strong.

One need not look far for telling examples of the association between sometimes radical, indeed, sometimes violent, left-wing political sympathies and philosophical sympathies of a positivist stripe. Surely the most notorious case is that of Friedrich Adler, son of Victor Adler, who was one of the founders of the Austrian Social Democratic Party. Friedrich Adler was the translator of Pierre Duhem's *La Théorie physique: Son objet et sa structure* (Duhem 1906, 1908), and was well enough known as a propagandist on behalf of Mach (F. Adler 1908, 1909, 1918) to draw the attention of Lenin, who devoted some paragraphs to an attack on Adler in his *Materialism and Empirio-Criticism* (see, for example, Lenin 1909, 46). In 1916, the younger Adler assassinated the Austrian prime minister, Count Sturgkh. Prominent intellectuals, such as Adler's student friend Albert Einstein, offered their support at the time of his trial (see Einstein to Adler, May 13, 1917, AE 6-023). Adler was sentenced to death, but the sentence was first postponed, reportedly as a result of his stirring indictment of wartime tyranny during his trial, and then he was pardoned by the Emperor Charles when the socialists took over at the end of World War I (Johnston 1972, 101; Ardelt 1984, 43; Blum 1985, 150–51, 202–5). Adler did not himself exert a profound or lasting influence on the development of logical empiricism in Vienna, partly because, truth be told, he was not that good a philosopher. But his highly visible role in the Austrian Social Democratic Party and his widely known championing of a nondialectical, Machian Marxism as a distinctive Austrian brand of Marxism played a crucial role in defining the local political and intellectual environment in which people like Neurath worked.

One could adduce still further examples of association between empiricism or positivism and socialism, including that of Mach himself, who was well known to be sympathetic to socialism though he was not very actively involved in politics beyond his support for workers' rights and workers' education (see F. Adler 1918, 27). More actively political than Mach was his contemporary, Eugen Dühring, whose many writings on the history and philosophy of science (Dühring 1873a, 1878) were complemented by his writings on social, political, and economic questions (Dühring 1871, 1873b). Dühring was a non-Marxist socialist whose scientific materialism influenced such thinkers and activists as Eduard Bernstein but drew the critical attention of Friedrich Engels in his *Anti-Dühring* (1878) because of its deviation from more orthodox dialectical materialism. Mention should also be made of Franz Mehring, Josef and Eugen Dietzgen, and the very original Viennese utopian social philosopher, Josef Popper Lynkeus (see Johnston 1972, 308–11), who inspired many people, including Mach,

Einstein, and Neurath. Depending upon how one conceives the intellectual origins of logical empiricism, one might also point to the tradition of neo-Kantians who expressed sympathy for socialism, ranging from Friedrich Albert Lange, Hermann Cohen, and Paul Natorp to Friedrich Adler's fellow Austrian Social Democrat (but not a relative), Max Adler (see M. Adler 1925).[8]

The association between empiricism and socialism is no accident. To my knowledge, no one has yet written a comprehensive history or analysis of this phenomenon, but some of the features of a full explanation are clear. At least since the time of the *Materialismusstreit* and the revolutions of 1848, materialism of one kind or another was a centerpiece of socialist theory (see Wittich 1971). What kind of socialism one represented was determined in no small measure by one's understanding of how science was related to the material world. Thus, being a socialist required having a philosophy of science. Relevant also is the fact that nascent, empiricist, nineteenth-century philosophy of science lived, for the most part, at the periphery of the academy. Even in Vienna in the 1920s and 1930s, empiricism and the philosophy of science occupied at best a precarious position at the University of Vienna, which was otherwise dominated by politically conservative, religiously oriented philosophical tendencies (see Stadler 1995). To this very day, the philosophy of science and analytic philosophy have been unable to establish themselves securely in German and Austrian universities owing to their politically liberal reputation (see Gadol 1982a; Dahms 2000; Fischer 2000; Heiss 2000). Whether this "outsider" cultural position of empiricist philosophers of science was in part a consequence of their socialist sympathies or a contributing factor in the cultivation of such sympathies is hard to say. But it is noteworthy that some of the loudest voices criticizing materialism, empiricism, and positivism after the middle of the nineteenth century were those of figures like Hermann von Helmholtz and Max Planck, who were securely established at the center of the German scientific establishment.

The example of Friedrich Adler points up a crucial feature of the Austrian and more broadly central-European political-philosophical context, namely, the blend of Marxism and Machian positivism that was the distinguishing trait of Austro-Marxism (see Blum 1985). That Viennese logical empiricism grew up in an atmosphere suffused with Machian Austro-Marxism is important for two reasons. First, it meant that Machian positivism and its descendants were almost universally understood to be liberal and progressive in their political implications. Empirical science was naturally seen as opposing the rationalist and a priorist ideologies that legitimated traditional claims to authority by a church and a state that were viewed as serving only a narrow class interest, especially when such science extended itself from physics and biology into the study of social and economic relations. Second, it meant that the socialism of Austrian social democrats, whether Marxists or non-Marxists, was universally understood to be compatible with objective, empirical natural science.

Given this background, it is not surprising that the dominant political orientation among the members of the Vienna Circle and allied groups such as the Gesellschaft für empirische Philosophie centered around Hans Reichenbach in Berlin was left-liberal. Opinion ranged from the rather quiescent liberalism of Moritz Schlick, which was nevertheless radical enough in the context of mid-1930s Austria to make him an object of suspicion in the eyes of the reactionary government of Engelbert Dolfuss, to the activist socialism of Neurath. Largely because of the accident of who survived to speak for logical empiricism and scientific philosophy after the war, we tend to remember logical empiricism as an apolitical philosophical movement. We assume that thinkers who argued for the cognitive meaninglessness of normative discourse could not, qua philosophers, take a political stance. Nothing could be further from the truth.

In the political context of 1920s and 1930s Vienna, Berlin, and Prague, an antimetaphysical scientific philosophy was perceived by its conservative opponents and was understood by its champions to pose a threat to a social and political order that depended upon metaphysical abstractions like *Volksgeist* for its legitimation (see Richardson 2000; Stadler 2000). One might pose as the defender of allegedly politically neutral science, but however sincere one's intention, one did so knowing full well that this pose gave one at least a rhetorical advantage, since one could argue that it was not mere personal political prejudices but *objective scientific criteria* that cut the ground from beneath conservative and reactionary social and economic theories. In this way, even the apolitical theoretical stance of someone like Schlick functioned politically in an environment where objective science was seen as a threat to resurgent reactionary political interests. It would have required extraordinary naiveté to think otherwise.

Some, however, were more explicit about their politics. Neurath was, without question, the most explicitly political member of the Vienna Circle. His political involvements were such as to lead to his being put on trial in 1919 with the leaders of the revolutionary Munich Räterepublik, which he served as president of the Central Economic Office, and to his having to flee Vienna after the conservative Dolfuss government put down a brief workers' uprising in Vienna in February 1934 and thereupon banned all communist and social democratic organizations and institutions, including the Social and Economic Museum, which Neurath had directed since his founding of it in 1925 (see P. Neurath 1994; Cartwright et al. 1996), and the Verein Ernst Mach, which Neurath had cofounded along with Rudolf Carnap, Hans

Hahn, and others in 1928 (Stadler 1982, 171–73, 175–205; Stadler 1997, 364–70). Neurath was never active in the electoral politics of the Austrian Social Democratic Party, but he was one of the most prominent intellectual figures associated with the party, well known as an early advocate of socialist economic planning and a program that he termed *"Vollsozialisierung"* (O. Neurath 1920), and he was deeply involved in many of the institutions of Austrian social democracy, including the very important adult education programs promoted by the party (see P. Neurath 1994; Stadler 1982, part 2) and the mentioned Social and Economic Museum, in connection with which Neurath developed his influential method of "ISOTYPE" for the graphical presentation of statistical information.

Like Neurath, Hahn made no secret of his politics. Among other things, for several years he was a leader of the Vereinigung sozialistischer Hochschullehrer (Stadler 1995, 53). But even most of those who did not draw an explicit connection between their politics and the philosophy of science were nevertheless politically of a cast of mind similar to Neurath's.

For example, at least early in his life, Hans Reichenbach stood nearly as far to the left as Neurath. A non-Marxist socialist, he was a leader of the progressive Freie Studentenschaft in Berlin before World War I. In the revolutionary days after Germany's defeat in November 1918, Reichenbach became the chair of the Sozialistische Studentenpartei Berlin, drafting its platform (H. Reichenbach 1918a) and various articles on socialism and education, including "Socializing the University" (H. Reichenbach 1918b) and "Student und Sozialismus" ("The Student and Socialism") (H. Reichenbach 1919).[9] His reputation for political activism was still strong enough in the mid-1920s to be an obstacle to his obtaining a hoped-for appointment in philosophy at the University of Berlin, which is what led Einstein and Planck to champion the creation of a new position for him in the physics department (see Hecht and Hoffmann 1982; Hoffmann 1993).

Not all of the members of the founder generation of logical empiricism and scientific philosophy were as radical in their politics as Neurath and the young Reichenbach. Schlick's political moderation is often remarked upon in the memoirs of others. A revealing episode is Schlick's effort to persuade the Dolfuss government not to dissolve the Verein Ernst Mach in February 1934: He argued that it was an "absolutely unpolitical" organization, even though it might have included a number of social democrats among its members and on its governing board. In a letter of March 2, 1934, to a "Hofrat Ganz" at the Bundespolizeidirektion, Schlick wrote,

> Personally, I would never allow myself to made head of an association that pursued political ends. In order to preserve the independence of my views as a philosopher under all circumstances, I have never in my life been a mem-

ber of a party, least of all the social democratic party. I am totally against all partisan political activity, and I must confess that I would be sick at heart if the association of which I am head would be dissolved precisely because of such activity. (As quoted in Stadler 1982, 198)

Neurath and Carnap are both reported to have opposed Schlick's argument (Stadler 1982, 199), no doubt because Neurath, especially, valued the explicitly intended role of the Verein Ernst Mach in the larger context of social democratic politics in Vienna—the social democratic workers' education movement and the socialist-dominated Freidenkerbund had particularly close ties to the members of the Verein Ernst Mach—and had long promoted the idea that the antimetaphysical scientific worldview championed by the Verein Ernst Mach was valuable because it furthered a socialist reorganization of the social and economic order along Marxist lines, as Neurath had argued in one of his own lectures to the Verein in 1930 on "Einheitswissenschaft und Marxismus" ("Unity of Science and Marxism") (O. Neurath 1930a; see also Stadler 1982, 184).

It was also in 1934, one year after Hitler's *Machtergreifung* and the beginning of the intellectual emigration, that Reichenbach declared a similar apolitical stance, this time on behalf of the journal *Erkenntnis*. The occasion was a published attack on the Vienna Circle and *Erkenntnis* by Hugo Dingler, who had long been famous for his conservative defense of Euclidean geometry against refutation by general relativity. Writing during the anti-Semitic, antirelativity campaign of the early 1920s, Dingler had argued that Euclidean geometry could not be refuted by means of measurements carried out with "Euclidean" instruments built in "Euclidean" machine shops, and he exploited a conventionalist strategy to reconcile observational evidence, such as Arthur Eddington's eclipse observations, with Euclidean geometry (see, for example, Dingler 1919, 1920, 1921).[10] At that time, Reichenbach had published what most regarded as a definitive critique of Dingler's arguments (H. Reichenbach 1921). But Dingler was not silenced. In 1933, he returned to the attack in a book on the foundations of geometry, in the foreword to which he wrote of

> the purely formalistic-computational mode of thinking that regards calculation not as a broadly useful auxiliary instrument but as the thing itself, as the absolute (Einstein, the so-called Vienna Circle, the Gesellschaft für wissenschaftliche Philosophie in Berlin, the circle around the journal "Erkenntnis" published by Felix Meiner as well as the journal *Die Naturwissenschaften* edited by Arnold Berliner) and that evinces such a strong analogy to the senseless absolutizing of organizational forms in political Bolshevism (as well as in socialist and personal directions).

He further declared that he would demonstrate the "complete barrenness and hollowness" of this mode of thinking, by contrast with which "truly productive human activity *[schaffenden Tun]* and creative human thinking *[schöpferishen Denken]*" would be given its complete due (Dingler 1933, as quoted in H. Reichenbach 1934, 75). By singling out all the prominent individuals and institutions associated with left-liberal tendencies in the philosophy of science, Dingler was obviously trying to curry favor with the new Nazi regime, in hopes of finally winning the kind of prestigious academic appointment he had long sought and finally got with a call to Munich in 1940 (Wolters 1987, 261–62).

Reichenbach's reply first points out the inanity of Dingler's charge that logical empiricism and scientific philosophy turn calculation into some kind of "Absolute." But then Reichenbach turns to the charge of bolshevism and socialism:

> Someone's trying to attack our scientific orientation by associating it with political bolshevism is a way of proceeding that all respectable scientists *[wissenschaftliche Anstand]* should oppose, above all when it concerns a journal that has demonstrated its political independence through the purely objectively technical *[sachliche]* choice of its contributors *[Mitarbeiter]*. (H. Reichenbach 1934, 76)

As Reichenbach notes, even Dingler had published in *Erkenntnis* (Dingler 1931).[11] As with Schlick's assertion of the apolitical nature of the Verein Ernst Mach, Reichenbach's insistence on the political neutrality of *Erkenntnis* might be justified from the point of view of explicit editorial policy and practice,[12] but it was, to say the least, a bit disingenuous from the point of view of the way he, Carnap, and especially Neurath and the other more avowedly political members of the Vienna Circle, such as Frank and Hahn, understood the larger social and political role played by scientific philosophy.

In the politically threatening situation of 1934, it would obviously have been expedient for Schlick and Reichenbach to proclaim an apolitical stance on behalf of the institutions of scientific philosophy. But there is something more subtle and more interesting going on here than a mere trimming of one's sails in perilous political weather, as there would be again in the 1950s in North America during the McCarthy period. For there were interesting differences of opinion among the members of the Vienna Circle as regards the place of politics in philosophy, and for some members of the Circle the adoption of an apolitical stance, or rather the celebration of the image of objective science, was, in fact, part of the political agenda of Viennese and Berlin philosophy of science.[13]

A good source for understanding this issue of political stance is Rudolf

Carnap, who was himself, like Reichenbach, a non-Marxist socialist but also someone who, again like Reichenbach, did not display his politics explicitly in his philosophical work. Carnap devotes several pages of his "Intellectual Autobiography" to the subject of the social and political engagements of the Vienna Circle "because very little has been written about this in earlier publications" (1963, 22). Because Carnap's remarks constitute a rare, honest, first-person recollection of these issues by one of the major figures in the Vienna Circle, I shall quote from them at length.

Carnap begins by noting that one of Neurath's important contributions to the life of the Vienna Circle "consisted in his frequent remarks on the social and historical conditions for the development of philosophical conceptions." Neurath criticized "the customary view . . . that a widespread acceptance of a philosophical doctrine depends chiefly on its truth" (22), stressing that certain historical and social situations were "favorable" to some philosophical views and "unfavorable" to others.[14] The time was ripe, in particular, for the acceptance of scientific philosophy:

> He shared our hopeful belief that the scientific way of thinking in philosophy would grow stronger in our era. But he emphasized that this belief is to be based, not simply on the correctness of the scientific way of thinking, but rather on the historical fact that the Western world at the present time . . . will be compelled for economic reasons to industrialize more and more. Therefore, in his view, on the one hand the psychological need for theological or metaphysical ways of thinking will decrease, and on the other hand the cultivation of the natural sciences will be strongly increased because they are needed by the technology of industrialization. Consequently the general cultural atmosphere will become more favorable toward the scientific way of thinking. (1963, 22)

How did the other members of the Vienna Circle react to Neurath's position? Carnap explains:

> Up to this point Neurath did not find much opposition. But he went further and often presented arguments of a more pragmatic-political rather than of a theoretical nature for the desirability or undesirability of certain logical or empirical investigations. All of us in the Circle were strongly interested in social and political progress. Most of us, myself included, were socialists. But we liked to keep our philosophical work separated from our political aims. In our view, logic, including applied logic, and the theory of knowledge, the analysis of language, and the methodology of science, are, like science itself, neutral with respect to practical aims, whether they are moral aims for the individual, or political aims for a society. Neurath criticized strongly this neutralist attitude, which in his opinion gave aid and comfort

to the enemies of social progress. We in turn insisted that the intrusion of practical and especially of political points of view would violate the purity of philosophical methods. (1963, 22–23)

One place where this difference in attitude made itself felt was in Neurath's arguments for a physicalist language as the framework for unified science. For Neurath, a major argument in favor of physicalism was that it provided an antidote to the then widely emphasized distinction between the *Naturwissenschaften* and the *Geisteswissenschaften,* a distinction that Neurath regarded as an obstacle to social and political progress, in part because of its reliance on what he saw as an obscure and possibly obscurantist notion of *Verstehen*. Carnap, who by the early 1930s had come to share Neurath's preference for a physicalist protocol language, comments as follows:

> We conceded that the acceptance of a physicalistic language might possibly have a positive correlation with social progress. But we thought it advisable to disregard this fact in our investigations, so as to avoid any prejudice in examining the possibility of a unified physicalistic language. Neurath rejected these doubts and warnings. He would deride those purist philosophers who sit on their icy glaciers and are afraid they might dirty their hands if they were to come down and tackle the practical problems of the world. (1963, 23)

Am I wrong to hear, especially in that last paraphrase of Neurath's views, at least some tension or ambivalence on Carnap's part about his own neutralist stance? The Carnap who, in the politically troubled mid-1950s, traveled to Mexico in a public gesture of support for jailed Communist philosophers[15] and who made no secret of his support for the farm workers in southern California surely does not fit the description of a philosopher sitting on his "icy glacier" refusing to dirty his hands by tackling the problems of this world. As Carnap makes clear, there was a rather large area of agreement between himself and Neurath:

> Of particular importance for me personally was his emphasis on the connection between our philosophical activity and the great historical process going on in the world: Philosophy leads to an improvement in scientific ways of thinking and thereby to a better understanding of all that is going on in the world, both in nature and in society; this understanding in turn serves to improve human life. (1963, 23–24)

It would appear that the "series of private discussions" with Carnap "and some younger members of the Circle," in which Neurath "explained the basic ideas of Marxism and showed their relevance to a better understanding of the sociological function of philosophy" (Carnap 1963, 24), left their mark on Carnap.

What did Neurath himself have to say about the question of the relation between politics and the philosophy of science?[16] Begin with the specific question raised by Carnap about the political neutrality of scientific and philosophical investigations. Neurath's position was quite clear and was often repeated in print: The promotion of objective science best served the interests of the proletariat. Here is how Neurath formulated the idea in his 1928 book, *Lebensgestaltung und Klassenkampf (Personal Life and Class Struggle)*:

> Nothing would be further amiss than to think that a Marxist-minded representative of the proletarian class struggle would respect only such scientific work which relates directly to the strategy of the class struggle. It is precisely Marxism that uncovers indirect relations and detours, and thus might ascertain that cultivating pure logic and the most general problems of mathematics and physics is especially favorable to revolutionary thinking. The Marxist will tend to regard it not as a mere accident that among the representatives of just these abstract disciplines ordinarily thought to be impractical, there are so many socialists as well as bourgeois in opposition, as for instance, the English logician and mathematician Bertrand Russell or the German physicist Albert Einstein. A cultivation of this kind of scientific thought seems almost a form of dissolution of metaphysical and half-theological thought, which under many disguises and masks is more alive today among the bourgeoisie than two generations ago. This is quite understandable, for bourgeois groups are closing ranks against the proletariat which has no traditions, and they must make their peace with the powers of yesterday, above all with clerical groups. . . . The cultivation of scientific, unmetaphysical thought, its application above all to social occurrences, is quite Marxist. (295)

Neurath went on to conclude a few pages later: "For the proletarian front, the technique of the struggle and the interests of propaganda coincide with high esteem for science and the overcoming of metaphysics" (297).

It must be remembered that it was this conception of the political role of science and scientific philosophy that Neurath had in mind when he composed the famous 1929 manifesto of the Vienna Circle, *Wissenschaftliche Weltauffassung: Der Wiener Kreis (The Scientific Conception of the World: The Vienna Circle)*. Recall the penultimate paragraph of this manifesto:

> The increase of metaphysical and theologizing leanings which shows itself today in many associations and sects, in books and journals, in talks and university lectures, seems to be based on the fierce social and economic

> struggles of the present: one group of combatants, holding fast to traditional social forms, cultivates traditional attitudes of metaphysics and theology whose content has long since been superseded; while the other group, especially in central Europe, faces modern times, rejects these views and takes its stand on the ground of empirical science. This development is connected with that of the modern process of production, which is becoming ever more rigorously mechanized and leaves ever less room for metaphysical ideas. It is also connected with the disappointment of broad masses of people with the attitude of those who preach traditional metaphysical and theological doctrines. So it is that in many countries the masses now reject these doctrines much more consciously than ever before, and along with their socialist attitudes tend to lean towards a down-to-earth empiricist view. In previous times, *materialism* was the expression of this view; meanwhile, however, modern empiricism has shed a number of inadequacies and has taken a strong shape in the *scientific world-conception*. (Neurath, Hahn, and Carnap 1929, 317)

That the language and ideas follow so closely Neurath's own earlier formulations is not surprising, since, as we know, Neurath drafted the text of the manifesto. But Hahn and Carnap edited the text, and many other members of the Circle had a hand in it as well (see editorial note 2 in Neurath 1973, 318; and Stadler 1997, 370–75).[17]

Three points should be made here. First, in a typically Marxist fashion, Neurath regarded Marxism as a preeminently *scientific* and objective way of understanding the world. Indeed, Neurath thought that he had gone Marx one better in this respect by substituting physicalism for materialism, physicalism being a more thoroughly scientific framework for Marxism because it avoided the metaphysics that still lurked in Marx's conception of the material:

> It is important to emphasize that *the scientific conception of the world with its physicalism as well as its unified science stand entirely on the side of "objectivism."* . . . Physicalism is the modern expression of this point of view which takes the place of the earlier "materialism" or "realism." (1931, 416–17)

Second, and again in typically Marxist fashion, Neurath believed that science and philosophy were part of the "ideological superstructure" determined at its basis by the mode of production (1930a, 75), but he saw in this no conflict with Marxism's claim to be objective science. The objectivity of bourgeois science may be questionable, but that of proletarian science is not, because of the unique social and economic position of the proletariat as an oppressed class. Here is how Neurath explains a point of view that is a first cousin to what today goes by the name of standpoint theory:

> The workers who lack a rich bourgeois education, can become superior to the bourgeois precisely in the field of social life in that they have a greater understanding for social connections and can apply even a smaller amount of knowledge more significantly. Marxism shows the proletarians who are engaged in the class struggle what is especially important to know; and it preserves adherents from the often disorganized educational endeavour of bourgeois enlightenment, which from the outset sees in merely increasing knowledge something worth striving for as such. (1928, 292–93)

A few pages later he concludes,

> Marxism makes it understandable why the bourgeoisie, conditioned by its class position, becomes ever more unscientific in the field of social theory.... To many bourgeois it may seem degrading, and an infringement of the dignity which is conceded to science, if one looks at it from the point of view of the class struggle. The proletariat appreciates science properly only as a means of struggle and propaganda in the service of socialist humanity. Many who came from the bourgeoisie are worried whether the proletariat will have some feeling for science; but what does history teach us? It is precisely the proletariat that is the bearer of science without metaphysics. (297)

Finally, Neurath believed that Marxism could claim the status of objective science in the service of the proletariat only by virtue of what would today be called its "reflexivity," its turning of the methods of critical science upon itself: "And so, Marxism, by applying its own method to itself, announces to the proletarian front that it has become the carrier of the scientific attitude" (297).

Neurath had an easier time than did more orthodox Marxists in explaining how a proletarian science that is part of an ideological superstructure could nevertheless be objective, this thanks to his holistic conception of theories and of the relation of theories to evidence and the associated thesis that considerations of logic and evidence alone underdetermine theory choice. For where considerations of logic and experience leave off, extralogical and extra-empirical factors may come into play, including, perhaps, social and political factors.

If anything, Neurath was even more radical in his holism and underdeterminationism than was Duhem, because Neurath believed that the holism and underdetermination went all the way down, as it were, there not even being a wholly determinate set of observational protocols. As Neurath argued so forcefully in the context of the crucial protocol-sentence debate of the early 1930s (Carnap 1932b; O. Neurath 1932; Carnap 1933; Schlick

1934; O. Neurath 1934), in the course of which he introduced the now-famous "Neurath's boat" metaphor (O. Neurath 1932, 92), even protocol sentences are subject to revision if that is the simplest way of restoring consistency between observation and theory:

> Science is *ambiguous—and is so on each level*. When we have removed the contradictory groups of statements, there still remain several groups of statements with differing protocol statements that are equally applicable; that are without contradictions in themselves but exclude each other. Poincaré, Duhem and others have adequately shown that even if we have agreed on the protocol statements, there is an unlimited number of equally applicable, possible systems of hypotheses. We have extended this tenet of the uncertainty of systems of hypotheses to all statements, including protocol statements that are alterable in principle. (O. Neurath 1934, 105 [translation corrected]; see also O. Neurath 1932, 94–95)

Protocols are subject to revision along with all of the other statements making up the total body of scientific theory because all comparison of theory with experience is just a comparison of one kind of statement with another kind of statement, and no report of experience that can be formulated propositionally—no matter how "basic" the experience might appear—can be distinguished qua statement from other statements with respect to its susceptibility to revision. There are no self-warranting observation reports (O. Neurath 1934, 111–14).

The corrigibility of protocols is one of Neurath's main arguments against Carnap's earlier preference for a phenomenalistic protocol language (see Carnap 1928b, 61–67) and against Schlick's conception of *Konstatierungen* (Schlick 1934), because phenomenalist protocols (Carnap's *"Elementarerlebnisse"* ["elementary experiences"]) and *Konstatierungen* (Schlick's "here-now-blue") invite the illusion of epistemic determinacy and certainty. Of course, that foundation in the seeming certainty of immediate experience is what Schlick and Carnap wanted. Indeed, it is what they thought a responsible, verificationist empiricism demanded, until Carnap, at least, was convinced otherwise by Neurath's critique. But Neurath's empiricism was of a different kind: "Thus for us striving after knowledge of reality is reduced to striving to establish agreement between the statements of science and as many protocol statements as possible. But this is very much; in this rests *empiricism*" (O. Neurath 1934, 109).

Neurath's antifoundationalist empiricism was also opposed to what Neurath dubbed the "pseudorationalism" of Karl Popper's falsificationist program (O. Neurath 1935b), since like all varieties of pseudorationalism, falsificationism indulges in the illusion that observation somehow yields a determinacy in theory choice, whereas Neurath's holism entails the essential underdetermination of theory choice by observation. As Neurath says, "*multiplicity* and *uncertainty* are essential" (1935a, 116); he adds that "multiplicity and uncertainty exist in all science . . . there is no *tabula rasa* for us that we could use as a safe foundation on which to heap layers upon layers. *The whole of science is basically* always under discussion" (118).

Especially when one views Neurath's holism and underdeterminationism in opposition to Carnap's earlier verificationism and Popper's falsificationism, one might wonder how, indeed, Neurath can maintain that science, on his view, is objective, all the more so when one takes the full measure of his insistence that the underdetermination goes all the way down to the level of the protocols. If the "whole of science," including the protocols, is "always under discussion," does it not follow that anything goes in science, that we have fallen into the worst kind of radical relativist coherentism?

That was precisely Schlick's charge against Neurath in the context of the protocol-sentence debate, at the very time in early 1934 when the Marxist Neurath had just been driven from "Red Vienna" by the Dolfuss government and when Schlick was trying to hold the Vienna Circle together in an ever more hostile political environment. Give up phenomenalism, give up the incorrigibility of protocols, reasoned Schlick, and one gives up the bedrock of certainty in experience that empiricism requires. Give up such strict control over theory choice by experience, and one is left with "only grounds of expediency" for choosing one among a set of empirically equivalent theories, whereas, according to Schlick, the purpose of science is "to provide a *true* account of the facts" (Schlick 1934, 374).

There is considerable irony in Schlick's charging Neurath with relativism and abandoning the aim of truth by virtue of his Duhemian underdeterminationist view of the relation between theory and experience, for in his earlier years Schlick was famous for having defended a view of the nature of truth according to which such empirically equivalent alternative theories are all equally true! The fundamental idea, as Schlick explained in his seminal 1910 paper on the nature of truth, is that concepts and propositions are nothing more than signs correlated with facts, from which it follows that truth is nothing more nor less than the unambiguous coordination of signs with facts (Schlick 1910). But that being the case, more than one theory can be true. Here is a typical formulation of the argument:

> The totality of our scientific propositions, in word and formula, is in fact nothing else but a system of symbols *correlated* to the facts of reality; and that is equally certain, whether we declare reality to be a transcendent being or merely the totality and interconnection of the immediately "given." The system of symbols is called "true" however, if the correlation is completely

unambiguous. Certain features of this symbol system are left to our arbitrary choice; we can select them in this way or that without damaging the unambiguous character of the correlation. It is therefore no contradiction, but lies, rather, in the nature of the matter, that under certain circumstances, several theories may be true at the same time, in that they achieve indeed a different, but each for itself completely unambiguous designation of the facts. (Schlick 1915, 149)

Schlick believed that one's choice among these alternative true theories was driven mainly by considerations of simplicity. Nevertheless, he conceded that there was an ineradicable element of convention in theory choice (Schlick 1917, 267). Starting in the early 1920s, when faced with the need to formulate a version of empiricism that was, by his lights, adequate to reply to neo-Kantian critiques of general relativity, Schlick gradually moved away from this position in the direction of the stricter verificationism he defended at the time of the protocol-sentence debate.[18] But Schlick's "multiple truths" theory of truth was so much a part of Vienna Circle philosophy of science that Neurath would have thought it quite natural to understand the objectivity of a science underdetermined by experience in such terms. Moreover, he must have been saddened, if not surprised, by Schlick's charge of relativism, since it was Schlick himself who had taught his Viennese colleagues that underdeterminationism entails no compromise with the aim of truth.

If, as Neurath says, multiplicity and uncertainty are essential, how do communities actually make a choice of one among a set of empirically equivalent theories, a choice underdetermined by logic and evidence?[19] Simplicity considerations obviously play some role, as they did for other conventionalists, such as Poincaré and the younger Schlick (O. Neurath 1934, 105). But for Neurath, that is by no means the whole story. What interests him more are the social and political determinants of theory choice. What, however, are those factors, and how do they make themselves relevant? And why are we so often unaware that we are, thus, making an underdetermined choice?

Underdetermination is the norm, both at the level of abstract theory and at the level of protocols:

This is how matters stand in every *"layer"* of scientific work, not only in the narrower sphere of systems of hypotheses, as Poincaré and Duhem have pointed out with such intensity. But these initiatives in multiplicity are constricted by life. A whole human lifetime is hardly long enough to immerse oneself in even a single view and to give full thought to its consequences. And how soon one senses the weakening effect of isolation. Thus one deserts the lonely, though perhaps auspicious, notions of an outsider to join in the work in a way of thought that enjoys more support and has therefore better chances of greater scientific achievement. In such ways it happens that not even too many possibilities are treated by several groups at the same time: through adaptation and selection a kind of assimilation of whole generations takes place—not to speak of the cases in which certain trains of thought are anathema, persecuted and suppressed. (O. Neurath 1935a, 117)

Not everyone is eager to accept this view of science, since it does not conform to the more traditional progressivist and accumulationist view of the history of science:

> This insight that a logically tenable multiplicity is reduced by life has little hope of response because it contradicts the usual view of a connection between achievement and "success." The representatives of a victorious doctrine are too much inclined to believe that their victory could be justified as it were by closer logical investigation. Many see the course of the history of science like that. (O. Neurath 1935a, 117)

But once one understands how external circumstance constrains theory choice, one can get on with the doing of science, assured that the integrity of science is in no way thereby compromised: "If, in spite of these comments on multiplicity and uncertainty, one sets unswervingly to the work that is seen as a common one, one can do so only because one knows how much the historical situation reduces the manifoldness *via facti*" (O. Neurath 1935a, 119).[20]

How external circumstance constrains theory choice is something for the historian, sociologist, psychologist, and even the evolutionary biologist to investigate. Herein lies perhaps the most important moment of Neurath's epistemological naturalism, a topic to which we will turn in a moment. But let us first return to the question of how objective science can further the interests of the proletariat, or to turn the question around, how a proletarian science can still be an objective science.

According to Neurath, theory choice is underdetermined by logic and evidence, with social influences of various kinds providing the explanation for how communities make a choice among empirically equivalent theories and for why we are often unaware of the community's thus making a choice. But there is no logical reason why social influence need work unconsciously or otherwise hidden from view. There is no logical reason why the community cannot consciously and deliberately make its choices for social and political reasons. Moreover, a community's allowing social and political considerations to play a conscious and deliberate role in theory choice—within the domain of empirical underdetermination—entails no compromise whatsoever with the objectivity of science. If it is a fact about science that extralogical, extra-empirical factors do play a role in theory choice, why is a science that is aware of and deliberately directing those social influences toward desirable social and political ends any less objective than a science in which those influences are at work but hidden from view? Indeed, one might argue that science is less objective to the extent that it allows the role of social influence to remain obscure, that a science unaware of the social influences at work in theory choice is more susceptible to being used to support reactionary social and political aims.

Neurath's preference is for the conscious and deliberate direction of theory choice in the service of progressive social and political aims. His blunt expression of this view is a bit shocking:

> The Marxist, as strict scientist, must admit that the course of history allows of various interpretations. But successful collaboration is possible only when those who act fix on one possibility, whether by agreement or propaganda. This choice is itself a matter of action and resolution, but that does not mean that such action has no scientific basis. (O. Neurath 1928, 293)

"Whether by agreement or propaganda"! One cannot accuse Neurath of being either reticent or disingenuous. Shocking though Neurath's blunt description of politically directed theory choice might be, we should not lose sight of the crucial feature of this view for our purposes, which is that even if the choice is a matter of propaganda, the objectivity of science is in no way compromised, as long as politically directed choice is constrained to operate within the domain of underdetermination. But at the same time, one must bear in mind the fact that, for Neurath, the domain of underdetermination is quite broad, given his inclusion of the physicalistic protocol statements within the sets of statements choice among which is underdetermined by logic and evidence.

Pierre Duhem, from whom Neurath learned his underdeterminationism, was not a socialist but, rather, a Catholic reactionary and critic of the French Third Republic's secular liberalism (see Jaki 1984; Martin 1991).[21] Nevertheless, he loomed large in the thinking of quite a few of the prominent thinkers who lived at the intersection of scientific philosophy and social democratic politics in Vienna, including not only Neurath but also Philipp Frank, Hans Hahn, and Richard von Mises from the philosophical side (see Frank 1949 and Haller 1985) and Otto Bauer and Gustav Eckstein from the political side (see Ardelt 1984, 298, n. 33). Of course, Friedrich Adler, who, like Neurath, was equally at home in both of these worlds, was the translator of the 1908 German edition of Duhem's *La Théorie physique: Son objet et sa structure* (1906).[22]

Part of the explanation for Duhem's popularity in Vienna was that he was widely seen at the time as being Mach's ally in the development of an antimetaphysical "new positivism" (see Frank 1949, 25–28). Today, we see Duhem's holism as being antithetical to the epistemological atomism implicit in Mach's phenomenalism, but that was not how their relationship was understood at the time. Mach himself endorsed Duhem's thesis of the theory-ladenness of observation (Mach 1906, 202, n. 3) as well as his holism and his critique of the concept of the *experimentum crucis* (Mach 1906, 244, n. 1; see also Howard 1990). Austro-Marxists would thus have seen the instrumentalist philosophies of science of Mach and Duhem as pointing in the direction of a more sophisticated alternative to the crude mechanism, materialism, and realism of Engels and Lenin, an alternative such as Neurath's physicalism. But it may well be that Duhem was also popular because his underdeterminationism could be put to use (as Neurath did put it to use), for the purpose of reconciling objectivity with political engagement in the image of science dear to scientific philosophers and Austro-Marxists alike.[23] We will find Frank making just this argument in the very different context of North American philosophy of science in the years right after World War II.

I mentioned above that naturalism was another feature of Neurath's theory of science, inasmuch as he holds that questions about the role of social influence in theory choice are questions of history and sociology (when they are not questions of practical, political action). This aspect of Neurath's view was agreed upon not only by all of the left wing of the Vienna Circle but even by less political types like Carnap, for recall that Carnap said of Neurath's view that the acceptance of a doctrine depended not only upon its truth but also upon the historical and social situation: "Up to this point Neurath did not find much opposition" (Carnap 1963, 22).[24]

Neurath's naturalism is important for two reasons. First, it helps us to understand why another, contemporary approach to the study of science with Marxist roots, namely, Karl Mannheim's program in the sociology of knowledge (Mannheim 1929), need not be viewed by thinkers of Neurath's persuasion as impugning the objectivity of science.[25] For the point is, again, that the domain of social influence is confined to the domain of underdetermination by logic and evidence. Second, naturalism and the image of a socially and politically engaged science constitute two of the main bridges between logical empiricism and the American pragmatism of Dewey that welcomed the refugees from Vienna, Prague, and Berlin upon their arrival in the United States in the 1930s.[26]

By highlighting the naturalism and political engagement of left-wing Viennese logical empiricism I mean, quite deliberately, to be stressing a facet of logical empiricism not normally emphasized in the histories of the movement that were written by those who came to represent logical empiricism and its legacy in North America in the 1950s and beyond. One of the main points that I want to make is that the neglect of this facet of logical empiricism in the history of the movement that came to be accepted in North America after the 1950s is, itself, a crucial part of the process whereby the social and political engagements of Viennese philosophy of science were read out of the discipline during the course of its institutionalization in North America in the 1950s. Whether this was a deliberate misrepresentation of the history of logical empiricism for the purpose of making possible the institutionalization of the philosophy science at a time of "the end of ideology" (Bell 1960) or whether it happened because the social and political aims of the left wing of the Vienna Circle simply seemed irrelevant in a very different cultural and political climate is something I cannot judge. That the history was written as it was, however, is a fact that cannot be ignored.

Naturalism and the image of a socially and politically engaged science could be reconciled with the objectivity of science in Neurath's fairly well worked out physicalist version of logical empiricism. But while these features of left-wing logical empiricism formed, as I claim, the bridge to scientific philosophy's new, North American home, the lack of a comparably detailed and persuasive theory of science ultimately made it impossible for pragmatism to manage the tensions inherent in the image of a politically engaged objective science.

Pragmatism and the North American Reception of Logical Empiricism and Scientific Philosophy

Pragmatism and positivism are too often compared by asking whether the pragmatic criterion of truth and the verificationist criterion of meaningfulness have anything in common. But even enthusiastic early mediators between the two traditions understood the difference between a theory of truth and a theory of meaning, and they saw the valorization of a scientific approach to life as the more important common element.

Perhaps the most influential of these mediators is Charles W. Morris. A product of the Chicago school of Dewey and George Herbert Mead, Morris spent his first sabbatical year, 1934–35, studying in Vienna and Prague with the remaining members of the Vienna Circle. He was later to be instrumental in bringing both Carnap and Reichenbach to the United States, and he became coeditor with Neurath and Carnap of the *International Encyclopedia of Unified Science*.[27] Morris's reaction to his initial encounter with logical empiricism is instructive.

In a series of papers written during and immediately after his year abroad,

Morris set out to explore the differences between pragmatism and logical empiricism and to effect a rapprochement between them (Morris 1935a, 1935b, 1936, 1937a). Given that he was later to become Neurath's close collaborator in editing the *International Encyclopedia of Unified Science,* it is ironic that his chief complaint against logical empiricism concerns its neglect of the social dimensions of science, both the social context within which science is done and the social implications of science.[28] An emphasis on the social dimensions of science was the centerpiece of Dewey's theory of science, as it was of Neurath's. Had Neurath not been forced to flee Vienna a few months before Morris's arrival, Morris might have formed a different picture of the dominant tendencies within the Vienna Circle. As it was, he found Schlick and Friedrich Waismann in a Vienna where the Social Democratic Party was now outlawed, and Carnap and Frank in Prague.

Morris's own most famous philosophical contribution to the joint movement that he dubbed "scientific empiricism" (Morris 1935a, 21) was his urging that a comprehensive semiotics include a third, pragmatic component of meaning alongside the syntactic and semantic components already studied by logical empiricists. But his reasons for urging this addition were not purely technical. A purely formal theory of meaning, one that neglected the pragmatic dimension, would not serve in a complete theory of science because it could not solve the problem of theory choice. Commenting on the shift from earlier verificationism to the purely linguistic conventionalism of Carnap's *Philosophy and Logical Syntax* (Carnap 1935), Morris wrote,

> The implication of this shift for the problem of truth is obvious: it involves passing from a correspondence view to a form of coherence view, i.e., a true proposition is simply one compatible with or unifiable with the accepted propositions of a science. . . . This shift is the price paid for neglecting other aspects of meaning than the formal when certain difficulties in the earlier empirical formulations of meaning were encountered. (Morris 1935a, 9, n. 4)

The neglected aspect of meaning is pragmatic meaning, consideration of which is Morris's way of taking account of the social dimension of science. Consideration of pragmatic meaning in the context of a Deweyan, pragmatist theory of science thus fills the same void that Neurath would fill by consideration of the social dimensions of empirically underdetermined theory choice.

As with Dewey and Mead, Morris's approach to the study of science is a species of naturalism, inasmuch as he takes semiotics, the framework within which a theory of science is to be constructed, to be an empirical science (see Morris 1935b). Also like Dewey, Morris emphasized a role for science in reshaping social values. In his very first paper on pragmatism and positivism he wrote,

> Dewey has been peculiarly sensitive to the instrumental relation of symbols to the life of the individual and the community. He has envisaged intelligence as a tool in the service of some value, and science as co-ordinated and institutionalized intelligence. . . . Here is science crowned by and ministering to social vision. Dewey has written: "One of the few experiments in the attachment of emotion to ends that mankind had not tried is that of devotion, so intense as to be religious, to intelligence as a force in social action." (Morris 1935a, 14–15, quoting from Dewey 1934, 79)

Dewey was, of course, not a Marxist like Neurath but a liberal social democrat (see Ryan 1995). Nevertheless, the image of science promoted by Morris on behalf of Deweyan pragmatism shared with Neurath's theory of science both a commitment to science as an agent of directed social change and the naturalist's belief that science is a part of culture and is therefore itself subject to study by empirical, scientific means.

One crucial place where the pragmatist theory of science differed from Neurath's version of logical empiricism was in its explication and defense of the objectivity of science. For Neurath, objectivity is secured by the fact that social influence and even conscious political choice works within the domain of underdetermination, leading to a choice of one among many empirically equivalent and equally true theories. For Morris and Dewey, objectivity is intersubjectivity, and scientific propositions are distinguished from mere subjective beliefs because they are those propositions that command the assent of the entire relevant community. A thinker like Quine might some years later succeed in incorporating Dewey's pragmatist conception of meaning and intersubjectivity within Neurath's underdeterminationist physicalism in a consistent and persuasive picture of objective science,[29] but the vague notion of intersubjectivity alone was not up to the task of fending off those who saw intersubjectivity as just an especially subtle rationalization for relativism.

Dewey's influence in American philosophy and American public intellectual life more generally was enormous when logical empiricism landed on American shores in the 1930s. Indigenous American philosophy of science was largely Deweyan in character, although several other influences cannot be overlooked, such as the critical realism of Roy Wood Sellars and Arthur O. Lovejoy, the operationalism of Percy W. Bridgman, and the scientific rationalism of Morris Raphael Cohen.[30] Special mention must be made of the influential mid-1930s textbook by Cohen and the young Ernest Nagel, *An Introduction to Logic and Scientific Method* (Cohen and

Nagel 1934). But their differences notwithstanding, all of these thinkers tended to agree on certain basic ideas about science, such as a biological-evolutionary naturalism and positivism in the old, Enlightenment sense of trusting only positive, scientific knowledge as a safeguard against dogmatic metaphysics.

There was, however, disagreement among indigenous North American philosophers of science in the crucial area of science and values, with some, like Sellars, defending a rather stricter separation of fact and value than Dewey was wont to do (see, for example, Sellars 1932, 475–77). Dewey's intention in blurring the distinction was to urge the extension of scientific method into domains traditionally regarded as the reserve of value. Ostensibly, the method was that of objective, politically neutral science. But Dewey was convinced that scientific method thus applied would lead to a betterment of the human condition by promoting an extension of social control over economic arrangements, such economic democracy being understood by Dewey as a necessary condition for genuine individual liberty. In this process, however, science itself as organized intelligence must undergo a transformation.

Prior to the rise of liberalism, science was conceived as a species of "Reason," which is to say, as "a remote majestic power that discloses ultimate truths" (Dewey 1935a, 20). The rise of liberalism went hand in hand with the rise of the traditional empiricism of thinkers like John Stuart Mill, but in the twentieth century, liberalism faced a "crisis" as it developed from its older, Benthamite form to a radical liberalism adequate to the challenge of extending social control over the means of production. Science was again implicated in this change. The science of the nineteenth century was still "immature" (Dewey 1948, xxv). That "partial and incomplete" science was perhaps adequate for the physical and biological domains, but it was not developed to the point where it could address human and social concerns (Dewey 1948, xxviii). As Dewey says,

> The conception of intelligence as something that arose from the association of isolated elements, sensations and feelings, left no room for far-reaching experiments in construction of a new social order. It was definitely hostile to everything like collective social planning. . . . The crisis of liberalism is connected with failure to develop and lay hold of an adequate conception of intelligence integrated with social movements and a factor in giving them direction. (Dewey 1935a, 43–44)

This point has special bearing on the social sciences:

> Social and historical inquiry is in fact a part of the social process itself, not something outside of it. The consequence of not perceiving this fact was that

the conclusions of the social sciences were not made (and still are not made in any large measure) integral members of a program of social action. (45)

This did not happen because nineteenth-century science was science under capitalism:

> The application of science, to a considerable degree, even its own growth, has been conditioned by the system to which the name of capitalism is given, a rough designation of a complex of political and legal arrangements centering around a particular mode of economic relations. Because of the conditioning of science and technology by this setting, the second and humanly most important part of Bacon's prediction has so far largely missed realization. The conquest of natural energies has not accrued to the betterment of the common human estate in anything like the degree he anticipated. (75)

The implication is that only a science resituated in the context of social control of the means of production, a context of economic democracy, can achieve this Baconian aim of the betterment of the human condition. It is by this latter standard, and not the traditional philosopher's notion of detached truth, that the success of science, as of any application of organized intelligence, is judged.

Science thus transformed still respects the claims of experience, but Dewey's theory of science is not an empiricism in which experience is some fixed datum exerting unidirectional control over theory choice. For at the center of Dewey's entire philosophical project is a radically new, enriched conception of experience itself. Commenting on the passive, tabula rasa associationism of the Hume-Mill tradition in empiricism, Dewey remarks, "Either natural sciences do not have the intimate dependence on experience that the enthusiasts on this subject think they have, or experience is a different sort of thing from what it had been analyzed as being either by the classic conception or by the eighteenth-century conception" (Dewey 1935b, 71; see also Dewey 1929b). Dewey's biological-evolutionary naturalism leads him to view experience as something other than the residue of an external world writing itself through sensation upon passive subjective consciousness. Experience is, instead, an activity of the organism, a process leading from a starting point to an end point. Such experience is not opposed to reason, just as facts are not opposed to ideas; these dualisms have no place in a biological view of experience (see Dewey 1916, 230–49). Reason and experience, fact and idea are mutually implicated in one another:

> The true "stuff" of experience is recognized to be adaptive courses of action, habits, active functions, connections of doing and undergoing; sensori-motor

co-ordinations. Experience carries principles of connection and organization within itself. . . . This organization intrinsic to life renders unnecessary a super-natural and super-empirical synthesis. It affords the basis and material for a positive evolution of intelligence as an organizing factor within experience. (Dewey 1948, 91)

Therefore, science as organized intelligence neither stands above experience, as an older rationalism might have held (Dewey 1948, 95), nor is it simply subordinated to experience, as an older empiricism might claim. Both rationalism and crude empiricism fail to take adequate account of experience as action:

> If we exclude acting upon the idea, no conceivable amount or kind of intellectualistic procedure can confirm or refute an idea, or throw any light upon its validity. How does the non-pragmatic view consider that verification takes place? Does it suppose that we first look a long while at the facts and then a long time at the idea, until by some magical process the degree and kind of their agreement become visible? Unless there is some such conception as this, what conception of agreement is possible except the experimental or practical one? And if it be admitted that verification involves action, how can that action be relevant to the truth of an idea, unless the idea is itself already relevant to the action? . . . The self-rectification of intellectual content through acting upon it in good faith is the "absolute" of knowledge, loyalty to which is the religion of intellect. (Dewey 1916, 241)

Perhaps the most important consequence of viewing experience as action is that the object of experience, and hence the object of knowledge, far from controlling that knowledge, is changed by its being the object of the activity of knowing:

> The organs, instrumentalities and operations of knowing are inside nature, not outside. Hence they are changes of what previously existed: the object of knowledge is a constructed, existentially produced, object. The shock to the traditional notion that knowledge is perfect in the degree in which it grasps or beholds without change some thing previously complete in itself is tremendous. (Dewey 1929a, 211)

Dewey's naturalism regards the human as both a biological and a social being, which entails that "social . . . organization enters into the formation of human experience" (Dewey 1948, 91). Dewey rejects the epistemological individualism of the Cartesian tradition. Ideas are not the private possessions of individual knowing subjects, because ideas live only through the expression they find in action, including communicative action, which is inherently social (see Dewey 1929b, chap. 5). For Dewey, objectivity is thus not the end point of an artificial construction starting from within the solipsistic predicament of individual subjective consciousness. Instead, intersubjectivity is the starting point, with the consequence that knowledge is through and through something social.

Naturalism also entails that values are not transcendent and that science does not aim simply for true pictures of an external reality bereft of value. Instead, all valuation is something intrinsic to experiencing and to the nature of which that experiencing is part, inasmuch as the ongoing selection of ends is part of experiencing as action. Science, as organized intelligence, is thus an instrument both for relating means to ends and for selecting those ends in the first place.[31] Moreover, valuation as the selection of ends is as much an integral part of the doing of science as it is of any action. One of many consequences of this integration of value in experience as part of nature is that there can be no distinction between judgments of fact and judgments of value: "Evaluations as judgments of practice are not a particular kind of judgment in the sense that they can be put over against other kinds, but are an inherent phase of judgment itself" (Dewey 1938, 179). For all these reasons, science, as organized intelligence in experience, "is inherently an instrument of critically determining what is good and bad in the way of acceptance and rejection" (Dewey 1929b, 423), and to the extent that knowledge and experience are something social, the ends thus selected will be social ends. The most important end that Dewey wants this new science to pursue is socialism:

> The only form of enduring social organization that is now possible is one in which the new forces of productivity are coöperatively controlled and used in the interest of the effective liberty and cultural development of the individuals that constitute society. . . . Organized social planning . . . is now the sole method of social action by which liberalism can realize its professed aims. Such planning demands in turn a new conception and logic of freed intelligence as a social force. (Dewey 1935a, 54–55)

By 1938, the emigration of logical empiricism and scientific philosophy from Europe to North America was nearly complete. Schlick had, of course, been assassinated in Vienna in 1936. Herbert Feigl had been in the United States since 1930, teaching at the University of Iowa since 1931; he moved to the University of Minnesota in 1940. Carnap had been teaching at the University of Chicago since 1936. Reichenbach started teaching at UCLA in 1938. Gustav Bergmann arrived at the University of Iowa in 1938, taking over Feigl's position in 1940. Carl Hempel had been Carnap's assistant at Chicago for the 1937–38 academic year, and he returned permanently to the United States in 1939, teaching briefly at City College in 1940 and then at Queens College until 1948.

It is important to notice, however, that virtually none of the prominent voices of the left wing of the Vienna Circle were included in the group present in the United States in 1938. Hahn had died in 1935. When forced to flee Vienna in 1934, Neurath first moved to the Netherlands and then fled again to England, where he died in 1945. Frank arrived in the United States only in 1938, settling in Cambridge in 1939. Not until 1941 did he secure an academic position, but even then only as a half-time lecturer (with tenure) in physics and mathematics at Harvard, where he continued to teach until his retirement in 1954.[32] Some of Neurath's vision for scientific philosophy found expression in the United States as Carnap and Morris became the coeditors, with Neurath, of the *International Encyclopedia of Unified Science,* which was published by the University of Chicago Press (see Reisch 1995). Then, after World War II, Frank reestablished in the United States the Institute for the Unity of Science, which he and Neurath had first established, with headquarters in the Netherlands in 1938.[33] But as we shall see, Frank did not succeed in establishing in North America the left-wing Vienna Circle version of the philosophy of science.

With the émigrés now established in their new North American home, the time was right for them to position themselves publicly with respect to the homegrown pragmatism that had prepared the ground for their reception. One of the earliest and most illuminating of the studies that sought to do this was Reichenbach's contribution to the Library of Living Philosophers volume on Dewey, "Dewey's Theory of Science" (Reichenbach 1939). In most respects, this essay can only be described as an essay in persuasive misreading. Reichenbach recognizes an affinity with Dewey's pragmatism in their common commitment to a form of empiricism, but otherwise differences loom larger, with Reichenbach faulting Dewey for his alleged nominalism(!) and his "nonrealistic interpretation of scientific concepts." Somehow Reichenbach also manages to convince himself that the analysis and justification of induction are as important for Dewey as they were for Charles Sanders Peirce, and he uses this as an excuse to expound upon his own recently published theory of induction (Reichenbach 1935).

Most importantly, however, Reichenbach recognizes the centrality of the theme of value in Dewey's theory of science:

> In restoring the world of everyday life as the basis of knowledge, Dewey does not only want to establish knowledge in a better and more solid form. What he intends, and perhaps to a greater extent, is establishing the sphere of values, of human desires and aims, on the same basis and in an analogous form as the system of knowledge. If concrete things as immediately experienced are the truly "real" world, if the scientific thing is nothing but an

auxiliary logical construction for better handling of the "real" things, then ethical and esthetical valuations are "real" properties of things as well as are the purely cognitive properties, and it is erroneous to separate valuations as subjective from cognitive properties as objective. In persuasive language and in ever renewed form Dewey insists upon this outcome of his theory, the establishment of which seems to be the motive force in the work of this eminently practical mind, "practical" to be taken in both its implications as "moral" and "directed toward action." (Reichenbach 1939, 162–63)

One is struck by the fact that Reichenbach does not pass judgment on this aspect of Dewey's thinking. He is content to quote Dewey himself on the question of whether or not such a view somehow compromises one's empiricism: "A philosophy of experience may be empirical without either being false to actual experience or being compelled to explain away the values dearest to the heart of man" (Reichenbach 1939, 164, quoting Dewey 1929a, 107).

Reichenbach's silence regarding the place of value in Dewey's theory of science is surprising because only one year earlier, in *Experience and Prediction,* Reichenbach had premiered his distinction between the "context of discovery" and the "context of justification" (Reichenbach 1938, 7, 382). Reichenbach's purpose had been to demarcate the context of justification as the proper domain for rational reconstruction in the epistemology of science, and he relegated all questions of value to the context of discovery along with the considerations of biology, psychology, and sociology that a naturalistic theory of science like Dewey's and Neurath's might deem important.[34] Reichenbach, like Schlick, had not followed Neurath and Carnap in the turn toward holism and underdeterminationism at the time of the protocol-sentence debate. For Reichenbach, the role of convention in science was confined to the conventional stipulation of correspondence rules or bridge principles, tying theoretical terms to observational primitives, the truth or falsity of all empirical propositions thereafter being unambiguously determined by experience. Empirically equivalent theories were to be regarded as mere notational variants for the expression of the same empirical content. For Reichenbach, as for Schlick, therefore, there is no room for social influence in theory choice in an objective science of nature. But even Reichenbach had to admit that the phenomenalism that undergirded the earlier strict verificationism of the Vienna Circle and the associated distinction between cognitively meaningful scientific discourse and cognitively meaningless normative discourse was untenable. The distinction between context of discovery and context of justification was his new way of attempting to preserve the epistemology of science from intrusions from the side of value and social influence.

Ronald Giere has pointed out that the discovery/justification distinction would be useful to exiled philosophers of science seeking to establish a new home in North America: "For the distinction says: Don't think about the fact that I am a German immigrant, or speak with an accent; just consider the validity of my ideas" (Giere 1996, 346). But when one looks ahead to the way in which an expressly politically neutral philosophy of science came to be institutionalized in North America after World War II, one must conclude that something more is going on here.

On a psychological level, Reichenbach and his fellow refugees from fascism must have been relieved to be working in a relatively much more tolerant American intellectual environment. Nothing could have been more welcome than a quiet haven in which to work on induction and the philosophy of physics, and the discovery/justification distinction worked to protect that haven from the intrusion of the political.[35] But at the same time, Reichenbach, the former socialist student leader, was surely astute enough to recognize that on this side of the Atlantic, religious and political conservatives were also unsettled by a science that respected no a priori claims to authority or privilege, even if that conservativism did not normally express itself in violent ways. Just as the pose of political neutrality had conferred at least a rhetorical advantage in the confrontation with reactionary political forces in Europe, so too it should confer the same advantage here. Of course, for Reichenbach this was not just a pose. As noted, his denial of Neurath's holism and underdeterminationism precluded any place for politics in theory choice.

American pragmatists were not yet ready to follow Reichenbach in so strictly separating justification from discovery and fact from value. An essay by C. I. Lewis on "Logical Positivism and Pragmatism" from 1941 makes this clear.[36] The essay provides, not surprisingly, a much more sympathetic and correct account of Dewey's theory of science than Reichenbach had given, and whereas Reichenbach had chosen to report Dewey's view on science and value without passing judgment, Lewis makes clear his own wholehearted sympathy with Dewey and his rejection of positivist distinctions between fact and value:

> The conception that determinations of correctness and incorrectness are subjective, and statement of them merely "expressive," or that they fall exclusively within the province of psychological and sociological description, is inadmissible, because such admission would erase the distinction between valid and invalid, and eventually between truth and untruth.... For the pragmatist, there can be no final division between "normative" and "descriptive." The validity of any standard of correctness has reference to some order of "descriptive facts"; and every determination of fact reflects some

judgment of values and constitutes an imperative for conduct. The validity of cognition itself is inseparable from that final test of it which consists in some valuable result of the action which it serves to guide. Knowledge—so the pragmatist conceives—is for the sake of action; and action is directed to realization of what is valuable. If there should be no valid judgments of value, then action would be pointless or merely capricious, and cognition would be altogether lacking in significance. (Lewis 1941, 111–12)

The distance between Lewis and Dewey, on the one hand, and Reichenbach, on the other hand, was considerable. As Lewis himself wrote, "It is with respect to problems of evaluation and of ethics that the contrast between logical positivism and pragmatism is strongest" (Lewis 1941, 107). But the distance between pragmatism and the left-wing logical empiricism of Neurath and Frank was much less. With Carnap's and Morris's bringing Neurath's *International Encyclopedia of Unified Science* to Chicago and Frank's bringing his and Neurath's International Institute for the Unity of Science to Harvard, one might have expected a pragmatist and left-wing logical empiricist common front to dominate North American philosophy of science. That did not happen, however, thanks partly to transformations brought about by World War II.

The War Years and Their Immediate Aftermath

During and immediately after World War II, science moved to center stage in American public life. This was partly due to the new cultural authority accruing to science as a result of the role played by the atomic bomb, radar, and other new technologies in winning the war for democracy. But science was also credited with playing an ideological role in the battle to preserve freedom and democracy from the threats of fascism and communism. Yet another threat that was frequently cited was resurgent religious fundamentalism and neo-Thomism, this following the debacle of Bertrand Russell's attempted appointment at City College, which was viciously opposed by a variety of religious interests (see Dewey and Kallen 1941).

A steady stream of books appeared with titles like *The Scientific Spirit and Democratic Faith* (Lindemann 1944), *Free Science* (Weaver 1945), *Science for Democracy* (Nathanson 1946), *Science and Freedom* (Bryson 1947), and *Modern Arms and Free Men: A Discussion of the Role of Science in Preserving Democracy* (Bush 1949). The rough consensus that emerged from these discussions was that science as a mode of inquiry and the institutions of science were essential to the survival of liberal democracy. The philosophy of science had a role to play as well, as was spelled out in an unsigned page-one editorial in the January 1949 issue of *Philosophy of Science,* under the title "Philosophy of Science and Liberalism." After noting the presence of both "conservative" and "revolutionary" tendencies within science and after observing that "the conservative physicist abhors the projected vacuum of a sociology of knowledge that threatens to judge modern physics as an aspect of present day class struggles," the anonymous editorialist then stated a role for the philosophy of science as a discipline to play and drew a surprising conclusion about how the discipline should conceive itself:

> This is the dilemma of modern liberalism: how to maintain a freedom for the introduction of revolutionary ideas, and simultaneously to preserve freedom to develop along conservative and well practiced lines. . . . Philosophy of science should play the liberal role within science today. It should enable the conservative to strengthen his position to the utmost, and it should enable the revolutionary to give expression to his conceptions that are not rigidly confirmed within the accepted mode. For this reason, the field and its journal, cannot and should not become "respectable" in the eyes of the "competent experts"; on the other side, it cannot and should not encourage ill-advised attacks on creeds already outworn. Its function is to keep alive the conflicts of general viewpoints that may give rise to a more powerful and fruitful science of the future. For this reason, philosophy of science is *not* professional philosophy, *nor* professional science, which are both in the main conservative in their outlook. Its aim is the liberalization of science and the scientification of liberalism. (*Philosophy of Science* 1949, 1–2)

The formulation is more moderate, but the sentiment is similar to that expressed by the previous editor of *Philosophy of Science,* William M. Malisoff, in a short essay five years earlier, "A Science of the People, by the People, and for the People," where he wrote, "No one will object to the 'dictatorship' of a true science nor to the 'dictatorship' of a true democracy. Neither can interfere with our freedom, since they serve to define it and the control and power derived from knowledge" (Malisoff 1946, 169; see also Malisoff 1944). One should not let one's surprise at finding *editorials* in the journal *Philosophy of Science* interfere with an appreciation of the position being defended.

The philosophy of science is supposed to promote the liberal cause in science, but a necessary condition for its doing so is held to be its *not* becoming a professionalized discipline, because professionalization implies a conservative stance. The journal's editorial policy under Malisoff, who died in November 1947, and his successor, C. West Churchman, who took over in 1951, reflected this view, as the journal welcomed contributions on a wide variety of topics, from many different political perspectives, authored by scientists and nonscientists, philosophers and nonphilosophers, academics and nonacademics alike. (Some might add that the quality varied as widely as the subject matter.) Over the course of the next decade, the journal and the discipline did professionalize themselves as the philosophy of science won for itself a secure institutional home in departments, centers, and programs on North American campuses, as a division within NSF, and elsewhere. The question is, in doing so, did it move to the right politically, as the 1949 editorial would imply it must have?

The Transition in the 1950s

One would have expected pragmatist philosophers of science in the tradition of Dewey to resist a drift to the right of an institutionalized philosophy of science. Dewey, himself, could no longer play a significant role in debates about the place of science in society and politics. He turned ninety in October 1949 and died on June 1, 1952. But many of Dewey's students, colleagues, and friends still occupied positions of considerable public influence. Many of them contributed to a symposium volume, *John Dewey: Philosopher of Science and Freedom,* that was assembled in honor of Dewey's ninetieth birthday and published in 1950.

Horace Kallen's lead article in the volume is typical of many in its expression of the traditional Deweyan faith. Kallen emphasizes the role of knowledge "in the remaking of the world from a jungle to flee from into a home to live in." About this remaking, he comments,

> Especially since the democratic revolution set reason free and confirmed the equal liberty of different human beings everywhere in the world, has this remaking—whose method is science and whose ends and means are the free cooperation of free men on equal terms—broken barrier after barrier until it has become the common working faith of the frontiersmen of the human spirit everywhere. (Kallen 1950, 45)

But a different note is sounded by some of the other contributors, perhaps signaling the first signs of the eclipse of Dewey's vision of a socially engaged and socially responsible science.

Most illuminating is Nagel's contribution, "Dewey's Theory of Natural Science." Far more sympathetic and understanding than Reichenbach's contribution to the 1939 Library of Living Philosophers volume on Dewey, Nagel's essay nevertheless bluntly faults Dewey's philosophy of science for various shortcomings when judged by the standards of the technical philosophy of science that Nagel was himself championing. Chief among these shortcomings is Dewey's failure to attend closely to science itself, by which Nagel means primarily physical science:

> The great William Harvey is reported to have said of Francis Bacon that he wrote about science like a Lord Chancellor. Of Dewey it can be said with equal justice that he writes about natural science like a philosopher, whose understanding of it, however informed, is derived from second-hand sources.... It is indeed curious that a thinker who has devoted so much effort to clarifying the import of science as has Dewey, should exhibit such a singular unconcern for the detailed articulation of physical theory. (Nagel 1950, 247)

Dewey is also faulted for being a "lone wolf in the formulation of his ideas on science." Nagel explains,

> His central views are in close agreement with conceptions that have been developed during the past half-century by eminent physicists concerned with the methodology of their discipline. Nevertheless, though he is obviously familiar with many of these analyses, he does not appear to have been strongly influenced by them, and he cites them only rarely. But what is more to the point, he does not use these specialized and expert studies to the best advantage in his own discussions. (Nagel 1950, 247–48)

Note that these are criticisms not of the content of Dewey's philosophy of science but of his way of being a philosopher of science. What we are witnessing here in these criticisms of Dewey is, precisely, an exercise in the professionalization of the philosophy of science as a discipline, something that was otherwise and elsewhere a major concern of Nagel's.[37]

What does Nagel have to say about Dewey's views on science and value, science and politics, science and society? In stark contrast to the warm endorsement of Dewey's position in Lewis's 1941 essay, Nagel treats the topic only under the heading of "external" reasons for why Dewey's views have failed to win "general assent." Along with the "notorious" difficulties of Dewey's literary style, there is, according to Nagel, the problem that

> his theory of logic is in effect and by implication a serious intellectual threat to social views whose chief support is tradition and authority; and it would be utopian to suppose that antecedent ideological commitments on the part of his readers have not played a role in their evaluation of his conception of science. (Nagel 1950, 246–47)

This remark is followed immediately by the just-quoted criticisms of Dewey's philosophy of science, which are designated as "less external reasons for the hesitations which even those in full sympathy with Dewey's aims and over-all conclusions have experienced with his account of natural science" (Nagel 1950, 247).

Nagel's attitude toward Dewey seems to me to be representative of the general opinion among those promoting the professionalization of institutionalized philosophy of science in the 1950s. When it comes to the specific question of the relation of science to social concerns, one finds occasional expressions of the Enlightenment belief that empirical science is inherently opposed to authoritarianism and dogma of any kind and that the extension of empirical knowledge to new domains virtually always tends to the betterment of humankind. Nagel sees the philosophy of science supporting a liberal political agenda of the kind advocated in the 1949 *Philosophy of Science* editorial, and he follows Dewey in advocating a naturalism that sees the human being as a part of nature and hence capable of being understood in purely scientific, naturalistic terms. But he does not follow Dewey in finding values as a part of nature, and so he cannot follow Dewey in viewing a science of nature as yielding positive conclusions about the future direction of the social and political order. Nagel writes of the philosophy of science,

> It has given vigorous support and expression to an attitude, at once critical and experimental, toward the perennial as well as the current issues of human life; and it has thereby been a champion of the central values of liberal civilization. The basis for a general outlook on the place of man in nature is supplied by detailed knowledge of the structure of things supplied by the special sciences—an outlook that contemporary philosophy of science has helped to articulate and defend. In the perspective of that outlook, the human creature is not an autonomous empire in the vast entanglement of events and forces constituting the human environment. . . . Moreover, in the perspective of that scientifically grounded outlook, human aspirations are expressions of impulses and needs which, whether these be native or acquired, constitute the ultimate point of reference for every justifiable moral judgment. The adequacy of such aspirations must therefore be evaluated in terms of the structures of human capacities and the order of human preferences. Accordingly, though the forces of nature may one day extinguish the human scene, those forces do not define valid human ideals, and they do not provide the measure of human achievement. But an indispensible condition for the just definition and the realization of those ideals is the employment and extension of the method of intelligence embodied in the scientific enterprise. . . . The cultivation of that intellectual temper is a fundamental condition for every liberal civilization. (Nagel 1954a, 307–8)

The scientific way of thinking is necessary for the "definition" and "realization" of human ideals, but since those ideals are not defined by the "forces of nature," science alone cannot determine them.

The growing popularity of such views among professional philosophers of science in the 1950s does not make them any less puzzling as regards the place of intelligence in the moral life. A thoroughgoing naturalism of the kind Nagel and Dewey professed regards the human being as a biological and social being, nothing more.³⁸ If human beings are thus part of nature in every aspect of their being, if they are not in touch with some transcendent realm of value, then all valuation must grow out of human activity in nature; it must be part of that activity. Dewey drew the conclusion that science as organized intelligence in experience had a role to play not just in relating means to ends but also in selecting ends, and not just from the point of view of feasibility. But Nagel hesitates when it comes to locating valuing in nature and so hesitates when it comes to ascribing to science a positive role in the selection of ends. Dewey would have criticized such hesitation as threatening to relinquish a positive role for intelligence in the moral life, leaving the selection of ends as merely a matter of subjective preference, something Dewey regarded as one of the lingering ill effects of the laissez-faire liberalism of the nineteenth century. But by 1954, Dewey was dead and so could not say these things.

One of the few remaining proponents of a positive social role for science and the philosophy of science in the early 1950s was Frank. His position as a lecturer in physics and mathematics at Harvard and his role as head of the Institute for the Unity of Science (which also counted Carnap and Morris among its directors, along with Reichenbach until his death in 1953) gave him some visibility and influence in the community of North American philosophers of science. For years he had defended essentially the same position as Neurath, albeit without quite the same stridently pro-socialist tone. He continued to defend this position throughout the 1950s, being if anything even more explicit than Neurath had been about the role of social and political considerations in theory choice.

Frank's argument, like Neurath's, was that logic and empirical evidence underdetermine theory choice, leaving a certain domain in which social and political considerations can affect our theory choice without impugning the objectivity of science. Here is how Frank stated the case in his last major systematic book on the philosophy of science in 1957: After reviewing several famous historical cases in which extrascientific factors clearly did play a role in theory choice, as in debates over the Copernican and Ptolemaic theories of the planetary system, he writes,

> Scientists and scientifically minded people in general have often been inclined to say that these "nonscientific" influences upon the acceptance of scientific theories are something which "should not" happen; but since they do happen, it is necessary to understand their status within a logical analysis of science. We have learned by a great many examples that the general principles of science are not unambiguously determined by the observed facts. If we add requirements of simplicity and agreement with common sense, the determination becomes narrower, but it does not become unique. We can still require their fitness to support desirable moral and political

doctrines. All these requirements together enter into the determination of a scientific theory. (Frank 1957, 355)

A few pages later he ties the point about social and political influences on theory choice to naturalistic understanding of such influences:

> The validity of a scientific theory cannot be judged unless we ascribe a certain purpose to that theory. The achievement of that purpose depends upon the degree to which the different criteria for the acceptance of a theory are satisfied, agreement with observed facts, simplicity and elegance, agreement with common sense, fitness to support desirable human conduct, etc. Hence, the validity of a theory cannot be judged by "scientific" criteria in the narrower sense: agreement with observations and logical consistency. After application of all these criteria, there remains often a choice among several theories. However, if we mean by "science" not only physical science, but also the sciences of human behavior (psychology and sociology), then we can decide which among several physical theories achieves a certain human purpose in the best way. . . . New lines of research arise for the scientist who wants to achieve a real understanding of his science. We are guided into the wide field which embraces science as a part of human behavior in general. We may speak of a "sociology of science" or of the "humanistic background of science" if we want to give these new fields a frame of reference in our traditional parlance. (Frank 1957, 359)

In his contribuition to the Carnap volume of the Library of Living Philosophers, Frank ties this point of view explicitly to Dewey. Frank's paper is curious in that it reproduces in its entirety the first published Soviet review, by one V. Brushlinsky, of Carnap's classic essay "The Elimination of Metaphysics through Logical Analysis of Language" (Carnap 1932a), this for the purpose of arguing that, without a pragmatics of language of the kind advocated by Morris one cannot really eliminate metaphysics entirely. If one neglects the pragmatic components in theory choice, then one is forced to say that underdetermined theory choice reflects just the whim of the individual scientist. But Frank, like Neurath, regards such "freedom of choice" as yet another species of metaphysics. What is the alternative? Frank writes,

> Not only the Soviet philosophers but also Western philosophers who follow the pragmatism of John Dewey would rather argue: If several choices are possible from the viewpoint of logic and physical epxerience, there are some of them which support desirable social effects and some support undesirable ones. The question whether and why hypotheses or principles like the existence of a god or the non-existence of the external world have desirable social effects can, of course, not be investigated without empirical research in psychology and sociology. (Frank 1963, 164)

Needless to say, Frank's view of the role of social and political factors in theory choice was growing ever more controversial in the 1950s. One occasion for a public airing of the debate was a conference, "The Validation of Scientific Theories," held in Boston in December 1953 in conjunction with the annual meeting of the American Association for the Advancement of Science (AAAS). The conference had an imposing list of cosponsors: Frank's Institute for the Unity of Science, the Philosophy of Science Association (PSA), Section L (History and Philosophy of Science) of the AAAS (PSA was in the habit of meeting every year together with Section L), the National Science Foundation, and the American Academy of Arts and Science, which was the home of the Institute for the Unity of Science.

Frank's paper "The Variety of Reasons for the Acceptance of Scientific Theories" (1953) opens the published conference volume. In it, he states the general argument for social and political factors in theory choice quoted above from his 1957 book (where his statements of 1953 are sometimes repeated verbatim). But in the 1953 statement, he expands a bit upon the actual mechanisms whereby the influence of such factors is mediated:

> The special mechanism by which social powers bring about a tendency to accept or reject a certain theory depends upon the structure of the society within which the scientist operates. It may vary from a mild influence on the scientist by friendly reviews in political or educational dailies to promotion of his book as a best seller, to ostracism as an author and as a person, to loss of his job, or, under some social circumstances, even to imprisonment, torture, and execution. The honest scientist who works hard in his laboratory or computation-room would obviously be inclined to say that all this is nonsense—that his energy should be directed toward finding out whether, say, a certain theory is "true" and that he "should not" pay any attention to the fitness of a theory to serve as an instrument in the fight for educational or political goals. This is certainly the way in which the situation presents itself to most active scientists. However, scientists are also human beings and are definitely inclined toward some moral, religious, or political creed. Those who deny emphatically that there is any connection between scientific theories and religious or political creeds believe in these creeds on the basis of indoctrination that has been provided by organizations such as churches or political parties. This attitude leads to the conception of a "double truth" that is not only logically confusing but morally dangerous. (Frank 1953, 21)[39]

Frank's paper was included with three others in a section entitled "The Acceptance of Scientific Theories." The last of these, by Barrington Moore Jr., takes Frank's view of the role of social factors in theory choice as its starting point and then explores the descriptive question of how such influences work with special reference to the situation of the sciences in the Soviet Union in the late 1940s and early 1950s (Moore 1953).[40] But I want to focus on the other two papers.

The paper immediately following Frank's was by C. West Churchman, then a dean of engineering at Case Institute of Technology, who had taken over as the editor of *Philosophy of Science* in 1951. Churchman asks what a pragmatic theory of induction would look like with the problem of induction conceived, pragmatically, "in terms of the relationship between evidence and decisions" (1953, 26). He distinguishes several such kinds of relationship on the basis of the number of aims and the varieties of uncertainty infecting the decisions. He then sketches three possible approaches to the problem of induction: (1) Science merely summarizes the evidence but does not evaluate possible policies. (2) Science also evaluates policies, but only relative to "given" values of various objectives, meaning that science does not evaluate the objectives. (3) Science itself also assigns values to objectives. Very much in the tradition of Dewey, Churchman declares himself for option (3), a more detailed version of which he had himself developed earlier in his *Theory of Experimental Inference* (1948). Some measure of the sophistication of Churchman's position is provided by Frederick Will's review of Churchman's book, which focuses on Churchman's interesting effort to defend his "experimentalism" against the charge of relativism by arguing that the existence of common purpose stops the relapse into relativism that such naturalistic theories of valuation might otherwise be thought to threaten (Will 1951).

A rather different perspective is found in the paper by Richard Rudner, who would take over from Churchman as editor of *Philosophy of Science* in 1959. Rudner declares himself sympathetic to the view that value judgments play an essential role in science, but in fact Rudner drastically restricts the scope of value judgments in science, locating them only in judgments that the strength of the evidence is sufficient in a specific situation to warrant acceptance or rejection of a hypothesis, given the risk associated with that being the wrong judgment. The examples he gives are telling: A high degree of confirmation is necessary if the hypothesis in question is that "a toxic ingredient of a drug was not present in lethal quantity," because the consequences of a mistake in this situation "are exceedingly grave by our moral standards" (Rudner 1953, 33). Needless to say, we are a long way here from Dewey's and Neurath's arguments on behalf of science in the service of socialism.

One gets the clear impression that this debate over the role of social factors in theory choice played a much larger role than the written record might betray in the evolving self-understanding of the discipline of the philosophy of science during the course of its institutionalization in the 1950s. For at this very time, major changes were afoot in the profession, taking it from a set of assumptions about the social role of the philosophy of science rather like those of Dewey, Neurath, and Frank to a rather different set of assumptions quite antithetical to that pragmatist and left-wing logical empiricist conception of the philosophy of science.

The nature of these changes was hinted at above in connection with Nagel's evaluation of Dewey's theory of science. Still within the framework of a basically naturalistic point of view, and still viewing science and the philosophy of science as allies of liberal democracy, there is nevertheless a gradual retreat from the typically strong claims of Dewey and Neurath about the role of social and political factors in theory choice, along with a retreat from strong claims about the role of values in science and of science in the making of value judgments. This tendency was reinforced with the publication in 1951 of Reichenbach's *The Rise of Scientific Philosophy,* which tells a story that politely ignores the Neurath-Hahn-Frank tradition in scientific philosophy, reaffirms the central importance for scientific philosophy of the distinction between discovery and justification (Reichenbach 1951, 231), and concludes, with respect to science and value, "The modern analysis of knowledge makes a cognitive ethics impossible: knowledge does not include any normative parts and therefore does not lend itself to an interpretation of ethics" (Reichenbach 1951, 277). As with Nagel, this retreat from the realm of value is connected with the rise of scientific philosophy as a professional discipline. The process begins in the nineteenth century:

> On the ground of the new science there arose a new philosophy. This new philosophy began as a by-product of scientific research. The mathematician, the physicist, or the biologist, who wanted to solve technical problems of his science, saw himself unable to find a solution unless he first could answer certain more general, philosophical questions. It was his advantage that he could look for these philosophical answers unburdened by preoccupation with a philosophical system. . . . And thus, carried along by the logic of the problems, he found answers unheard of in the history of philosophy. (Reichenbach 1951, 118–19)

In this century, such scientific philosophy has become a specialization:

> It was not until our generation that a new class of philosophers arose, who were trained in the technique of the sciences, including mathematics, and who concentrated on philosophical analysis. These men saw that a new distribution of work was indispensible, that scientific research does not leave a man time enough to do the work of logical analysis, and that conversely logical analysis demands a concentration which does not leave time for scientific work—a concentration which because of its aiming at clarification

rather than discovery may even impede scientific productivity. The professional philosopher of science is the product of this development. (Reichenbach 1951, 123)

One might add that such concentration on logical analysis also leaves no time for the political action that was so much a part of the "professional" life of Neurath and Dewey.[41]

The best evidence for the changing character of the philosophy of science as a discipline undergoing professionalization and institutionalization is found in the changing editorial policy of the journal *Philosophy of Science*. The journal was founded in 1934 by the biochemist William M. Malisoff, who served as its editor until his death in 1947. Interim arrangements that included Philipp Frank sustained the editorial operation for a few years. Then in 1951, as mentioned, the editorship of the journal was taken on by C. West Churchman, who was broadly sympathetic to the social interests of Frank and Dewey.[42] By 1956, some changes seem already to have been stirring, with Gustav Bergmann, Arthur Burks, Henry Margenau, Ernest Nagel, and Chester Ruddick joining Churchman as editors of the journal. Then, starting with volume 26 in 1959, Richard Rudner became the first editor institutionally identifiable as a professional philosopher of science, and in short order he turned the journal into the publication we know today.

The editorial changes that started in 1959 are stunning, all the more so given that there was no published statement of a change in editorial policy, only a short notice of Rudner's appointment, simultaneous with a reorganization of the governance of the Philosophy of Science Association, the adoption of a new constitution, and the election of Nagel to the newly created post as vice president of the association (Curt Ducasse was president at time). All of these changes resulted from the work of a committee appointed in the fall of 1956 by then-president Margenau to draw up proposals for a reorganization of the association (see Ducasse 1959).

When one compares the contents of the journal *Philosophy of Science* before 1959 and after, one notices the almost total and almost immediate disappearance of at least four genres of literature that had been common in the journal before 1959. Let me briefly describe them.

Essays on Science and Value

This was a theme evidently close to the hearts of both Malisoff and Churchman, as it was to everyone schooled in the Deweyan tradition. Essays in this category include:

- William M. Malisoff, "Virtue and the Scientist," 6 (1939): 127–36.
- R. W. Gerard, "A Biological Basis for Ethics," 9 (1942): 92–120.
- James Feibleman, "The Mythology of Science," 11 (1944): 117–21.
- Emmanuel G. Mesthene, "On the Need for a Scientific Ethics," 14 (1947): 96–101.
- Carl F. Butts, "Science and Social Responsibility," 15 (1948): 100–103.
- Elgin Williams, "Can We Save Science?" 15 (1948): 333–41.
- Reed Bain, "Natural Science and Value Policy," 16 (1949): 182–92.
- George Simpson, "The Scientist—Technician or Moralist?" 17 (1950): 95–108.
- Robert S. Hartman, "Is a Science of Ethics Possible?" 17 (1950): 238–46.
- George Simpson, "Science as Morality," 18 (1951): 132–43.
- D. W. Gottschalk, "Value Science," 19 (1952): 183–92.
- Abraham Edel, "Concept of Value in Contemporary Philosophical Value Theory," 20 (1953): 198–207.
- L. Haworth and J. S. Minas, "Concerning Value Science," 21 (1954): 54–61.
- W. A. Koistro, "Discussion: Moral Judgments and Value Conflict," 22 (1955): 54–57.
- Donald Davidson, J. C. C. McKinsey, and Patrick Suppes, "Outlines of a Formal Theory of Value," 22 (1955): 140–60.
- K. E. Boulding, "Some Contributions of Economics to the General Theory of Value," 23 (1956): 1–14.
- A. Bachem, "Ethics and Esthetics on a Biological Basis," 25 (1958): 157–62.

As late as 1958, the journal published Robert S. Hartman's "Value, Fact, and Science" (25 [1958]: 97–108), a long critical essay review of Everett W. Hall's *Modern Science and Human Values* (Hall 1956). Hartman's main criticism focused on Hall's doubts about the capacity of science to bring us knowledge of value. Hartman holds out hope for the development of a value theory as sophisticated in the knowledge of value as science is in the knowledge of facts.

Explicitly Ideological Essays or Essays Explicitly concerning Matters of Political Ideology

This tradition started in volume 1, issue 1, with Dirk Struik's review ([1934]: 122–23) of Engels's *Dialectics of Nature*. Other articles in this category include:

- A. Emery, "Dialectics versus Mechanics: A Communist Debate on Scientific Method," 2 (1935): 9–38.

- John Somerville, "Soviet Science and Dialectical Materialism," 12 (1945): 23–29.

- John Somerville, "Commentary. Ethics and Social Science: Case History of a Sharp Practice," 14 (1947): 345–47. (This concerns an alleged effort by *The Nation* to censor Somerville's reply to a hostile review by Sidney Hook of Somerville's book *Soviet Philosophy* [Somerville 1946].)

- Lewis S. Feuer, "Dialectical Materialism and Soviet Science," 16 (1949): 105–24.

- John Somerville, "A Key Problem of Current Political Philosophy: The Issue of Force and Violence," 19 (1952): 156–65.

- Edward G. Ballard, "On the Nature and Use of Dialectic," 22 (1955): 205–13.

- Dale Riepe, "Flexible Scientific Naturalism and Dialectical Fundamentalism," 25 (1958): 241–48.

Others will no doubt be as surprised as I was to find the journal publishing as late as 1956 and 1957 articles by Hans Freistadt with titles like "Dialectical Materialism: A Friendly Interpretation" (23 [1956]: 97–110) and "Dialectical Materialism: A Further Discussion" (24 [1957]: 25–40).

Essays on Science Planning, Science Policy, and Related Topics

This too was evidently a major interest of both Malisoff and Churchman. Articles include:

- T. Swann Harding, "The Mass Production of Research," 6 (1939): 98–105.

- Oscar Kaplan, "Concerning the Mass Production of Research," 6 (1939): 374–77.

- T. Swann Harding, "Exploiting the Creators," 8 (1941): 385–90.

- Walter Rautenstrauch, "What Is Science Planning?" 12 (1945): 8–18.

- Jerome Frank, "The Place of the Expert in a Democratic Society," 16 (1949): 3–24.
- Robert K. Merton, "The Role of Applied Social Science in the Formation of Policy: A Memorandum," 16 (1949): 161–81.
- Philip M. Hauser, "Social Science and Social Engineering," 16 (1949): 209–18.
- E. A. Shils, "Social Science and Social Policy," 16 (1949): 219–42.
- Gerard Hindrichs, "Toward a Philosophy of Operations Research," 20 (1953): 59–66.
- Fred H. Blum, "Action Research—A Scientific Approach?" 22 (1955): 1–7.
- Walter E. Cashen, "War Games and Operations Research," 22 (1955): 309–20.

Special mention is reserved for Herbert A. Shepard's "Basic Research and the Social System of Pure Science" (23 [1956]: 48–57), which played an important role in establishing the concept of "basic research" in the context of debates about the funding of scientific research in the 1950s (see Hollinger 1990).

Essays on the Sociology of Knowledge

One is not surprised to find this genre well represented in a journal still feeling the influence of Dewey's conception of science. Articles include:

- Sidney Ratner, "Evolution and the Rise of the Scientific Spirit in America," 3 (1936): 104–22.
- Robert K. Merton, "Science and the Social Order," 5 (1938): 321–37.
- Frank E. Hartung, "The Social Function of Positivism," 12 (1945): 120–33.
- Frank E. Hartung, "Sociological Foundations of Modern Science," 14 (1947): 68–95.
- Elgin Williams, "Sociologists and Knowledge," 14 (1947): 224–30.
- Virgil Hinshaw Jr., "Epistemological Relativism and the Sociology of Knowledge," 15 (1948): 4–10.
- Walter P. Metzger, "Ideology and the Intellectual: A Study of Thorstein Veblen," 16 (1949): 125–33.

- Frank E. Hartung, "Problems of the Sociology of Knowledge," 19 (1952): 17–32.
- Lewis S. Feuer, "Sociological Aspects of the Relation between Language and Philosophy," 20 (1953): 85–100.
- Frank E. Hartung, "Cultural Relativity and Moral Judgments," 21 (1954): 118–26.
- Gideon Sjoberg, "Science and Changing Publication Patterns," 23 (1956): 90–96.

After 1959, virtually all of this literature disappears from *Philosophy of Science* within a very short time. There is one batch of articles on Marxism and science, by Paul Mattick, Donald Clark Hodges, and Barrows Dunham, all published together in the fourth issue of volume 29 in October 1962 (333–68), and one article by Laird Addis on Marxism and freedom in volume 33 ([1966]: 101–17). There is a paper by Leonard Goodwin on social science and social problem solving in volume 29 ([1962]: 377–92), and one paper by T. J. Gordon and M. J. Raffensperger on basic research planning in volume 36 ([1969]: 205–18). Finally, in volume 38 ([1971]: 395–412), Enrico Cantore published a paper under the title "Humanistic Significance of Science: Some Methodological Considerations."

Otherwise, these four formerly prominent genres of literature simply disappear. I can find only three traces. First, in volume 29 (1962), there is a noteworthy article by Howard J. Ehrlich entitled "Some Observations on the Neglect of the Sociology of Science," in which the author ascribes the lack of interest in the sociology of knowledge in the United States to "the prestige position of the sociologist and the clarity of his status with regard to natural science" (Ehrlich 1962, 369). Second is an article in volume 31 (1964) by Irving Louis Horowitz, "Discussion: Professionalism and Disciplinarianism. Two Styles of Sociological Performance," which looks at the rise of positivism in sociology and compares what Horowitz calls the "pragmatic" and "positivistic" views of the field (Horowitz 1964).

Most interesting, by far, however, is an exchange over Frank's philosophy of science. It begins with a critical analysis of Frank's recently published book, *Philosophy of Science* (Frank 1957), by Charles W. Kegley, a theologian otherwise best known for his editions of the works of Rudolf Bultmann (Kegley 1966), Reinhold Neibuhr (Kegley and Bretall 1956), and Paul Tillich (Kegley 1952). Whatever else one might think about the changes wrought by Rudner as editor of *Philosophy of Science,* one must admit that the overall intellectual quality of the journal rose dramatically. It is all the more surprising, then, to find the journal publishing in this case what can only be described as a second-rate hatchet job by a figure utterly peripheral to the philosophy of science. Kegley focues, as one might expect, on Frank's view of the role of social factors in theory choice, charging him with failing to distinguish was "*does* determine" the acceptance of a theory and what "*ought* to" (1959, 36). But he so thoroughly misunderstands Frank that he winds up accusing him of being sympathetic to Thomism and of "obsequiousness" with respect to Aristotle and Aquinas (37). It fell to F. James Rutherford to reply on Frank's behalf a little over a year later, patiently unpacking Kegley's distortions and confusions but also bluntly challenging the ad hominem character of Kegley's attack (Rutherford 1960, 184).

Kegley's assault on Frank appeared in the first issue of 1959, having been received in February of 1958. It was thus in the editorial pipeline before Rudner took over. But it is hard to figure out why Churchman, who was sympathetic to Frank, would not have insisted on changes at least for the sake of accuracy, unless he was no longer closely attending to editorial business. And if Churchman was not paying attention, what about the other editors? One does not want to believe that eminently fair-minded scholars like Nagel would tolerate such an insult to Frank, however much they might disagree with his philosophy. Whatever the story behind the publication of Kegley's article on Frank, the published article stands there in the journal as a permanent marker of the changes that had taken place in the discipline of the philosophy of science in North America by the end of the decade of the 1950s.

How are we to understand these changes? Why did they occur? As with any such phenomenon, this one, too, is surely overdetermined. We must remember that the 1950s, a period of red-baiting and McCarthyism, would not have been hospitable to public celebrations of a leftist political agenda for the philosophy of science. This was a time when people like Robert Cohen were being fired from teaching jobs for political reasons (Cohen was fired from his job at Wesleyan). Gerald Holton recalls Frank's being the recipient of a worrisome visit by FBI agents.[43] And as late as 1961, Robert G. Colodny, a historian of science at the University of Pittsburgh, became the target of attacks in the local press and in the Pennsylvania state legislature for his support of left-wing groups, such as the Fair Play for Cuba Committee and the Committee for a Sane Nuclear Policy, as well as for his having fought in the Lincoln Brigade during the Spanish Civil War (Alberts 1986). Moreover, as John McCumber has shown, the pressures of the McCarthy period definitely had a chilling effect on the expression of left-liberal political sentiments by members of the broader professional philosophical community in the United States (McCumber 2001). Leaders in the nascent field of the philosophy of science, anxious to secure public and private funding for new departments, programs, and institutes, could not be faulted for being cautious about the political reputation of the philosophy of science in an era when an official at the Rockefeller Foundation wondered about "a faint tinge of pink" in a proposal that Charles Morris developed for an "Institute for the Study of Man" (see Reisch 1995, chap. 6, 6).

Rudner is remembered as a staunch anticommunist in the tradition of Sidney Hook.[44] But for every anticommunist in the PSA, one could find an influential, long-time leftist like Carnap. Moreover, in some settings the more openly leftist members of the profession continued to play a crucial role in the establishment of the very institutions one might have expected to exert rightward political pressure, as in the case of Frank's having advised Raymod J. Seeger in his setting up of the National Science Foundation's Program on the History and Philosophy of Science (see Seeger 1965, xxvii–xxviii).

Real though the pressures of McCarthyism were, it would be a mistake to attribute the shift to a politically disengaged philosophy of science solely or even primarily to a wish to trim sail in politically troubled weather, just as it was a mistake to read Schlick's protestations of the political neutrality of the Verein Ernst Mach that way in 1934, because to do so requires discounting the sincerity of those protestations of neutrality, when in fact there is every reason to believe that the individuals involved were utterly sincere.

Not only are such simplistic political explanations insulting to the participants, but they also overlook the many other factors at work. One such factor that surely must be mentioned in a complete explanation is the logic of professionalization in the philosophy of science in the 1950s. A more avowedly political posture is to be expected from a discipline that stands on the margins of the academy and the cultural establishment, as did logical empiricism and scientific philosophy in Europe in the 1920s and 1930s. (Whether this description also fits Deweyan pragmatism in the 1930s and 1940s is more difficult to say in the absence of a detailed social and institutional history of the movement, especially in the area of the philosophy of science.) A drift to the political center and toward political neutrality often accompanies a discipline's establishing itself among the cultural institutions of the day, not so much because the members of the discipline change their views as because a discipline at the cultural center attracts relatively more of its new members from the politically and culturally less alienated segments of the population. Professionalization can have a similar effect but for a different reason. Professionalization establishes clearer criteria of community membership and thereby makes easier the exclusion of those whose opinions stray too much from the core commitments that define community membership, with the result that opinion tends to cluster closer to that core.

The felt need to professionalize the philosophy of science in the 1950s was acute. I mentioned above that Nagel and Reichenbach stressed the development of the philosophy of science as a profession. A most interesting expression of such sentiments is found in the article "Some Remarks on Problems and Methods in the Philosophy of Science," which Patrick Suppes, a student of Nagel's, published in *Philosophy of Science* in 1954. Suppes writes,

> Philosophers of science need be neither journalists of science nor acute men of common sense eternally restricted to contemplating the general meaning of such notions as those of mind, free will, cause and determinism.... The main purpose of this paper is to outline a partial program for ... a hard core of studies in the philosophy of science. (1954, 242)

Ron Giere has drawn our attention to the fact that Dewey and his followers failed to train new Ph.D.'s in the philosophy of science in anything like the numbers we see in the case of those who stood closer to the neopositivist core commitments of emergent professional philosophy of science, such as Reichenbach at UCLA, Carnap at Chicago and later UCLA, Feigl at Minnesota, Hempel at Princeton, and Nagel at Columbia (Giere 1996, 349–50). This happened partly because, as Giere points out, Dewey's pragmatism seems to have played itself out by the early 1950s as a program for research, whereas neopositivism was something fresh and new in the postwar American context, with a rich fund of problems and projects waiting to be tackled and something of a consensus on the method to be followed in those investigations. Moreover, the central prominence of the physical sciences in logical empiricism and scientific philosophy would have given it added lustre in an era when the atom bomb had won the war and our nuclear deterrent was keeping us safe from the Red Menace. The human sciences about which Dewey had always cared more were trying hurriedly to bring their methodologies up to the high standard of the physical sciences.

The full story of the growth and prominence of the new departments, centers, and institutes for the history and philosophy of science and of the discipline's prestige within existing philosophy departments has yet to be told. The sheer intellectual excitement associated with the philosophy of science and the widespread sense that it was tackling some of the most important problems of the day go a long way toward explaining this phenomenon. But external factors, such as the establishment of the

Program on the History and Philosophy of Science at the National Science Foundation, are surely also relevant. The role of private foundation money must be examined as well.

We must also remember that the decade of the 1950s was the era of the "end of ideology" (Bell 1960) in many arenas other than the philosophy of science. With fascism defeated and the advance of communism checked both in Europe and Asia, one could glimpse a day when liberal democracy and corporate capitalism would no longer need an ideological justification because their authority would be unquestioned thanks to the absence of any serious rivals. Objective science was widely viewed as the natural companion of political democracy and economic freedom. Of course the "end of ideology" was itself an ideology, but the prevalence of such attitudes would help make the academy more receptive to a depoliticized philosophy of science.

But all of these externalist and sociological explanations for the political disengagement of the philosophy of science in the 1950s leave one unsatisfied, because they neglect the interesting internal intellectual developments within the philosophy of science that for decades had been tending in the same direction. As I have tried to indicate, even the most explicitly political left-wing Vienna Circle logical empiricists and Deweyan pragmatists marched under the banner of objective science. This is the enduring historical fact that ties together a strident Austro-Marxist like Otto Neurath and a quiet liberal like Ernest Nagel and that, in the end, defines the historical trajectory of the philosophy of science from the 1920s to the 1960s. What changes is the manner in which the image of objective science finds expression in different decades, on different continents, and in dramatically different social, cultural, and political circumstances. For those of us who want to understand the political disengagement of the philosophy of science during the decade of the 1950s, the principal historical-philosophical task is, therefore, to understand how it came to pass that scientific objectivity was once thought to be wholly compatible with political engagement, both by left-wing logical empiricists and Deweyan pragmatists, but later came to be regarded as fatally compromised by the intrusion of social and political concerns.

Conclusion

In the 1950s and 1960s the philosophy of science was one of the most exciting places to be in the academy. Not just within philosophy departments but from many other places on the typical North American campus, the philosophy of science was seen as one of the most important centers of intellectual activity. In the broader public intellectual sphere, as well, the philosophy of science was one of those fields that set the tone. To mention, for the moment, only the most obvious outward signs of the high regard in which the field was held, no philosophy department, even a small one on a small campus, could be credible without one or more philosophers of science on its faculty. Deans and department chairs were eager to spend money to buy a reputation in the philosophy of science. It was a time when many schools established fields of concentration, programs, centers, and departments of history and philosophy of science. New journals and book series were launched, conference series were inaugurated, and government funding of research in the philosophy of science was expanding.

The times, however, have changed. I have the impression that today, on more and more campuses, the philosophy of science is moving toward the periphery. While we may be reluctant to admit it, and while there are important exceptions in areas like the foundations of physics and biology, the plain fact is that the philosophy of science is no longer so widely seen as a discipline that sets the agenda for others to follow.

The marginalization of the philosophy of science is occurring both within philosophy departments and elsewhere on campus. The problem within philosophy is well illustrated by a conversation I had some time ago, in which the current chair of a prestigious philosophy department explained to me that the department had decided to replace its only philosopher of science, who had resigned to take a position elsewhere, with someone just completing a dissertation in the philosophy of mind, because—this is a close paraphrase—after long and careful deliberation, the department concluded that the philosophy of science no longer needed to be represented specifically among a faculty of twelve, since it was no longer one of those important centers of philosophical activity from which significant new lines of thinking emerge, its place having been taken by the philosophy of mind and analytic metaphysics and epistemology. If such is the opinion of the philosophy of science on a bellwether campus, then small wonder that our students are having such difficulty finding jobs elsewhere.

Outside of philosophy departments, the marginalization of the philosophy of science is made evident by the fact that when our colleagues in other departments ask themselves the kinds of questions about science currently at the forefront of interest, they turn, more and more, not to philosophers of science but to sociologists, historians, literary theorists, and faculty members specializing in such areas as gender studies and African-American studies for the kinds of answers that they find relevant and illuminating. If a "traditional" philosopher of science is invited to participate in such discussions, as often as not it is to represent a retrograde, "positivist" conception of science, a point of view to be pitied or pilloried rather than seriously engaged as a significant alternative.

This change in the standing of the philosophy of science troubles me, not just for selfish professional reasons but because I believe that philosophers of science can and should be making vital contributions to all manner of current discussions, from policy debates over funding the superconducting supercollider and protocols for clinical trials of drugs designed for the terminally ill to the science front in the culture wars.[45] I believe that many of the problems with which we have wrangled are of continuing relevance, from questions about the relation between theory and evidence to the realism/instrumentalism debate, controversies over intertheoretical reduction, and puzzles about conceptual change. Most importantly, I believe that, if asserted with tact rather than condescension, the high standard of careful scholarship that we have set for ourselves can and should be a model for other disciplines and discourses instead of being derogated as, at best, a symptom of some intellectual neurosis or, at worst, a form of intellectual imperialism.

So I want to understand why this transformation in the standing of the philosophy of science has taken place. It would be a mistake simply to blame our colleagues for ignoring us because they are too stupid to do the kind of careful work that we demand of ourselves or because they are driven by a partisanship incompatible with scholarly independence and integrity. Balm though it might be to our wounded pride, such defensive dismissal of those who ignore us is undignified and it obscures the real causes of the change in our fortunes.

The marginalization of the philosophy of science is, no doubt, a complex phenomenon. Partly to blame are changes in the funding of research and higher education, changes that affect many other fields as well. Partly to blame are broader changes in the cultural, sociological, and economic role of the university. Simple change in intellectual fashion surely plays a role as well. But we, as philosophers of science, bear little direct responsibility for such changes as these.

We alone are to blame, however, for the fact that no new paradigm has come to the fore to define the field following the demise of neopositivism. This is a big part of the problem, for the result is a directionless drift, as we fiddle with various naturalisms, neo-Kantianism, neopragmatism, conventionalism, coherentist and constructive empiricisms, and the natural ontological attitude, all with ever less passion and a growing sense of despair over our ever again pretending to give large answers to large questions. This is not to say that it is right to assume that we should be pretending to give large answers to large questions. But we certainly have not had much fun lately. Amid all this ennui, one of the few encouraging developments is the recent surge of interest in the history of the philosophy of science, but although historical studies might help us understand how we got ourselves into this cul-de-sac, history alone cannot point the way to a new conception of the field. That will require creative imagination and an enthusiasm for the challenge that is lacking in an era when arrogant, ironic, postmodern disdain for "naive," "hegemonizing" projects is the norm among fashion-conscious, intellectual sophisticates.

My own preference is for us to take up again the naturalism of Neurath and Dewey. I should think that the example of Quine would have taught us that the blending of Neurath and Dewey need not produce the conceptual muddle that some might imagine would result. On the contrary, the holism of Neurath and the naturalism of Dewey can be made to yield a coherent framework for the study of science, in which there is a place for everything from the most technical studies in the foundations of physics to the most externalist sociology of knowledge.[46] But this naturalist program must also open itself, as Neurath and Dewey would have wanted, to an unblushing engagement with questions of value and the pressing social and political problems of the day.[47] In this way, it can provide the much needed left-liberal empiricist philosophy of science that is missing in contemporary public intellectual life.

Important though the lack of a successor paradigm might be in explaining the diminishing fortunes of the philosophy of science today, that is not the cause that I want to emphasize. I want to argue that the single most important factor in the decline of the philosophy of science in the academy and the public intellectual sphere is our loss of the sense of a cultural, social, and political mission. One cannot discount the intrinsic interest of the problems we have addressed over the years nor the intrinsic pleasure we take in trying to solve them. After all, I have always thought that Plato and Aristotle were close to the mark in holding that intellectual pleasure was the greatest human good. But Plato, at least, believed that the highest kind of wisdom was associated with a knowledge of the form of the good, and that after glimpsing the form of the good, the philosopher was obliged to go back in the cave. I would argue that our settling for anything less is morally and socially irresponsible.[48]

Notes

This paper was first written, under the title "Philosophy of Science and Social Responsibility," for presentation at a symposium, "Science and Social Responsibility," organized by Helen Longino, at the 1996 Biennial Meeting of the Philosophy of Science Association in Cleveland. Parts of that earlier version were the basis for the talk "The End of the Science and Values Debate in 1950s Philosophy of Science," which I gave at the conference on Logical Empiricism in North America at Harvard in May 1998. Different versions of the paper were also presented in talks to the philosophy departments at the University of Minnesota, the University of Toronto, and Northwestern University. A number of helpful comments and criticisms were offered in discussion at each of these venues. Joe Pitt, George Reisch, Alan Richardson, John Stachel, and Tom Uebel have read the manuscript with care and, I would like to think, sympathetic interest, offering more extended critical reactions. To them I extend my sincere thanks. Special thanks are owed to Robert Cohen and Adolf Grünbaum for lengthy and extremely helpful conversations about North American philosophy of science in the 1950s and 1960s. I am similarly indebted to Paul Meehl for observations about the situation at the University of Minnesota in the 1950s. Finally, a note of thanks to W. Gerald Heverly and the staff at the Archive for Scientific Philosophy at the University of Pittsburgh for their assistance during a rushed but, thanks to them, very productive day spent working with the Carnap and Reichenbach papers.

    1. In brief, I would argue that, as with other cultural domains, an understanding of the historical context is essential for an understanding of science and philosophy if only because scientific and philosophical problems have histories of such kind that one does not even know what the problem is unless one understands that history. Which aspects of the context are relevant and the extent of contextual determination of content varies widely from one cultural domain to another and even within a given domain, such as physics. In some cases, it is only the immediate intellectual context that is relevant; in others, one must take technology, economics, or politics into consideration as well. In some cases, context explains much; in others, it explains little. In virtually no cultural domain does context explain everything. Thus, although I believe that social factors can and often do influence theory choice, I would deny that the content of physical theory is simply to be understood as social construction. I am all the more skeptical of radical social constructivist approaches to science studies because of their principled embrace of relativism at the metatheoretical level and because of the inherent difficulty one faces in attempting to tell a convincing microcausal story about the influence of large-scale factors like class interest. The better the causal account of external influence on science, the more the large-scale factors are obscured behind the immediate causal efficacy of very local, small-scale influences, such as peer pressure, careerism, and the like. Finally, I would oppose to the social constructivist model the picture of social influence promoted by such early logical empiricists as Otto Neurath and Philipp Frank. Their holistic, underdeterminationist variety of conventionalism provides a robustly empiricist, nonrelativist understanding of the role of social factors in theory choice, one in which theory choice is constrained by evidence—but not up to the point of uniqueness—leaving a space within which both the freely creative imagination of the scientist and social factors can play a role.

    2. Critical theory and left-wing Vienna Circle logical empiricism had been in conversation with one another in the 1930s; see Dahms 1994.

    3. Koertge 1998 represents an attempt to have the mainstream philosophy of science community engage the contemporary critical science studies literature. But this is still not the argument for which I hope, because what is missing are, not surprisingly, the voices of left-wing Vienna Circle logical empiricism and Deweyan pragmatism. Even the interesting contribution by Philip Kitcher (Kitcher 1998b) is disappointing because of Kitcher's misrepresentation, in my opinion, of the lessons learned about issues like truth and objectivity through the work of professional philosophers of science in the latter half of the twentieth century; see, especially, his list of "uncontroversial" theses in the "Realist-Rationalist Cluster" (34) and the "Socio-Historical Cluster" (36). The crucial question concerns the role of

social and political values in theory choice and hence in shaping the content of science. Kitcher is right to note that the claim of the underdetermination of theory by evidence "has been dramatically overblown by some historians and sociologists who have contended that it shows that the world can have no bearing on what scientists accept" (40). But Neurath and Dewey did not make such "overblown" claims. They strove to combine respect for objective, empirical science with respect for the social influences on the content of science. That such a *via media* is lost in the polarized contemporary debate is precisely the circumstance that I lament.

4. Four particularly helpful recent studies are Hollinger 1995b, Ryan 1995, Campbell 1995, and Menand 2001; see also the essays collected in Hollinger and Depew 1995.

5. A good start on a history of pragmatist theories of science, with special emphasis on Dewey, is to be found, however, in Wilson 1995.

6. These movements were curiously eager to tell their own history right from the very start; see, for example, Kraft 1950 and Reichenbach 1951. More recent histories have begun to correct the unavoidable distortions of these earlier partisan histories. Among the many items in what has been, in recent years, a rapidly growing literature, see especially Runggaldier 1984, Proust 1986, Cirera Duocastella 1990, Coffa 1991, Uebel 1992, Haller 1993, Richardson 1998a, Friedman 1999 and 2000, and the essays collected in Uebel 1991, Stadler 1993, Giere and Richardson 1996, and Nemeth and Stadler 1996. An important role in the telling of a more objective history of logical empiricism and scientific philosophy has been played by the various centennial conferences; see, for example, Gadol 1982b, Haller 1982b, Haller and Stadler 1993, and Salmon and Wolters 1994. Mention should also be made of the Vienna Circle Collection, published by Reidel (Kluwer) and edited by Henk L. Mulder, Robert S. Cohen, and Brian McGuinness, which has so far presented selections from the writings of O. Neurath (1973, 1983), Hans Reichenbach (1978), and Moritz Schlick (1979), along with the writings of a number of other figures associated with the Vienna Circle, including Mach, Ludwig Boltzmann, Herbert Feigl, Hans Hahn, Belá Juhos, Eino Kaila, Felix Kaufmann, Victor Kraft, Karl Menger, Josef Schächter, and Friedrich Waismann. Indeed, it was Michael Friedman's review of the Schlick volumes in the Vienna Circle Collection that marked the beginning of serious, scholarly, English-language studies of the history of the movement (Friedman 1983), and Friedman's own later work continued to set the standard for this scholarship; see, for example, Friedman 1987, 1988, 1991, 1992. A comparable leading role has been played in the German-language historical literature by Rudolf Haller; see, for example, Haller 1982a, 1985. Another helpful contribution is the new, six-volume series edited by Sahotra Sarkar and published by Garland, Science and Philosophy in the Twentieth Century: Basic Works of Logical Empiricism.

7. Among the more helpful studies are Nemeth 1981, Stadler 1982, Cartwright et al. 1996, and the papers collected in Uebel 1991, P. Neurath and Nemeth 1994, and Stadler 1997.

8. A helpful discussion of the link between socialism and neo-Kantianism is van der Linden 1988, especially chap. 6, "Hermann Cohen: From Social Ethics to Socialist Ethics," and the appendix, "A Historical Note on Kantian Ethical Socialism."

9. For an overview of Reichenbach's political activities as a student during the 1910s, see M. Reichenbach 1978.

10. For more on Dingler's critique of relativity, including his role in the forging of an "antirelativity" preface to Mach's posthumously published book on optics (Mach 1921), see Wolters 1987.

11. What Reichenbach fails to note is that Dingler's 1931 article was sandwiched between critical notes about Dingler's work by Erich von Aster (1931) and Reichenbach himself (1931).

12. Among the regular contributors to *Erkenntnis* whose politics were well to the right of those of Neurath, Carnap, and Reichenbach was Pascual Jordan, who rather easily accommodated himself to the strictures of intellectual life during the Nazi period.

13. Here I differ with the interpretation of logical positivist protestations of value neutrality in Proctor 1991, chap. 12.

14. See the letter from Carnap to Neurath of October 7, 1928 (Rudolf Carnap Collection, Archive for Scientific Philosophy, University of Pittsburgh 029-16-01) in which he agrees on the importance of work on historical and sociological problems and promises to work on them himself. This letter to Neurath is noteworthy also because Carnap quotes from a letter from Frank about the recently published *Scheinprobleme in der Philosophie* (Carnap 1928a) in which Frank emphasizes the convergence between logical empiricism and pragmatism.

15. Robert S. Cohen, personal communication.

16. A helpful discussion of the political side of Neurath's philosophy of science is Wartofsky 1982; see also Nemeth 1981; Stadler 1982; P. Neurath 1994; and Cartwright et al. 1996.

17. Was Neurath's argument for an objective science in the service of the proletariat just a pose behind which hid a dogmatic, Marxist ideologue? I think not. More revealing of Neurath's true feelings are remarks directed not to a broader public audience but to his fellow Austrian socialists in the journal *Der Kampf*. In a 1923 reply to a series of articles on the role of money in a socialist economy by Helene Bauer (wife of the effective leader of the Austrian Social Democratic Party, Otto Bauer), Neurath wrote, "Like everything that is accomplished in a scientific way, the view I defended regarding the capitalist money economy, on the one hand, and the socialist planned economy must be grounded in direct investigations. What is correct is not any more correct because Marx also taught it and is not false because Marx was of another opinion. But it is important for the worker's movement that the agreement of correct propositions with Marx be pointed out. . . . Marxism is not the carrying out of what Marx taught, it is instead a world view that was historically introduced by Marx and in broad measure established by him. Marxism can criticize Marx and deviate from him" (O. Neurath 1923, as quoted in P. Neurath 1994, 77).

18. For a discussion of this shift in Schlick's thinking, with special attention to the attendant changes in Schlick's conception of the role of conventions in science, see Howard 1994.

19. It should be noted that for a holist like Neurath, such a choice among empirically equivalent alternative theories is a real choice, in the sense that these alternative theories say something different about the world. Indeed, as Neurath said in an already quoted remark, empirically equivalent alternative theories may even be so different as to "exclude each other" (O. Neurath 1934, 105). The theoretical and semantic holism to which Neurath was committed blocks the inference from empirical equivalence to semantic equivalence, an inference required by the stricter verificationism and epistemological atomism championed by Reichenbach (1928) and Schlick (1935), according to which empirically equivalent theories say the same thing about the world, being related to one another like notational variants or the expression of the same proposition in two different languages. I have discussed this difference between holistic conventionalism and purely linguistic conventionalism in Howard 1994.

20. Neurath's remarks on the manner in which history and social circumstance constrain theory choice and even disguise the fact of its existence should be compared to Einstein's similar contrast between underdetermination in principle and determination in practice; see Einstein 1918 and Howard 1990.

21. If the example of Duhem is not enough to deter one from thinking that the under-

determinationism espoused by Neurath is inherently some kind of left-leaning philosophy of science, then one should consider as well the example of W. V. O. Quine, whose holism and underdeterminationism were as radical as Neurath's (see, for example, Quine 1951 and 1960) but whose politics were diametrically opposed to Neurath's.

22. Philipp Frank was the translator of Duhem's other major work that was influential in these circles, his *L'Evolution de la mécanique* (1903; German edition 1912).

23. Friedrich Adler sought to defend the objectivity of science differently than did Neurath. In a somewhat confused discussion in his book *Ernst Machs Ueberwindung des mechanischen Materialismus*, Adler argued that the relations of production do not "determine" the intellectual superstructure, which is what he took Marx to have claimed, but instead merely "condition" the superstructure by constraining the range of possibilities, relating this alternative view to Mach's economic-biological understanding of the "adaptation" of thoughts to objects (F. Adler 1918, 169–77). In particular, he ridiculed the views of some more-orthodox contemporary Marxists who sought to relate the development of non-Euclidean geometry to changes in the relations of production consequent upon the discovery of the Americas, or who held that what was crucial about Kant's philosophy was that Kant himself came out of a petit bourgeois background and worked under an absolute monarchy (1918, 180–81). He concluded that, in fact, the natural sciences and mathematics are not really part of the superstructure of class relations, that they belong instead, like all tools, to that part of the basis or infrastructure that consists in the technological relations: "The natural sciences is no reflex, no ideology of the class relations, as are, perhaps, many parts of philosophy. Their foundations are much more solid, they do not change with changes in class relations, but only as part of the totality of the technical conditions of production" (1918, 185).

24. Carnap's at least lukewarm sympathy for Neurath's naturalism as regards the social determinants of theory choice should be kept in mind when one tries to understand in what sense Carnap saw the choice of a linguistic framework to be a pragmatic question (Carnap 1950) and when one tries to understand his not unsympathetic reaction to Thomas Kuhn's *Structure of Scientific Revolutions* (see Reisch 1991).

25. For more on the relation between logical empiricism and the sociology of science in the first half of the twentieth century, see Uebel 1998 and Richardson 1998b.

26. In characterizing naturalism as a bridge from Neurath to Dewey, I do not mean to imply that they were naturalists of the same stripe. "Naturalism" is notoriously vague as a designation for a philosophical position. Neurath's naturalism is epistemological, inasmuch as it accords a place within the philosophy of science for sociological and historical studies of the social, cultural, and political influences on theory choice. Dewey's naturalism owes a more explicit debt to Charles Darwin and emphasizes the biological roots of human cognition. What they have in common is the idea that scientific theorizing exists in a context (human, historical, biological, social, cultural, and political), that this context makes a difference in the content of science, and that the study of this context is a proper subject for the philosopher of science.

27. Much helpful information on Morris and his role as mediator is to be found in Reisch 1995. Albert Blumberg surely also deserves mention as an early mediator of the reception of logical empiricism in North America (see Blumberg and Feigl 1931), as do Quine and Ernest Nagel, both of whom, like Morris, traveled to Vienna in the early 1930s to learn more about the philosophy of the Vienna Circle.

28. See Reisch 1995, chap. 3, for more on Morris's critique of logical empiricism.

29. See Quine 1960, chap. 2, sec. 10, where "observation sentences" are defined as occasion sentences with a high degree of "observationality," the latter being defined in turn as "degree of constancy of stimulus meaning from speaker to speaker" (43). Quine

freely acknowledges his debts to both Neurath (see, for example, Quine 1969b, 84–85) and Dewey (see Quine 1969a, 26–27).

30. An informative snapshot of North American philosophy of science in the early 1930s is given in Morris 1935c. For more general background, see Hollinger 1975.

31. Here is an important point of difference between Neurath and Dewey, for Neurath denied any role for science in the selection of ends.

32. Helpful information on the migration of logical empiricism and scientific philosophy to North America can be found in Feigl 1969 and Dahms 1995.

33. See Holton 1992 for a more detailed discussion of Frank's move to Harvard and his activities on behalf of reconstituting the institutional structures of Vienna Circle philosophy of science.

34. Reichenbach mentions the distinction between context of discovery and context of justification in his essay on Dewey's theory of science, but this only in a footnote on Peirce's concept of "abduction" (Reichenbach 1939, 188, n. 28).

35. One does well to recall just how horrible the situation in Europe was throughout much of the 1920s and 1930s. Let just two facts suffice as reminder: In 1922, Albert Einstein had already been forced to flee Germany for a few months in the wake of the assassination of Walther Rathenau, because rumor had it that he too was targeted for assassination (see Clark 1971, 292–94). And in 1927, anti-Semitic riots at the University of Vienna left ninety people dead (see Stadler 1995, 53–54).

36. As noted by the editors of the *Collected Papers of Clarence Irving Lewis,* this essay was intended for publication in the *Revue Internationale de Philosophie* but was not published because of the German invasion of Belgium. The version that appears in the *Collected Papers* was completed in 1941. I thank Paul Taylor for bringing this essay to my attention.

37. See, for example, Nagel 1943 and the introduction to Nagel 1954b, where Nagel writes, "The philosophy of science is a difficult branch of intellectual analysis. It requires for its effective cultivation a rare combination of solidly founded substantive knowledge, analytical and constructive skill, and—not least—a sharp sense for relevance. . . . There are, indeed, no uniform standards of competent workmanship which control analyses in the philosophy of science, and which are also binding upon those who attempt to evaluate the analyses of others. But the best work in this domain of philosophy, so it seems to me, has been done by men who have sought to understand human reason by examining its operations in controlled inquiry, and who have interpreted the meaning of theoretical constructions in terms of their manifest functions in identifiable contexts" (1954b, 16).

38. For more on Nagel's naturalism, see his 1954 presidential address to the Eastern Division of the American Philosophical Association (Nagel 1954c).

39. When reading this passage one might want to ponder the effect that hearing it might have had on the young Kuhn, who was most probably in the audience. More work is needed on the question of Frank's possible influence on Kuhn. Frank was an important figure in the history and philosophy of science community in Cambridge during Kuhn's years at Harvard, and there are many points of contact between his philosophy of science and Kuhn's, especially as regards the role of contextual factors in theory choice. Steve Fuller's recent study of Kuhn barely mentions Frank though it pretends to tell in detail the story of Kuhn at Harvard (Fuller 2000).

40. Likewise, Alexandre Koyré reports himself in "perfect agreement" with Frank, adding only that Frank "did not go far enough" because he neglected to include philosophical trends among the extrascientific factors influencing theory choice (Koyré 1953, 177).

41. For more on the cultural significance of Reichenbach's *The Rise of Scientific Philosophy,* see Hollinger 1995a.

42. A more thorough study of the history of the journal is needed and would prob-

ably yield additional helpful insight. The interim editorial committee assembled in 1948 had equal representation from a faction more closely associated with the founding editor, Malisoff—this faction included Churchman—and a faction that Frank termed "our group," meaning the Philosophy of Science Association and the Institute for the Unity of Science. In a letter to Reichenbach of January 6, 1948, Frank writes that "some people suspected that our group wanted to take over the Journal by a kind of coup d'état" (Hans Reichenbach Collection, Archive for Scientific Philosophy, University of Pittsburgh, 18-34-63; quoted by permission), and it appears that some thought had been given to the institute's starting its own journal, perhaps with the University of California Press. To some extent, what separated these factions was a different attitude toward professionalizing the philosophy of science, an issue discussed further later in the text. But there are hints of a more straightforwardly political difference as well. The political spectrum spanned by the core members of the Vienna Circle and by Dewey and his followers ranged from centrist liberalism through non-Marxist democratic socialism to Austro-Marxism. The politics of those close to Malisoff seem to have been still further to the left. Recently opened Soviet-era intelligence archives suggest that Malisoff might have been passing information to Soviet intelligence agents (George Reisch, personal communication).

43. The agents' suspicions that Frank might have been a spy are said to have been allayed by Frank's showing them the passages where he is attacked by name in Lenin's *Materialism and Empirio-Criticism* (Holton 1992, 43–44).

44. Robert S. Cohen, personal communication.

45. One encouraging sign of a reengagement of philosophers of science with questions of broader public concern is Philip Kitcher's recent book on the human genome project, *The Lives to Come* (1996); see also Kitcher and Cartwright 1996.

46. Philip Kitcher's recent call for a revival of the sociology of science (Kitcher 1998a) is an encouraging sign, but Kitcher has in view a more modest conception of the field than that suggested by the perspectives of Neurath and Dewey. The essential difference stems from Kitcher's reluctance to acknowledge a role for social and political values in theory choice.

47. For an instructive recent exercise in constructing a theory of science that preserves a place for considerations of value alongside rational criteria of theory choice, see Longino 1990. Especially edifying are Longino's remarks on the way in which scholars outside of mainstream philosophy of science have "filled the void left by our silence" (Longino 1990, 7). For a somewhat less ambitious attempt to bring the philosophy of science back into the political arena, and one not at all shy about the legacy of left-wing Vienna Circle logical empiricism, see Reisch 1998.

48. A few years ago, C. West Churchman came out of retirement, as it were, to express a similar view; see Churchman 1994.

References

Adler, F. 1908. "Die Entdeckung der Weltelemente. (Zu Ernst Machs 70. Geburtstag.)" *Der Kampf. Sozialdemokratische Monatsschrift* 1: 231–40.

———. 1909. "Die Einheit des physikalischen Weltbildes." *Naturwissenschaftliche Wochenschrift* 8: 817–22.

———. 1918. *Ernst Machs Ueberwindung des mechanischen Materialismus.* Vienna: Verlag der Wiener Volksbuchhandlung Ignaz Brand & Co.

Adler, M. 1925. *Kant und der Marxismus. Gesammelte Aufsätze zur Erkenntniskritik und Theorie des Sozialen.* Berlin: E. Laub.

Alberts, R. C. 1986. "The Colodny Case." In *Pitt: The Story of the University of Pittsburgh, 1787–1987*, 291–96. Pittsburgh: University of Pittsburgh Press.

Archive for Scientific Philosophy (ASP). University of Pittsburgh. Rudolph Carnap Collection and Hans Reichenbach Collection.

Ardelt, R. G. 1984. *Friedrich Adler. Probleme einer Persönlichkeitsentwicklung um die Jahrhundertwende*. Vienna: Österreichischer Bundesverlag.

Bell, D. 1960. *The End of Ideology: On the Exhaustion of Political Ideas in the Fifties*. New York: Free Press.

Blum, M. E. 1985. *The Austro-Marxists, 1890–1918: A Psychobiographical Study*. Lexington: University Press of Kentucky, 1985.

Blumberg, A. E., and H. Feigl. 1931. "Logical Positivism: A New Movement in European Philosophy." *Journal of Philosophy* 28: 281–96.

Bryson, L. 1947. *Science and Freedom*. New York: Columbia University Press.

Bush, V. 1949. *Modern Arms and Free Men: A Discussion of the Role of Science in Preserving Democracy*. New York: Simon and Schuster.

Campbell, J. 1995. *Understanding John Dewey: Nature and Cooperative Intelligence*. Chicago and LaSalle, Ill.: Open Court.

Carnap, R. 1928a. *Scheinprobleme in der Philosophie. Das Fremdpsychische und der Realismusstreit*. Berlin-Schlachtensee: Weltkreis-Verlag.

———. 1928b. *Der logische Aufbau der Welt*. Berlin-Schlachtensee: Weltkreis-Verlag.

———. 1932a. "Überwindung der Metaphysik durch logische Analyse der Sprache." *Erkenntnis* 2: 219–41.

———. 1932b. "Die physikalische Sprache als Universalsprache der Wissenschaft." *Erkenntnis* 2: 432–65.

———. 1933. "Über Protokollsätze." *Erkenntnis* 3: 215–28.

———. 1935. *Philosophy and Logical Syntax*. London: Kegan Paul, Trench, Trubner, and Co.

———. 1950. "Empiricism, Semantics, and Ontology." *Revue internationale de philosophie* 4, no. 11: 20–40.

———. 1963. "Intellectual Autobiography." In *The Philosophy of Rudolf Carnap*, ed. Paul A. Schilpp, 3–84. LaSalle, Ill.: Open Court.

Cartwright, N., J. Cat, L. Fleck, and T. E. Uebel. 1996. *Otto Neurath: Philosophy between Science and Politics*. Ideas in Context, vol. 38, ed. Quentin Skinner, L. Daston, Wolfheppenies, R. Rorty, J. B. Schneewind. Cambridge: Cambridge University Press.

Churchman, C. W. 1948. *Theory of Experimental Inference*. New York: Macmillan.

———. 1953. "A Pragmatic Theory of Induction." In *The Validation of Scientific Theories*, ed. P. G. Frank, 26–31. Boston: Beacon Press. Reprint, New York: Collier Books, 1956.

———. 1994. "What Is Philosophy of Science?" *Philosophy of Science* 61: 132–41.

Cirera Duocastella, R. 1990. *Carnap i el Cercle de Veina: Empirisme i sintaxi lògica*. Barcelona: Anthropos.

Clark, R. W. 1971. *Einstein: The Life and Times*. New York and Cleveland: World Publishing Company.

Coffa, J. A. 1991. *The Semantic Tradition from Kant to Carnap: To the Vienna Station*, ed. Linda Wessels. Cambridge: Cambridge University Press.

Cohen, M. R., and E. Nagel. 1934. *An Introduction to Logic and Scientific Method*. New York: Harcourt, Brace, and Company.

Dahms, H.-J. 1994. *Positivismusstreit. Die Auseinandersetzungen der Frankfurter Schule mit dem logischen Positivismus, dem amerikanischen Pragmatismus, und dem kritischen Realismus*. Frankfurt: Suhrkamp.

———. 1995. "The Emigration of the Vienna Circle." In *Vertreibung der Vernunft: The*

Cultural Exodus from Austria, ed. F. Stadler and P. Weibel, 57–79. Vienna and New York: Springer-Verlag, 1995.

———. 2000. "The Absence of the 'Scientific World Conception' from Middle Europe after 1945: Causes and Consequences." Paper delivered at HOPOS 2000: Third International History of Philosophy of Science Conference, Vienna, July 6–9.

Dewey, J. 1916. *Essays in Experimental Logic.* Chicago: University of Chicago Press.

———. 1929a. *The Quest for Certainty.* New York: Minton, Balch, and Company.

———. 1929b. *Experience and Nature.* 2nd ed. New York: W. W. Norton.

———. 1934. *A Common Faith.* New Haven, Conn.: Yale University Press.

———. 1935a. *Liberalism and Social Action.* New York: G. P. Putnam's Sons.

———. 1935b. "An Empirical Survey of Empiricisms." In *Studies in the History of Ideas,* ed. Department of Philosophy, Columbia University, 3–22. New York: Columbia University Press. Page numbers from the reprinting in John Dewey, *On Experience, Nature, and Freedom,* ed. Richard J. Bernstein, 70–87. Indianapolis and New York: Bobbs-Merrill, 1960.

———. 1938. *Logic: The Theory of Inquiry.* New York: Holt, Rinehart, and Winston.

———. 1948. *Reconstruction in Philosophy.* Enlarged ed. Boston: Beacon Press.

Dewey, J., and H. M. Kallen, eds. 1941. *The Bertrand Russell Case.* New York: Viking Press.

Dingler, H. 1919. *Die Grundlagen der Physik. Synthetische Prinzipien der mathematischen Naturphilosophie.* Berlin and Leipzig: Vereinigung wissenschaftlicher Verleger (Walter de Gruyter).

———. 1920. "Kritische Bemerkungen zu den Grundlagen der Relativitätstheorie." *Physikalische Zeitschrift* 21: 668–75.

———. 1921. *Physik und Hypothese. Versuch einer induktiven Wissenschaftslehre nebst einer kritischen Analyse der Fundamente der Relativitätstheorie.* Berlin and Leipzig: Vereinigung wissenschaftlicher Verleger (Walter de Gruyter).

———. 1931. "Über den Aufbau der experimentellen Physik." *Erkenntnis* 2: 21–38.

———. 1933. *Die Grundlagen der Geometrie. Ihre Bedeutung für Philosophie, Mathematik, Physik, und Technik.* Stuttgart: Ferdinand Enke.

Ducasse, C. J. 1959. "A Statement from the President to the Members of the Philosophy of Science Association." *Philosophy of Science* 26: 171.

Duhem, P. 1903. *L'Evolution de la mécanique.* Paris: A. Joanin.

———. 1906. *La Théorie physique: Son objet et sa structure.* Paris: Chevalier & Rivière.

———. 1908. *Ziel und Struktur der physikalischen Theorien.* Trans. Friedrich Adler. Foreward by Ernst Mach. Leipzig: Johann Ambrosius Barth.

———. 1912. *Die Wandlungen der Mechanik und der mechanischen Naturerklärung.* Trans. Philipp Frank, with the collaboration of E. Stiasny. Leipzig: Johann Ambrosius Barth.

Dühring, E. 1871. *Kritische Geschichte der Nationalökonomie und des Socialismus.* Berlin: T. Grieben; 2nd ed., 1875; 3rd ed., Leipzig: Fues's Verlag (R. Reisland), 1879; 4th ed., Leipzig: C. G. Naumann, 1900.

———. 1873a. *Kritische Geschichte der allgemeinen Principien der Mechanik.* Berlin: T. Grieben; 2nd ed., Leipzig: Fues's Verlag (R. Reisland), 1877; 3rd ed., 1887.

———. 1873b. *Cursus der National- und Socialökonomie.* Berlin: T. Grieben; 2nd ed., Leipzig: R. Reisland, 1876; 3rd ed. Leipzig: O. R. Reisland, 1892; 4th ed. Leipzig: Reisland, 1925.

———. 1878. *Logik und Wissenschaftstheorie.* Leipzig: Fues's Verlag (R. Reisland); 2nd ed., Leipzig: T. Thomas, 1905.

Ehrlich, H. J. 1962. "Some Observations on the Neglect of the Sociology of Science." *Philosophy of Science* 29: 369–76.

Einstein, A. 1918. "Motive des Forschens." In *Zu Max Plancks sechzigstem Geburtstag.*

Ansprachen, gehalten am 26 April 1918 in der Deutschen Physikalischen Gesellschaft, 29–32. Karlsruhe: C. F. Müller.

Engels, F. 1878. *Herr Eugen Dühring's Umwälzung der Wissenschaft.* Leipzig: Druck und Verlag der Genossenschafts-Buchdruckerei. English translation, *Herr Eugen Dühring's Revolution in Science (Anti-Dürhing).* Chicago: Charles H. Kerr, 1935.

Feigl, H. 1969. "The *Wiener Kreis* in America." In *The Intellectual Migration: Europe and America, 1930–1960,* ed. D. Fleming and B. Bailyn, 630–73. Cambridge, Mass.: Harvard University Press. Reprinted in *Inquiries and Provocations: Selected Writings, 1929–1974,* by H. Feigl, 57–94. Dordrecht, Boston, and London: Reidel, 1980.

———. 1980. *Inquiries and Provocations: Selected Writings, 1929–1974,* Ed. R. S. Cohen. Dordrecht, Boston, and London: D. Reidel.

Fischer, K. R. 2000. "Philosophy in Austria and the United States since 1945." Paper delivered at HOPOS 2000: Third International History of Philosophy of Science Conference, Vienna, July 6–9.

Frank, P. G. 1949. "Historical Background." In *Modern Science and Its Philosophy,* Cambridge, Mass.: Harvard University Press; reprint, New York: Collier Books, 1961, 13–61.

———. 1953. "The Variety of Reasons for the Acceptance of Scientific Theories." In *The Validation of Scientific Theories,* ed. P. G. Frank, 13–26. Boston: Beacon Press. Reprint, New York: Collier Books, 1956.

———, ed. 1956. *The Validation of Scientific Theories.* Boston: Beacon Press. Reprint, New York: Collier Books, 1961.

———. 1957. *Philosophy of Science: The Link between Science and Philosophy.* Englewood Cliffs, N.J.: Prentice-Hall.

———. 1963. "The Pragmatic Components in Carnap's 'Elimination of Metaphysics.'" In *The Philosophy of Rudolf Carnap,* ed. P. A. Schilpp, 159–64. La Salle, Ill.: Open Court.

Friedman, M. 1983. Review of *Philosophical Papers,* by M. Schlick. *Philosophy of Science* 50: 498–514. Reprinted in *Reconsidering Logical Positivism,* by M. Friedman, 17–34. Cambridge: Cambridge University Press, 1999.

———. 1987. "Carnap's *Aufbau* Reconsidered." *Nous* 21: 521–45. Reprinted in *Reconsidering Logical Positivism,* by M. Friedman, 89–113. Cambridge: Cambridge University Press, 1999.

———. 1988. "Logical Truth and Analyticity in Carnap's *Logical Syntax of Language.*" In *History and Philosophy of Modern Mathematics,* ed. W. Aspray and P. Kitcher, 82–94. Minneapolis: University of Minnesota Press. Reprinted as "Analytic Truth in Carnap's *Logical Syntax of Language*" in *Reconsidering Logical Positivism,* by M. Friedman, 165–76. Cambridge: Cambridge University Press, 1999.

———. 1991. "The Re-evaluation of Logical Positivism." *Journal of Philosophy* 88: 505–19.

———. 1992. "Epistemology in the *Aufbau.*" *Synthese* 93: 15–57. Reprinted in *Reconsidering Logical Positivism,* by M. Friedman, 114–52. Cambridge: Cambridge University Press, 1999.

———. 1999. *Reconsidering Logical Positivism.* Cambridge: Cambridge University Press.

———. 2000. *A Parting of the Ways: Carnap, Cassirer, and Heidegger.* Chicago and La Salle, Ill.: Open Court.

Fuller, S. 2000. *Thomas Kuhn: A Philosophical History for Our Times.* Chicago and London: University of Chicago Press.

Gadol, E. T. 1982a. "Philosophy, Ideology, Common Sense, and Murder—The Vienna of the Vienna Circle Past and Present." In *Rationality and Science: A Memorial Volume for Moritz Schlick in Celebration of the Centennial of His Birth,* ed. E. T. Gadol, 1–35. Vienna and New York: Springer-Verlag, 1982.

———, ed. 1982b. *Rationality and Science: A Memorial Volume for Moritz Schlick in Celebration of the Centennial of His Birth*. Vienna and New York: Springer-Verlag.

Giere, R. N. 1996. "From Wissenschaftliche Philosophie to Philosophy of Science." In *Origins of Logical Empiricism,* ed. R. N. Giere and A. Richardson, 335–54. Minneapolis: University of Minnesota Press, 1996.

Giere, R. N., and A. Richardson, eds. 1996. *Origins of Logical Empiricism*. Minnesota Studies in the Philosophy of Science, vol. 16. Minneapolis: University of Minnesota Press.

Gross, P. R., and N. Levitt. 1994. *Higher Superstition: The Academic Left and Its Quarrels with Science*. Baltimore and London: The Johns Hopkins University Press.

Hall, E. W. 1956. *Modern Science and Human Values*. New York: Van Nostrand.

Haller, R. 1982a. "New Light on the Vienna Circle." *Monist* 65: 25–37.

———, ed. 1982b. *Schlick und Neurath: Ein Symposion*. Amsterdam: Rodopi.

———. 1985. "Der erste Wiener Kreis." *Erkenntnis* 22: 341–58.

———. 1993. *Neopositivismus. Eine historische Einführung in die Philosophie des Wiener Kreises*. Darmstadt: Wissenschaftliche Buchgesellschaft.

———. 1995. "Philosophy: Tool and Weapon." In *Vertreibung der Vernunft: The Cultural Exodus from Austria,* ed. F. Stadler and P. Weibel, 80–87. Vienna and New York: Springer-Verlag, 1995.

Haller, R., and F. Stadler, eds. 1993. *Wien—Berlin—Prag. Der Aufstieg der wissenschaftlichen Philosophie*. Vienna: Hölder-Pichler-Tempsky.

Hecht, H., and D. Hoffmann, 1982. "Die Berufung Hans Reichenbachs an die Berliner Universität." *Deutsche Zeitschrift für Philosophie* 30: 651–62.

Heiss, G. 2000. "Philosophy at the University of Vienna from the First to the Second Austrian Republic." Paper delivered at HOPOS 2000: Third International History of Philosophy of Science Conference, Vienna, July 6–9.

Hoffmann, D. 1993. "Die Berliner 'Gesellschaft für empirische/wissenschaftliche Philosophie." In *Wien—Berlin—Prag: Der Aufstieg der wissenschaftlichen Philosophie,* ed. R. Haller and F. Stadler, 386–401. Vienna: Hölder-Pichler-Tempsky, 1993.

Hollinger, D. A. 1975. *Morris R. Cohen and the Scientific Ideal*. Cambridge and London: MIT Press.

———. 1990. "Free Enterprise and Free Inquiry: The Emergence of Laissez-Faire Communitarianism in the Ideology of Science in the United States." *New Literary History* 21: 897–919.

———. 1995a. "Science as a Weapon in *Kulturkämpfe* in the United States during and after World War II." *Isis* 86: 440–54. Reprinted in *Science, Jews, and Secular Culture: Studies in Mid-Twentieth-Century American Intellectual History,* 155–74. Princeton, N.J.: Princeton University Press, 1996.

———. 1995b. "The Problem of Pragmatism in American History: A Look Back and a Look Ahead." In *Pragmatism: From Progressivism to Postmodernism,* ed. R. Hollinger and D. Depew, 3–18. Westport, Conn., and London: Praeger, 1995.

Hollinger, R., and D. Depew, eds. 1995. *Pragmatism: From Progressivism to Postmodernism*. Westport, Conn., and London: Praeger.

Holton, G. 1992. "Ernst Mach and the Fortunes of Positivism in America." *Isis* 83: 27–60. Reprinted in *Science and Anti-Science*. Cambridge, Mass., and London: Harvard University Press, 1993, 1–55.

Hook, S., ed. 1950. *John Dewey: Philosopher of Science and Freedom*. New York: Dial Press.

Horowitz, I. L. 1964. "Discussion: Professionalism and Disciplinarianism: Two Styles of Sociological Performance." *Philosophy of Science* 31: 275–81.

Howard, D. 1990. "Einstein and Duhem." *Synthese* 83: 363–84.

———. 1994. "Einstein, Kant, and the Origins of Logical Empiricism." In *Language, Logic, and the Structure of Scientific Theories: The Carnap-Reichenbach Centennial,*

ed. W. Salmon and G. Wolters, 45–105. Pittsburgh: University of Pittsburgh Press; Konstanz: Universitätsverlag.
Jaki, S. L. 1984. *Uneasy Genius: The Life and Work of Pierre Duhem*. Dordrecht: Martinus Nijhoff.
Johnston, W. M. 1972. *The Austrian Mind: An Intellectual and Social History, 1848–1938*. Berkeley, Los Angeles, and London: University of California Press.
Kallen, H. M. 1950. "John Dewey and the Spirit of Pragmatism." In *John Dewey: Philosopher of Science and Freedom*, ed. S. Hook, 3–46. New York: Dial Press, 1950.
Kegley, C. W., ed. 1952. *The Theology of Paul Tillich*. New York: Macmillan.
———. 1959. "Reflections on Philipp Frank's Philosophy of Science." *Philosophy of Science* 26: 35–40.
———. 1966. *The Theology of Rudolf Bultmann*. New York: Harper and Row.
Kegley, C. W., and R. W. Bretall, eds. 1956. *Reinhold Niebuhr: His Religious, Social, and Political Thought*. New York: Macmillan.
Kitcher, P. 1996. *The Lives to Come*. New York: Simon and Schuster.
———. 1998a. "Reviving the Sociology of Science." *PSA 98: Proceedings of the 1998 Biennial Meeting of the Philosophy of Science Association. Part II, Symposia Papers*, ed. D. Howard. *Philosophy of Science* S67: S33–S44.
———. 1998b. "A Plea for Science Studies." In *A House Built on Sand: Exposing Postmodernist Myths about Science*, ed. N. Koertge, 32–56. New York: Oxford University Press, 1998.
Kitcher, P., and N. Cartwright. 1996. "Science and Ethics: Reclaiming Some Neglected Questions." *Perspectives on Science: Historical, Philosophical, Social* 4: 145–53.
Koertge, N., ed. 1998. *A House Built on Sand: Exposing Postmodernist Myths about Science*. New York: Oxford University Press.
Koyré, A. 1953. "Influence of Philosophic Trends on the Formulation of Scientific Theories." In *The Validation of Scientific Theories*, ed. P. G. Frank, 177–87. Boston: Beacon Press. Reprint, New York: Collier Books, 1956.
Kraft, V. 1950. *Der Wiener Kreis: Der Ursprung des Neopositivismus. Ein Kapitel der jüngsten Philosophiegeschichte*. Vienna: Springer-Verlag.
Lenin, V. I. 1909. *Materializm i empiriokrititsizm*. Moscow: Zveno. Page numbers are cited from the English translation: *Materialism and Empirio-Criticism: Critical Comments on a Reactionary Philosophy*. Moscow: Foreign Languages Publishing House, 1952.
Lewis, C. I. 1941. "Logical Positivism and Pragmatism." In *Collected Papers of Clarence Irving Lewis*, ed. John D. Goheen and John L. Mothershead Jr., 92–112. Stanford, Calif.: Stanford University Press, 1970.
Lindeman, E. G., ed. 1944. *The Scientific Spirit and Democratic Faith*. New York: King's Crown Press.
Longino, H. E. 1990. *Science as Social Knowledge: Values and Objectivity in Scientific Inquiry*. Princeton, N.J.: Princeton University Press.
Mach, E. 1906. *Erkenntnis und Irrtum. Skizzen zur Psychologie der Forschung*. 2nd ed. Leipzig: Johann Ambrosius Barth.
———. 1921. *Die Prinzipien der physikalischen Optik. Historisch und erkenntnispsychologisch entwickelt*. Leipzig: Johann Ambrosius Barth.
Malisoff, W. M. 1944. "Philosophy of Science after Ten Years." *Philosophy of Science* 11: 1–2.
———. 1946. "A Science of the People, by the People, and for the People." *Philosophy of Science* 13: 166–69.
Mannheim, K. 1929. *Ideologie und Utopie*. Bonn: F. Cohen.

Martin, R. N. D. 1991. *Pierre Duhem: Philosophy and History in the Work of a Believing Physicist.* La Salle, Ill.: Open Court.

McCumber, J. 2001. *Time in the Ditch: American Philosophy and the McCarthy Era.* Evanston, Ill.: Northwestern University Press.

Menand, L. 2001. *The Metaphysical Club: A Story of Ideas in America.* New York: Farrar, Straus, and Giroux.

Moore, B., Jr. 1953. "Influence of Political Creeds on the Acceptance of Theories." In *The Validation of Scientific Theories,* ed. P. G. Frank, 35–41. Boston: Beacon Press. Reprint, New York: Collier Books, 1956.

Morris, C. W. 1935a. "Philosophy of Science and Science of Philosophy." *Philosophy of Science* 2: 271–86. Page numbers and quotations taken from the reprinting in *Logical Positivism, Pragmatism, and Scientific Empiricism,* by C. W. Morris, 6–21. Paris: Hermann, 1937.

———. 1935b. "The Relation of the Formal and Empirical Sciences within Scientific Philosophy." *Erkenntnis* 5: 6–14. Reprinted in *Logical Positivism, Pragmatism, and Scientific Empiricism,* by C. W. Morris, 46–55. Paris: Hermann, 1937.

———. 1935c. "Some Aspects of Recent American Scientific Philosophy." *Erkenntnis* 5: 142–49.

———. 1936. "Semiotic and Scientific Empiricism." In *Actes du Congrés International de Philosophie Scientifique.* Actualités Scientifique et Industrielles, vol. 388. Section 1, *Philosophie Scientifique et Empirisme Logique,* 42–56. Paris: Hermann. Reprinted in *Logical Positivism, Pragmatism, and Scientific Empiricism,* by C. W. Morris, 56–71. Paris: Hermann, 1937.

———. 1937a. "The Concept of Meaning in Pragmatism and Logical Positivism." In *Logical Positivism, Pragmatism, and Scientific Empiricism,* by C. W. Morris, 22–30. Paris: Hermann, 1937.

———. 1937b. *Logical Positivism, Pragmatism, and Scientific Empiricism.* Actualités Scientifique et Industrielles, vol. 449. Paris: Hermann.

Nagel, E. 1943. "Malicious Philosophies of Science." *Partisan Review* 10 (January–February 1943). Reprinted in *Sovereign Reason and Other Studies in the Philosophy of Science,* by E. Nagel, 17–35. Glencoe, Ill.: Free Press, 1954.

———. 1950. "Dewey's Theory of Natural Science." In *John Dewey: Philosopher of Science and Freedom,* ed. S. Hook, 231–48. New York: Dial Press, 1950.

———. 1954a. "The Perspectives of Science and the Prospects of Men." *Perspectives USA,* no. 7 (Spring). Reprinted in *Sovereign Reason and Other Studies in the Philosophy of Science,* by E. Nagel, 296–308. Glencoe, Ill.: Free Press, 1954.

———. 1954b. *Sovereign Reason and Other Studies in the Philosophy of Science.* Glencoe, Ill.: Free Press.

———. 1954c. "Naturalism Reconsidered." In *Proceedings and Addresses of the American Philosophical Association,* vol. 28. Reprinted in *Logic without Metaphysics and Other Essays in the Philosophy of Science,* 3–18. Glencoe, Ill.: Free Press, 1956.

Nathanson, J., ed. 1946. *Science for Democracy.* New York: King's Crown Press.

Nemeth, E. 1981. *Otto Neurath und der Wiener Kreis. Revolutionäre Wissenschaftlichkeit als Anspruch.* Frankfurt and New York: Campus.

Nemeth, E., and F. Stadler, eds. 1996. *Encyclopedia and Utopia: The Life and Work of Otto Neurath (1882–1945).* Vienna Circle Institute Yearbook, vol. 4. Dordrecht, Boston, and London: Kluwer.

Neurath, O. 1920. "Vollsozialisierung." In *Deutsche Gemeinwirtschaft,* no. 15. Jena: Eugen Dietrichs.

———. 1923. "Geld und Sozialismus." *Der Kampf* 16: 145–57.

———1928. *Lebensgestaltung und Klassenkampf.* Berlin: E. Laub. Page numbers and translations from the excerpt translated as "Personal Life and Class Struggle" in *Empiricism and Sociology,* by O. Neurath, 249–98. Dordrecht and Boston: Reidel, 1973.
———. 1930a. "Einheitswissenschaft und Marxismus." *Erkenntnis* 1: 75.
———. 1930b. "Wege der wissnschaftliche Weltauffassung." *Erkenntnis* 1: 106–25.
———. 1931. *Empirische Soziologie. Der wissenschaftliche Gehalt der Geschichte und Nationalökonomie.* Schriften zur wissenschaftlichen Weltauffassung, vol. 5, ed. P. Frank and M. Schlick. Vienna: Julius Springer. Page numbers and translations from the excerpt translated as "Empirical Sociology" in *Empiricism and Sociology,* by O. Neurath, 319–421. Dordrecht and Boston: Reidel, 1973.
———. 1932. "Protokollsätze." *Erkenntnis* 3: 204–14. Page numbers and translations from the excerpt translated as "Protocol Sentences" in *Philosophical Papers, 1913–1946,* by O. Neurath, 91–99. Dordrecht, Boston, and Lancaster: Reidel, 1983.
———. 1934. "Radikaler Physikalismus und 'wirkliche Welt.'" *Erkenntnis* 4: 346–62. Page numbers and translations from the excerpt translated as "Radical Physicalism and the 'Real World'" in *Philosophical Papers, 1913–1946,* by O. Neurath, 100–14. Dordrecht, Boston, and Lancaster: Reidel, 1983.
———. 1935a. "Einheit der Wissenschaft als Aufgabe." *Erkenntnis* 5: 16–22. Page numbers and translations from the excerpt translated as "The Unity of Science as a Task" in *Philosophical Papers, 1913–1946,* by O. Neurath, 115–20. Dordrecht, Boston, and Lancaster: Reidel, 1983.
———. 1935b. "Pseudorationalismus der Falsifikation." *Erkenntnis* 5: 353–65.
———. 1973. *Empiricism and Sociology.* Ed. Marie Neurath and Robert S. Cohen. Vienna Circle Collection, vol. 1. Dordrecht and Boston: D. Reidel.
———. 1983. *Philosophical Papers, 1913–1946.* Ed. and trans. Robert S. Cohen and Marie Neurath. Vienna Circle Collection, vol. 16. Dordrecht, Boston, and Lancaster: D. Reidel.
Neurath, O., H. Hahn, and R. Carnap. 1929. *Wissenschaftliche Weltauffassung: Der Wiener Kreis.* Vienna: Artur Wolf. Page numbers and translations from the translation as "The Scientific Conception of the World: The Vienna Circle" in *Empiricism and Sociology,* by O. Neurath, 299–318. Dordrecht and Boston: Reidel, 1973.
Neurath, P. 1994. "Otto Neurath (1882–1945): Leben und Werk." In *Otto Neurath oder die Einheit von Wissenschaft und Gesellschaft,* ed. P. Neurath and E. Nemeth, 13–96. Vienna, Cologne, and Weimar: Böhlau, 1994.
Neurath, P., and E. Nemeth, eds. 1994. *Otto Neurath oder die Einheit von Wissenschaft und Gesellschaft.* Monographien zur österreichischen Kultur- und Geistesgeschichte, vol. 6, ed. P. Kampits. Vienna, Cologne, and Weimar: Böhlau.
Philosophy of Science. 1949. "Philosophy of Science and Liberalism" (editorial). *Philosophy of Science* 16: 1–2.
Popper, K. 1974. "Autobiography of Karl Popper." In *The Philosophy of Karl Popper,* ed. Paul Arthur Schilpp, 3–181. The Library of Living Philosophers, vol. 14. La Salle, Ill.: Open Court.
Proctor, R. N. 1991. *Value-Free Science? Purity and Power in Modern Knowledge.* Cambridge, Mass., and London: Harvard University Press.
Proust, J. 1986. *Questions de forme: Logique et proposition analytique de Kant à Carnap.* Paris: Artheme Fayard. English translation: *Questions of Form: Logic and the Analytic Proposition from Kant to Carnap.* Trans. A. A. Brenner. Minneapolis: University of Minnesota Press, 1989.
Quine, W. V. O. 1951. "Two Dogmas of Empiricism." *Philosophical Review* 60: 20–43. Reprinted in *From a Logical Point of View,* by W. V. O. Quine, 20–46. Cambridge, Mass.: Harvard University Press, 1953.

———. 1960. *Word and Object*. Cambridge: MIT Press.
———. 1969a. "Ontological Relativity." In *Ontological Relativity and Other Essays,* 26–68. New York and London: Columbia University Press, 1969.
———. 1969b. "Epistemology Naturalized." In *Ontological Relativity and Other Essays,* 69–90. New York and London: Columbia University Press, 1969.
———. 1969c. *Ontological Relativity and Other Essays*. New York and London: Columbia University Press.
Reichenbach, H. 1918a. "Programm der Sozialistischen Studentenpartei." Original source and publication have not been established. English translation: "Platform of the Socialist Students' Party." In *Selected Writings, 1909–1953,* by H. Reichenbach, ed. M. Reichenbach and R. S. Cohen. Trans. E. H. Schneewind, vol. 1, 132–35. Dordrecht, Boston, and London: Reidel, 1978.
———. 1918b. "Die Sozializierung der Hochschule." Unpublished manuscript. English translation: "Socializing the University." In *Selected Writings, 1909–1953,* by H. Reichenbach, ed. M. Reichenbach and R. S. Cohen. Trans. E. H. Schneewind, vol. 1, 136–80. Dordrecht, Boston, and London: Reidel, 1978.
———. 1919. "Student und Sozialismus." *Die Aufbau* (Flugblätter der Jugend, Berlin), no. 5.
———. 1921. "Erwiderung auf H. Dinglers Kritik an der Relativitätstheorie." *Physikalische Zeitschrift* 22: 379–84.
———. 1928. *Philosophie der Raum-Zeit-Lehre*. Berlin: Julius Springer.
———. 1931. "Schlußbemerkung." *Erkenntnis* 2: 39–41.
———. 1934. "In eigener Sache." *Erkenntnis* 4: 75–78.
———. 1935. *Wahrscheinlichkeitslehre. Eine Untersuchung über die logischen und mathematischen Grundlagen der Wahrscheinlichkeitsrechnung*. Leiden: A. W. Sijthoff.
———. 1938. *Experience and Prediction: An Analysis of the Foundations and the Structure of Knowledge*. Chicago: University of Chicago Press.
———. 1939. "Dewey's Theory of Science." In *The Philosophy of John Dewey,* ed. P. A. Schilpp, 159–92. Evanston, Ill., and Chicago: Northwestern University Press.
———. 1951. *The Rise of Scientific Philosophy*. Berkeley and Los Angeles: University of California Press.
———. 1978. *Selected Writings, 1909–1953*. 2 vols. Ed. Maria Reichenbach and Robert S. Cohen. Trans. Elizabeth Hughes Schneewind. The Vienna Circle Collection, vol. 4. Dordrecht, Boston, and London: Reidel.
Reichenbach, M. 1978. "The Student Years: Introductory Note to Part I." In *Selected Writings, 1909–1953,* by H. Reichenbach, ed. M. Reichenbach and R. S. Cohen. Trans. E. H. Schneewind, vol. 1, 91–101. Dordrecht, Boston, and London: Reidel, 1978.
Reisch, G. 1991. "Did Kuhn Kill Logical Empiricism?" *Philosophy of Science* 58: 264–77.
———. 1995. "A History of the *International Encyclopedia of Unified Science*." Ph.D. diss., University of Chicago.
———. 1998. "Pluralism, Logical Empiricism, and the Problem of Pseudoscience." *Philosophy of Science* 65: 333–48.
Richardson, A. 1998a. *Carnap's Construction of the World: The Aufbau and the Emergence of Logical Empiricism*. Cambridge: Cambridge University Press.
———. 1998b. "Science as Will and Representation: Carnap, Reichenbach, and the Sociology of Science." *PSA 98: Proceedings of the 1998 Biennial Meeting of the Philosophy of Science Association. Part II, Symposia Papers,* ed. Don Howard. *Philosophy of Science* S67: S151–S162.
———. 2000. "Tolerance, Internationalism, and Scientific Community in Philosophy: Political Themes in the Philosophy of the Vienna Circle and Their Contemporaries." Paper

delivered at HOPOS 2000: Third International History of Philosophy of Science Conference, Vienna, July 6–9.
Rudner, R. 1953. "Value Judgments in the Acceptance of Theories." In *The Validation of Scientific Theories,* ed. P. G. Frank, 31–35. Boston: Beacon Press. Reprint, New York: Collier Books, 1956.
Runggaldier, E. 1984. *Carnap's Early Conventionalism: An Inquiry into the Historical Background of the Vienna Circle.* Amsterdam: Rodopi.
Rutherford, F. J. 1960. "Discussion: Frank's Philosophy of Science Revisited." *Philosophy of Science* 27: 183–86.
Ryan, A. 1995. *John Dewey and the High Tide of American Liberalism.* New York and London: W. W. Norton.
Salmon, W., and G. Wolters, eds. 1994. *Logic, Language, and the Structure of Scientific Theories: Proceedings of the Carnap-Reichenbach Centennial, University of Konstanz, 21–24 May 1991.* Pittsburgh, Pa.: University of Pittsburgh Press; Konstanz: Universitätsverlag.
Schlick, M. 1910. "Das Wesen der Wahrheit nach der modernen Logik." *Vierteljarhsschrift für wissenschaftliche Philosophie und Soziologie* 34: 386–477.
———. 1915. "Die philosophische Bedeutung des Relativitätsprinzips." *Zeitschrift für Philosophie und Philosophische Kritik* 159: 129–75. Page numbers and quotations from the translation in *Philosophical Papers,* ed. H. L. Mulder and B. F. B. van de Velde-Schlick. Trans. P. Heath, vol. 1, 153–89. Dordrecht and Boston: Reidel, 1979.
———. 1917. *Raum und Zeit in den gegenwärtigen Physik. Zur Einführung in das Verständnis der allgemeinen Relativitätstheorie.* Berlin: Julius Springer. Page numbers and quotations from the translation in *Philosophical Papers,* ed. H. L. Mulder and B. F. B. van de Velde-Schlick. Trans. P. Heath, vol. 1, 190–269. Dordrecht and Boston: Reidel, 1979.
———. 1934. "Ueber das Fundament der Erkenntnis." *Erkenntnis* 4: 79–99. Page numbers and quotations from the translation in *Philosophical Papers,* ed. H. L. Mulder and B. F. B. van de Velde-Schlick. Trans. P. Heath, vol. 2, 370–87. Dordrecht and Boston: Reidel, 1979.
———. 1935. "Sind die Naturgesetze Konventionen?" In *Actes du Congrès International de Philosophie Scientifique, Paris 1935,* vol. 4, *Induction et Probabilité,* 8–17. Actualités Scientifique et Industrielles, vol. 391. Paris: Hermann, 1936.
———. 1979. *Philosophical Papers.* 2 vols. Ed. H. L. Mulder and B. F. B. van de Velde-Schlick. Trans. P. Heath. Vienna Circle Collection, vol. 11. Dordrecht and Boston: Reidel.
Seeger, R. J. 1965. "Philipp G. Frank, Physicist Extraordinaire." In *In Honor of Philipp Frank,* ed. Robert S. Cohen and Marx W. Wartofsky, xxvi–xxviii. Boston Studies in the Philosophy of Science, vol. 2. New York: Humanities Press.
Sellars, R. W. 1932. *The Philosophy of Physical Realism.* New York: Macmillan.
Somerville, J. 1946. *Soviet Philosophy: A Study of Theory and Practice.* New York: Philosophical Library.
Stadler, F. 1982. *Vom Positivismus zur "Wissenschaftlichen Weltauffassung." Am Beispiel der Wirkungsgeschichte von Ernst Mach in Österreich von 1895 bis 1934.* Vienna and Munich: Löcker Verlag.
———, ed. 1993. *Scientific Philosophy: Origins and Developments.* Vienna Circle Institute Yearbook, vol. 1. Dordrecht, Boston, and London: Kluwer.
———. 1995. "The Vienna Circle and the University of Vienna." In *Vertreibung der Vernunft: The Cultural Exodus from Austria,* ed. F. Stadler and P. Weibel. 44–55. Vienna and New York: Springer-Verlag, 1995.

———. 1997. *Studien zum Wiener Kreis. Ursprung, Entwicklung und Wirkung des Logischen Empirismus im Kontext.* Frankfurt: Suhrkamp. English translation: *The Vienna Circle: Studies in the Origins, Development, and Influence of Logical Empiricism.* Vienna and New York: Springer-Verlag, 2001.

———. 2000. "On the Political Meaning and Cultural Context of Logical Empiricism." Paper delivered at HOPOS 2000: Third International History of Philosophy of Science Conference, Vienna, July 6–9.

Stadler, F., and P. Weibel, eds. 1995. *Vertreibung der Vernunft: The Cultural Exodus from Austria.* Vienna and New York: Springer-Verlag.

Suppes, P. 1954. "Some Remarks on Problems and Methods in the Philosophy of Science." *Philosophy of Science* 21: 242–48.

Uebel, T. E., ed. 1991. *Rediscovering the Forgotten Vienna Circle: Austrian Studies on Otto Neurath and the Vienna Circle.* Boston Studies in the Philosophy of Science, vol. 133. Dordrecht, Boston, and Lancaster: Kluwer.

———. 1992. *Overcoming Logical Positivism from Within: The Emergence of Neurath's Naturalism in the Vienna Circle's Protocol Sentence Debate.* Amsterdam and Atlanta, Ga.: Rodopi.

———. 1998. "Logical Empiricism and the Sociology of Knowledge: The Case of Neurath and Frank." *PSA 98: Proceedings of the 1998 Biennial Meeting of the Philosophy of Science Association. Part II, Symposia Papers,* ed. Don Howard. *Philosophy of Science* S67: S138–S150.

Van der Linden, H. 1988. *Kantian Ethics and Socialism.* Indianapolis, Ind.: Hackett.

Von Aster, E. 1931. "Kritische Bemerkungen zu Hugo Dinglers Buch 'Das Experiment.'" *Erkenntnis* 2: 1–20.

Wartofsky, M. 1982. "Positivism and Politics: The Vienna Circle as a Social Movement." In *Schlick und Neurath. Ein Simposion,* ed. R. Haller, 79–101. Amsterdam: Rodopi.

Weaver, W. 1945. *Free Science.* Society for Freedom in Science, Occasional Pamphlet, no. 3. Oxford: Potter Press.

Will, F. L. 1951. "Relativism and Experimental Inference." *Philosophy of Science* 18: 155–69.

Wilson, D. J. 1995. "Fertile Ground: Pragmatism, Science, and Logical Positivism." In *Pragmatism: From Progressivism to Postmodernism,* ed. R. Hollinger and D. Depew, 122–41. Westport, Conn., and London: Praeger, 1995.

Wittich, D. 1971. "Einleitung des Herausgebers." In *Vogt, Moleschott, Büchner. Schriften zum kleinbürgerlichen Materialismus in Deutschland,* ed. Dieter Wittich, v–lxxxii. Berlin: Akademie-Verlag.

Wolters, G. 1987. *Mach I, Mach II, Einstein und die Relativitätstheorie. Eine Fälschung und ihre Folgen.* Berlin and New York: Walter de Gruyter.

Michael Friedman

3
Hempel and the Vienna Circle

I first met Carl Hempel when I was an undergraduate at Queens College in New York. Hempel had taught there in the years 1940–48, as his first regular position at an American university, and he had now returned (I believe it was in the academic year 1967–68) for a period of two weeks as a Distinguished Visitor. I had just become seriously interested in philosophy of science, and I vividly remember the sense of profound excitement I felt while attending the variety of talks, seminars, and discussions Hempel held during this visit. His clarity and acuity of mind, wide-ranging knowledge and interests, and singular kindness and enthusiasm were exhilarating, and I resolved then and there to apply for graduate study at Princeton, where Hempel taught from 1955 until 1975. I attended Princeton in the years 1969–72 and took several inspiring seminars from Hempel. I remember one, in particular, when Hempel was working out his ideas on the problem of "provisoes": We in attendance were simply enthralled by the experience of witnessing a major philosopher fundamentally change his mind, and in the most open and relaxed way imaginable, about a central question of philosophical methodology that had essentially shaped much of his previous work.

I will here explore the intellectual roots of Hempel's thought—including the fundamental change of mind just mentioned—in his earlier encounter with the Vienna Circle. We will see, perhaps somewhat surprisingly, that virtually all the seeds of his philosophical development can be found there, including his later turn away from what has come to be identified as the characteristic mode of philosophizing of logical empiricism to what he himself calls a more pragmatic and naturalistic approach.

Hempel received the bulk of his philosophical education at the University of Berlin, which he attended from 1925 to 1934, when he completed a doctoral dissertation, principally under the direction of Hans Reichenbach, on the logical analysis of the concept of probability. Hempel had become acquainted with Reichenbach when the latter arrived at the university in 1926, whereupon Reichenbach became one of Hempel's most important influences and teachers at Berlin, along with the methodologically oriented psychologists Wolfgang Köhler and Kurt Lewin. Hempel was a member of Reichenbach's Society for Empirical Philosophy, which included the philosopher-logicians Walter Dubislav and Kurt Grelling and often attracted a number of other distinguished visitors, such as, in particular, the logician Paul Bernays, with whom Hempel had earlier studied in Göttingen. In 1928, Hempel read Rudolf Carnap's *Der logische Aufbau der Welt* (1928a) and *Scheinprobleme in der Philosophie* (1928b) and immediately decided to study in Vienna for a term—a decision that was reinforced, so he reports, by meeting Carnap in person at the first Tagung für die Erkenntnislehre der exakten Wissenschaften at Prague in September 1929. Hempel visited the University of Vienna in the winter term 1929–30, where he attended lectures and seminars given by Moritz Schlick, Carnap, and Friedrich Waismann. Most importantly, with a letter of introduction from Reichenbach, he was invited to attend the discussions of Schlick's Philosophical Circle, which included Carnap, Schlick, Waismann, Otto Neurath, Hans Hahn, Herbert Feigl, Kurt Gödel, and Karl Menger.[1]

Hempel's visit to Vienna came at a propitious and portentous moment, during the beginnings of a central debate then developing within the Vienna Circle—what we now know as the protocol-sentence debate—wherein a fundamental split developed between Schlick and Waismann, on the one side, and Neurath, followed quickly by Carnap, on the other. The two sides can be seen as adopting opposing stances towards the *Aufbau*, with the Schlick-Waismann camp (often referred to as the "right wing" of the Circle) pushing in a foundationalist-subjectivist direction and the Neurath-Carnap camp (often referred to as the "left wing" of the Circle) pushing in a "physicalist" or intersubjectivist direction. Beginning in 1928, with his review in the socialist journal *Der Kampf,* Neurath criticized the *Aufbau* for its reliance on "methodological solipsism," whereas Schlick can be read as objecting to the *Aufbau* from the opposite direction, on behalf of the need, in the end, for purely ostensive reference to individual subjective experience.[2] It is clear from Hempel's own reports of his visit that the Circle philosophers he was most impressed by, and felt most attracted to, were Neurath and Carnap, and also that he was very much interested in the emerging protocol-sentence debate.[3] So it is by no means surprising that Hempel's very first publications, appearing in *Analysis* in the years 1935–36 (Hempel 1935a, 1935b, 1936), are central contributions to this debate, in which Hempel weighs in clearly and explicitly, at a more public and developed stage of the debate, on the side of the "radical physicalism" of the Neurath-Carnap camp.

Nevertheless, it is important to see that the Neurath-Carnap camp was far from a united front. Neurath, in "Soziologie im Physikalismus" (1932a,

§1) begins the public phase of the debate by arguing that metaphysics is best definitively overcome by rejecting any point of view outside the language of unified science—including that of *Tractarian* "elucidations" purporting to relate language to "experience as a whole," "the world," or "the given" from some "not yet linguistic" standpoint, "as Wittgenstein and certain representatives of the 'Vienna Circle' seek to do." On the contrary, metaphysics can only be finally overcome by resolutely remaining entirely within the linguistic world of unified science itself:

> Unified science formulates statements, corrects them, makes predictions; but it cannot itself anticipate its future state. There is not a *"true" system of statements* over and above the present system of statements. It is meaningless to speak of such a thing, even as a limiting concept. *We can only ascertain that we are operating today with the space-time system to which that of physics corresponds,* and thereby achieve successful predictions. This system of statements is that of unified science—this is the standpoint which may be designated as *physicalism* (cf. [Neurath 1931], 2). . . .
>
> The one unified science peculiar to a definite historical period, as physicalism, proceeds, remote from all meaningless sentences, from statement to statement, which are combined together in a self-consistent system as a tool for successful predictions, and thus life. (1932a, 397–98)

Neurath thus emphasizes in the strongest possible terms that "physicalism," for him, means precisely this essentially interlinguistic standpoint. And from this standpoint, moreover, as Neurath explains, the concept of "successful prediction" also involves no reference whatsoever to an extralinguistic "given" (1932a, §2). On the contrary, it entails only that we compare the statements of unified science with other such statements—those that express the percipient state of an observer and are themselves expressible in neurophysiological or behavioristic terms within the very same physicalistic language. Accordingly, Neurath proceeds to criticize the *Aufbau* for its reliance on a "phenomenal language" and what Carnap there calls "methodological solipsism": "One cannot scientifically formulate this thesis of 'methodological solipsism'—as even Carnap would likely concede—but [also] one cannot even use it any longer to indicate a definite standpoint, which is an alternative to some other standpoint, because there exists only *one* physicalism. Everything scientifically formulable is contained within it" (1932a, 401–2).

Carnap, in "Die physikalische Sprache als Universalsprache der Wissenschaft" (1932a) and "Über Protokollsätze" (1932b), takes himself to be responding to, and indeed accepting, Neurath's criticisms. Unlike Neurath, however, Carnap envisions two alternative forms for the total language of science. In the first alternative, there is a separate language of protocol sentences distinct from the "system language" or physical language. In this case, in order to test sentences in the system language, there must be rules of translation of the latter into the protocol language. In the second alternative, by contrast, protocol sentences belong to the system language, either in Neurath's special linguistic form involving personal names or, following a suggestion by Karl Popper, having any linguistic form whatsoever. And, particularly if we follow Popper's suggestion, it is then entirely clear that protocol sentences are in no way epistemologically privileged. For this reason, above all, the Neurath-Popper form of language is actually preferable; for, as Carnap puts it, this language is indeed most effective in overcoming the "residue of idealistic absolutism . . . in the logical positivism of our Circle"—the residue of "an absolutism of the 'given', of 'experience', of the 'immediate phenomena'"—and Neurath, in particular, deserves full credit for this antimetaphysical move, since, within the Vienna Circle, "Neurath was the first to turn decisively against this absolutism, in that he rejected the unrevisability of protocol-sentences" (Carnap 1932b, 228).

In regard to this preference or choice of the Neurath-Popper form of language, moreover, Carnap appeals to his own antimetaphysical standpoint of "tolerance," a standpoint he is simultaneously developing in his more general theory of logical syntax or metalogic:

> Not only the question whether protocol-sentences occur outside or inside the system language, but also the further question of their more exact characterization, is to be answered, it seems to me, not by an assertion, but rather by a convention *[Festsetzung]*. Although I earlier ([Carnap 1932a], 438) left this question open and only indicated a few possible answers, I now think that the different answers do not contradict one another. They are to be understood as suggestions for conventions; the task consists in investigating these various possible conventions with regard to their consequences and in testing their practical utility. (1932b, 216)[4]

And it is here that the fundamental difference between Carnap's antimetaphysical standpoint and Neurath's first becomes fully explicit. For Carnap, metaphysics is overcome by adopting the metalogical standpoint of logical syntax, the discipline he soon comes to call *Wissenschaftslogik* (logic of science), relative to which a plurality of alternative forms for the total language of science is possible and legitimate. For Neurath, as we have observed, we can only overcome metaphysics by deliberately confining ourselves to a single language—the "universal-slang" of physicalism.

It is here, too, that we see the full force of Neurath's naturalism. There is only the language of unified empirical science. So the discipline Carnap calls *Wissenschaftslogik* must itself belong to empirical science, that is, to the psychological-sociological study of the actual linguistic behavior of empirically and historically given scientists as they continuously fashion and refashion the "ship" of knowledge (that is, the totality of currently accepted statements) without ever being in a position, in Neurath's famous words, "to build it anew out of the best constituents"—so that there is also a limit to the logical precision we can require or attain, since "imprecise, unanalyzed terms *('Ballungen')* . . . are always in some way constituents of the ship" (1932b, 206). For Carnap, by contrast, *Wissenschaftslogik* is a fully precise and rigorous subdiscipline of mathematical logic, where our task is not to describe actual linguistic behavior (which, of course, always remains less than fully precise) but, rather, to investigate in a fully precise way the consequences of adopting one or another *proposal* for the logical form of the total language of science. Carnap thus hopes to overcome traditional metaphysics by reinterpreting its "theses" as logico-linguistic proposals. Neurath, by contrast, will have none of this reinterpretive project; rather, he aims at a complete dismissal of the metaphysical tradition on behalf of empirical science.

We know that the fundamental contrast between Neurath's naturalism and Carnap's conception of *Wissenschaftslogik* is central to understanding Hempel's own later turn from the latter point of view to a version of the former one. (I was privileged to have heard two of Hempel's last public statements of this contrast: the first at a conference in honor of Thomas Kuhn at the Massachusetts Institute of Technology in May 1990; the second at the first biannual Pittsburgh-Konstanz Colloquium in the Philosophy of Science, on the occasion of Hempel's receipt of an honorary degree from the University of Konstanz, in May 1991. On both occasions Hempel joyfully described his own conversion from the point of view of Carnapian "explication" or "rational reconstruction" to the point of view of Kuhnian historical and sociocultural naturalism as a return to Neurath's original conception of the ineliminable necessity of *"Ballungen."*)[5] In order fully to understand the roots of Hempel's later development, however, it is first necessary to consider the remainder of the protocol-sentence debate, beginning with the exchange between Neurath and Schlick that precipitated Hempel's first official involvement in the pages of *Analysis*.

Schlick's "Über das Fundament der Erkenntnis" (1934) is a response to both Neurath's "Protokollsätze" (1932b) and Carnap's "Über Protokollsätze" (1932b). Schlick's main complaint, as is well known, is that the holistic conception represented by the latter two papers, according to which all sentences of science are revisable through confrontation with other such sentences, leads to a version of the "coherence theory of truth." Any logically consistent system of sentences, on this conception, is thus an equally good candidate for the "true" system of sentences.

But what empiricism requires instead, according to Schlick, is that some particular class of statements, those reporting the "raw facts" of immediate experience, be absolutely fixed in the process of empirical testing. And Schlick finds such fixed assertions, as is also well known, in his *"Konstatierungen,"* assertions that, according to Schlick, "express a *present* fact of one's own 'perception' or 'experience'" (1934, 89).

Neurath's "Radikaler Physikalismus und 'Wirklicher Welt'" (1934) is a point-by-point rebuttal of Schlick. Neurath, of course, rejects the demand for fixed and absolutely certain assertions against which all others are to be tested. And in accordance with his (and Carnap's) earlier papers, he rejects all talk of a comparison between sentences, on the one side, and "reality" or "experience," on the other: What takes place in unified science is simply a comparison, and consequent mutual adjustment, of sentences with one another. But what is of perhaps most interest is the way in which Neurath responds to Schlick's problematic of the "coherence theory of truth" and the possibility, in particular, that any logically consistent system of sentences may count as "true." Neurath argues that "[t]he terms 'sentence,' 'language,' etc. must be historically-sociologically defined" (1934, 356). Moreover, when one takes such a historical-sociological perspective on science, one sees, according to Neurath, that "[t]he practice of life reduces all the ambiguity [arising from alternative mutually consistent systems of hypotheses] very quickly," so that in real life "the individual scarcely has the power properly to carry out *one* system, to speak nothing of several systems" (352). In science as an actual, historically given social system—in other words, as opposed to any mere logical collection of propositions imagined by philosophers—Schlick's problem posed by the threat of mere logical coherence as our "criterion of truth" simply does not arise.

In January 1935 Hempel was invited by Susan Stebbing to present a lecture in London on the latest developments within the Vienna Circle and in particular on the exchange between Neurath and Schlick that had just appeared in the pages of *Erkenntnis.* Hempel's "On the Logical Positivists' Theory of Truth" (1935a) is a "hurriedly condensed" version of this lecture[6]—prepared, as is indicated at the end of the published version, in December 1934. Hempel accepts Schlick's characterization of "the logical positivists' theory of truth" (that is, the theory of Neurath and Carnap) as a "coherence-theory," although of a "restrained" kind (1935a, 49). Hempel then sets out to describe the historical evolution of this theory from the ideas of Wittgenstein's *Tractatus,* understood as a "correspondence-theory." A decisive step in this evolution, and one that finally enables us to clear up the confusions surrounding the notion of a correspondence or comparison between sentences and "facts," is Carnap's introduction of the distinction between the formal and the material modes of speech:

> As Carnap has shown, each non-metaphysical consideration of philosophy belongs to the domain of Logic of Science, unless it concerns an empirical question and is proper to empirical science. And it is possible to formulate each statement of Logic of Science as an assertion concerning certain properties and relations of scientific propositions only. So also the concept of truth may be characterized in this formal mode of speech, namely, in a crude formulation, as a sufficient agreement between the system of acknowledged protocol-statements and the logical consequences which may be deduced from the statement and other statements which are already adopted. (Hempel 1935a, 54)

Hempel thus appeals to Carnap's conception of *Wissenschaftslogik,* in particular, in formulating an explicit statement—admittedly "crude"—of the coherence theory of truth.

It is precisely this characterization of the concept of truth, however, that allows us to respond to Schlick's main objection—the specter of alternative, internally consistent but mutually incompatible "true" systems of sentences:

> As Carnap and Neurath emphasize, there is indeed no formal, no logical difference between the two compared systems, but an *empirical* one. The system of protocol statements which we call true, and to which we refer in every day life and science, may only be characterized by the historical fact, that it is the system which is actually adopted by mankind, and especially by the scientists of our cultural circle; and the "true" statements in general may be characterized as those which are sufficiently supported by that system of actually adopted protocol statements. (Hempel 1935a, 57)

It is theoretically possible, Hempel continues, that "the protocol statements produced by different men would not admit the construction of one unique system of scientific statements," but fortunately this abstract possibility is not realized in fact, since "by far the greater part of scientists will sooner or later come to an agreement, and so, as an empirical fact, a perpetually increasing and expanding system of coherent statements and theories results from their protocol statements" (57). In this way, by synthesizing Carnap's conception of *Wissenschaftslogik* with Neurath's historical-sociological naturalism, Hempel hopes to provide a definitive clarification of the Neurath-Schlick debate.

By September of that same year, however, at the International Congress for Scientific Philosophy held in Paris, the Neurath-Carnap synthesis has become fully and explicitly unraveled. Carnap, motivated largely by his recent acceptance of Alfred Tarski's semantical definition of truth, presented two papers (published in 1936 in the proceedings of the congress) in which he draws a pair of fundamental distinctions effectively dissolving the protocol-sentence debate.[7] In "Warheit und Bewährung" (1936b) Carnap introduces a sharp distinction between the concepts "true," on the one side, and "confirmed" or "scientifically accepted," on the other (18). The former is a timeless, logical concept, whose unobjectionable formal definition within metalogic has just been achieved by Tarski. The latter, by contrast, is a temporally relative, nonlogical concept, whose adequate articulation, according to Carnap, requires "not a logical, but rather a (psychological-sociological) presentation belonging to empirical science" (19). From this point of view, the Neurath-Hempel characterization of "true," as applying to those sentences currently accepted by the community of scientists, is thus seen to rest on a fundamental confusion. And it then follows, in addition, that there are importantly correct elements in Schlick's opposing conception. For, from the point of view of the (psychological-sociological) theory of confirmation, we must certainly admit a procedure or operation Carnap calls confrontation of a sentence with observation. In this case, by Tarski's theory of truth, there is a perfectly good sense in speaking of a comparison between facts and propositions after all. As Carnap explains, "[w]hen one sees a key, one may accept the sentence, 'here lies a key,'" and "[a]t this point the definition of the concept of truth enters into the problem of confirmation" (1936b, 21).

Carnap 1936b thus emphasizes a sharp distinction between truth and confirmation; it characterizes truth as a logical concept, whereas confirmation, by contrast, is counted as a nonlogical, "psychological-sociological" concept. The main point of Carnap's "Von der Erkenntnistheorie zur Wissenschaftslogik" (1936a) is that this second sharp distinction between logical and psychological considerations is absolutely central to fully clarifying the nature of properly scientific philosophy, especially as it has been practiced within the Vienna Circle:

> It seems to me that *epistemology, in the form it has taken until now, is an unclear mixture out of psychological and logical constituents.* That holds also for the works of our Circle, not excepting my own earlier works. Many unclarities and misunderstandings result from this. Thus, a short time ago, for example, an article in *Erkenntnis* called forth a great variety of hesitations and objections and lively discussions through its apparently logical theses, until the author finally explained that his statements are not meant as logical, but rather as psychological analysis. From this we see how important it is, in the case of every so-called epistemological discussion, to make clear and explicit whether logical or psychological questions are meant. (1936a, 36–37)

There can be very little doubt, moreover, that the article in question is precisely Schlick's 1934 "Über das Fundament der Erkenntnis," and that the hesitations and objections in question are precisely those then expressed by Neurath and Hempel.[8] Carnap's present point of view, therefore, is that the entire Schlick-Neurath-Hempel debate rests on a fundamental unclarity about the distinction between logical and psychological questions—an unclarity which is fully resolved, in Carnap's mind, by his own "Wahrheit und Bewährung" (1936b).

Carnap's first major publication in English was "Testability and Meaning" (1936–37). This famous paper further develops and applies the basic point of view of the earlier works (Carnap 1936a, 1936b). Carnap repeatedly insists, in particular, on the importance of distinguishing psychological and logical investigations. Indeed, this distinction determines the very structure of the paper, the heart of which consists of §2, "Logical Analysis of Confirmation and Testing," and §3, "Empirical Analysis of Confirmation and Testing." In describing this two-part structure in the introductory §1, Carnap remarks that "[t]he difficulties in discussions of epistemological and methodological problems are, it seems, often due to a mixing up of logical and empirical questions; therefore it seems desirable to separate the two analyses as clearly as possible" (1936–37, 421). And later in §1, he repeats the warning from "Von der Erkenntnistheorie zür Wissenschaftslogik" almost word for word: "In fact, however, epistemology in the form it usually takes—including many of the publications of the Vienna Circle—is an unclear mixture of psychological and logical components. We must separate it into its two kinds of components if we wish to come to clear, unambiguous concepts and questions" (1936–37, 429).

Here the fundamental tension between Carnap's conception of *Wissenschaftslogik* and Neurath's has become intolerable. According to Neurath's naturalistic understanding of this discipline, there is only the single unified language of empirical science. There is no room, therefore, for a metalanguage or syntax language describing the process of empirical testing from some idealized point of view outside the language of empirical science itself. Neurath's conception of *Wissenschaftslogik* is thus a historical-sociological one, wherein we describe how science, considered as an actual social system, operates with empirically and factually given real sentences and utterances (as opposed to mere "serial-structures" belonging to "pure syntax").[9] Moreover, from this historical-sociological point of view, as we have seen, there is a limit to the precision we can require or attain in the actual historical-social process. We can certainly introduce logical precision into our actual scientific methods, by axiomatization, for example, but it makes no sense either to represent or to replace our actual procedures by a fully precise logical version. There is an ineliminable residue of "imprecise, unanalyzed terms (*'Ballungen'*) . . . [that] are always in some way constituents of the ship."

In a letter to Carnap of October 27, 1935, sent along with his own comments on the manuscript version of Carnap 1936–37, comments that "are mostly of a purely technical nature," Hempel replies to a question posed to him by Carnap concerning whether Carnap should attempt to rewrite the manuscript in response to some fundamental objections from Neurath:

> Although I understand N[eurath']s tendencies in this connection, and would go along with them to a certain extent, I still have certain hesitations about recommending this radical procedure in the present situation: first, it would likely be very tedious to modify the entire article from this point of view, and second, it seems to me difficult to carry through the formal consideration concerning the testability of sentences with operators [that is, quantifiers] and the like with reference to such vague initial concepts. N[eurath] of course rages when he hears such arguments, but I say to him on the other side: if one presupposes that all concepts and sentences of the language about which one undertakes logical investigations are as vague and smeary *[schmierig]* as N[eurath] would like, then there is properly speaking no longer a point of entry for the least amount of logical analysis. (The language admitted by N[eurath] would be syntactically characterized thus: the sentences of the language are *schmierige Ballungen,* and no transformation rule [is] without exception.) (In this connection it seems to me that N[eurath] is entirely correct in his remark that one may shut oneself off from many insights by taking unsuitable models as one's basis. The question now is: Is logic in the form you presuppose an unsuitable model, and what can such an assertion, strictly taken, mean?) (Archives for Scientific Philosophy [hereafter ASP], 102-14-40)

Hempel thus appears to be clearly torn here between Neurath's naturalism, on the one side, and Carnap's conception of *Wissenschaftslogik,* on the other.

In an excited letter to Neurath of September 22, 1938, by contrast, Hempel seems strongly inclined in Neurath's direction—stimulated, this time, by a surprising source. Since this letter sheds considerable light, I believe, on Hempel's later development, I will here quote a substantial piece of it:

> Your remarks about discussion with Tarski interest me very much. On the whole I had a very pleasing impression, not only because T[arski] is in general very sharp and stimulating, but also more specifically in reference to the questions of empiricism. Among others, a conversation with him about the logic of testing empirical hypotheses made a very great impression on me. T[arski] thought, of course, that the Wittgensteinian idea of complete verifiability for empirical hypotheses is entirely naive; but also that, in his opinion, Carnap's logical theory in Test[ability] and Meaning, based on much more liberal principles, did *not* achieve what was desired: in fact he

is acquainted with no single example of a reduction-sentence that actually reduces a concept, say of physical theory, to concepts of the observation-language in materially correct fashion (i.e., so that the empirical investigator would agree). All examples known to him, e.g., C[arnap']s example "soluble," are schematizations, which the empirical [investigator] must view as inappropriate: in fact it can happen that a material is put in water, does not disappear, and yet is soluble. And no matter how many additional conditions and clauses one may add, "exceptions" are still always thinkable. . . . —In his opinion we so far have no theory that erects in materially adequate fashion a logical connection between theory, on the one side, and the realm of observations, on the other. . . . —Tarski's hesitations are shared (and are perhaps in part stimulated) by Wundheiler and Poznanski. However, whereas the latter, as T[arski] indicated, are still hopeful about the search for a logical bridge (as Carnap and, e.g., I, in his opinion, still take one to be constructible), it appears to him that it is at the very least not excluded that no such bridge can be forged in an adequate manner, and that the judgement of theories by means of observations perhaps follows instinctively, as it were, without being able to deduce theoretically the first predictions in the form of observation-sentences with the help of reduction-sentences, etc. (ASP 102-46-17)

Tarski is suggesting, then, that logic, in the form Carnap presupposes, is indeed an "unsuitable model." Whereas Carnap had appealed to Tarski's semantical conception in opposing, among other things, the sociological naturalism represented by Neurath's and Hempel's account of truth as community-wide agreement, a different, more skeptical side of Tarski can now be enlisted on behalf of precisely the *"schmierige Ballungen"* favored by Neurath.

What I find most remarkable here is that it is essentially Tarski's argument, although in a clearer and more explicit form, that constitutes the centerpiece of Hempel's "Provisoes" (1988). And this article, published, appropriately enough, in *Erkenntnis,* is in turn the centerpiece of Hempel's later conversion, noted several times above, from the Carnapian program of logical "explication" or "logical reconstruction" to a more naturalistic emphasis on historical, sociological, and other broadly "pragmatic" factors.[10]

Hempel raises the question of the logical relationship between two sentences: "*b* is a metal bar to which iron filings are clinging" and "*b* is a magnet" (1988, 148). The problem of provisoes shows us, surprisingly, that there is no deductive route from the second sentence to the first, even in the context of the whole of the theory of magnetism. For the theory of magnetism does not rule out the presence of disturbing factors (perhaps other counteracting magnetic forces, for example) that prevent the iron filings from clinging to the magnet in this case. We can arrive at the result that the iron filings will in fact cling only by adding a ceteris paribus clause, which, according to Hempel, remains necessarily "vague and elusive" (156). In particular, we cannot use such a ceteris paribus clause even to generate a determinate *probability* that the filings will cling, for, as Hempel explains, "surely, the theory of magnetism contains no sentences of this kind [probabilistic laws connecting our two sentences]; it is a matter quite beyond its scope to state how frequently air currents, disturbing further magnetic fields, or other factors will interfere with the effect in question" (153). And, more generally, he continues since "[a] scientific theory propounds an account of certain kinds of empirical phenomena, but it does not pronounce on what other kinds there are" (158), there is no way precisely to formulate a proviso ruling out *all* disturbing factors within any particular such theory.

This problem of provisoes is of course identical to the problem Tarski had raised for Carnap's account of theoretical terms in "Testability and Meaning." Here "magnetic" plays the role of "soluble," and "iron filings cling" is the observational counterpart of "dissolves in water." Just as Tarski, according to Hempel's 1938 letter to Neurath, had pointed out that a material may be soluble without dissolving in water, and that "no matter how many additional conditions and clauses one may add, 'exceptions' are still always thinkable," Hempel here points out that the very same situation holds with respect to the logical relationship between his two sentences. Accordingly, Hempel makes it perfectly clear how his problem bears on Carnap's account of reduction sentences: "The foregoing considerations show in particular that when a theory contains interpretive sentences in the form of explicit definitions or of Carnapian reduction chains based on the antecedent vocabulary, the applicability of these sentences is usually subject to the fulfillment of provisoes; they cannot be regarded as unequivocal complete or partial criteria of applicability for theoretical expressions" (1988, 151). Hempel's rejection in 1988 of the Carnapian program of "Testability and Meaning" is therefore entirely of a piece with the Tarskian rejection of this same program that he had earlier reported to Neurath some fifty years before.

It is by no means surprising, then, that the conclusion of Hempel's "Provisoes" points in the same naturalistic direction that had been most strongly defended, in the context of the earlier discussions, by Neurath:

> There is a distinct affinity, I think, between the perplexing questions concerning the appraisal of provisoes in the application of scientific theories and the recently much discussed problems of theory choice in science.
>
> As Kuhn in particular has argued in detail, the choice between competing

theories is influenced by considerations concerning the strength and the relative importance of various desirable features exhibited by the rival theories; but these considerations resist adequate expression in the form of precise explicit criteria. The choice between theories in the light of those considerations, which are broadly shared within the scientific community, is not subject to, nor learned by means of, unambiguous rules. Scientists acquire the ability to make such choices in the course of their professional training and careers, somewhat in the manner in which we acquire the use of our language largely without benefit of explicit rules, by interaction with competent speakers.

Just as, in the context of theory choice, the relevant idea of superiority of one theory to another has no precise explication and yet its use is strongly affected by considerations shared by scientific investigators, so in the inferential application of theories to empirical contexts, the idea of the relevant provisoes has no precise explication, yet it is by no means arbitrary and its use appears to be significantly affected by considerations akin to those affecting theory choice. (1988, 162)

That Hempel's later turn in a strongly Kuhnian direction may at the same time be viewed as a return to the conception earlier represented within the Vienna Circle by Neurath is thus perhaps even more true than Hempel himself ever realized.

The episodes we have been reviewing form a set of brackets, as it were, around the most productive and influential phase of Hempel's career, which consists, as is well known, of a series of papers on confirmation and explanation—beginning with Hempel 1943, Hempel 1945, Hempel and Oppenheim 1945, and Hempel and Oppenheim 1948—which dominated discussion in the discipline now known as philosophy of science for the better part of two decades.

These papers show Hempel as the master of Carnapian "explication," dedicated, above all, to finding a precise and explicit characterization in purely formal-logical terms of the crucial relationship between scientific theory, on the one side, and observational statements, on the other. Whether we look at this relationship in terms of confirmation (of theory by observational evidence) or explanation (of observational statements by theory), the central ambition is that it be reconstructed as perfectly precise and explicit and, therefore, as "objective." This point stands out most clearly, perhaps, in the conclusion of §2 of Hempel's "Studies in the Logic of Confirmation" (1945):

> Perhaps there are no objective criteria of confirmation; perhaps the decision as to whether a given hypothesis is acceptable in the light of a given body of evidence is no more subject to rational, objective rules than is the process

of inventing a scientific hypothesis or theory; perhaps, in the last analysis, it is a "sense of evidence," or a feeling of plausibility in view of the relevant data, which ultimately decides whether a hypothesis is scientifically acceptable. This view is comparable to the opinion that the validity of a mathematical proof or of a logical argument has to be judged ultimately by reference to a feeling of soundness or convincingness; and both theses have to be rejected on analogous grounds: they involve a confusion of logical and psychological considerations. . . . A rational reconstruction of the standards of scientific validation cannot, therefore, involve reference to a sense of evidence; it has to be based on objective criteria. In fact, it seems reasonable to require that the criteria of empirical confirmation, besides being objective in character, should contain no reference to the specific subject matter of the hypothesis or of the evidence in question; it ought to be possible, one feels, to set up purely formal criteria of confirmation in a manner similar to that in which deductive logic provides purely formal criteria for the validity of deductive inference. (8–9)

A more explicit argument in favor of Carnapian "explication," phrased in just the terms that had most sharply divided Carnap from Neurath (and also, at least in part, from Hempel himself) in the middle to late 1930s, could hardly be constructed or imagined.

What is the explanation for this striking turn of events? Why does Hempel now find himself firmly in the Carnapian camp, without even a hint, it seems, of the fiery opposition such a purely logical perspective on science had once provoked within the Vienna Circle (and for which he himself had once expressed considerable sympathy)? We are here faced, I believe, with one of the many consequences of the full-scale intellectual migration that moved the majority of "scientific philosophers" from Europe to the United States in the middle to late 1930s. This cataclysmic political and geographical shift, as recent scholarship is just beginning to reveal, had the most fundamental intellectual consequences as well. In the case of the logical empiricist movement, in particular, it entailed a process of thoroughgoing adjustment and accommodation to a new cultural and political climate, a new language, and a dramatically different intellectual environment.

The Vienna Circle was founded in the midst of the cultural, political, and intellectual turmoil of the Weimar period. Neurath, in fact, viewed the Vienna Circle as the counterpart, within philosophy, of the movement for a *neue Sachlichkeit* (new soberness, objectivity) typified by the Dessau Bauhaus—and, indeed, as providing the philosophical underpinnings for a radical form of Marxian socialism. In this perspective, he was quickly joined by Carnap, although the latter, to be sure, had a much less explicitly politicized conception of philosophy than did the former. In opposition to the more genteel and individualistic liberalism championed by Schlick, however, both Neurath and Carnap viewed their enterprise as part of a revolutionary intellectual movement, deeply intertwined with the other mass movements of the time. And it is in this context that both Neurath and Carnap proposed radical transformations of the discipline of philosophy itself—the former, in the direction of a historical-sociological naturalism with fundamentally Marxist revolutionary ambitions; the latter, in the direction of a logico-linguistic pluralism with deep affinities to the currently popular movements for an international language. In this sense, the opposition between a "right wing" and a "left wing" of the Vienna Circle that surfaced during the protocol-sentence debate had very explicit cultural and political significance.[11]

The fall of Weimar in 1933 set off a massive intellectual migration to the New World. But the complex cultural struggles dominating the intellectual life of central Europe during the Weimar period could not be easily transplanted onto U.S. soil. In the case of the logical empiricists, in particular, the revolutionary context and rhetoric of radical philosophical transformation, especially in light of its explicitly Marxist overtones, had to be quickly forgotten, as the erstwhile "scientific philosophers" from central Europe were embraced by the more down-to-earth and pragmatically minded American logicians and philosophers of science Charles W. Morris, W. V. O. Quine, Nelson Goodman, and Ernest Nagel. Moreover, key contributors to the old "scientific philosophy" were left behind, for neither Schlick nor Neurath made it to the New World: The former was murdered by a deranged student at the University of Vienna in 1936, the latter died in exile in England in 1945.

Carnap, with Morris's help, settled at the University of Chicago in 1936. There, in a philosophical environment that was generally hostile to him, he began to work out his ideas on semantics, the logic of modalities, and the logical foundations of probability in the late 1930s and early 1940s. And it was at this time that he first developed the idea of philosophy as "explication," the process of reconstructing initially vague and imprecise concepts in terms of fully exact, logically analyzed ones. This idea, to be sure, was certainly implicit in much of Carnap's earlier work—in the "logical reconstructions" of the *Aufbau,* for example, and in the construction of various types of formal languages (each intended to represent one or another philosophical position) characteristic of the logical syntax period. In the 1940s, however, the idea was extracted, as it were, from the radical programmatic setting of the earlier period, and this allowed Carnap, among other things, to forge deliberate links with the much less threatening and politically loaded notion of "analysis" deriving from British philosophy. As Ronald Giere points out, when Carnap introduces the idea of "explication" in §2 of his 1947 book *Meaning and Necessity,* for example, he adds a footnote to C. H. Langford's article on "analysis" in the G. E. Moore volume of the Library of Living Philosophers, published in 1942 (Giere 1996, 340).

With financial help from the Rockefeller Foundation, Carnap brought Hempel and his close friend Olaf Helmer to the University of Chicago for the academic year 1937–38. As Carnap explains, "[t]he three of us talked often about logical problems, mainly those of semantics, which I was trying to develop systematically" (1963, 35). Another close friend of Hempel's, Paul Oppenheim, with whom Hempel had worked intensively in Brussels in the years 1935–37, emigrated to Princeton, New Jersey, in 1939. And it was in close cooperation with both Helmer and Oppenheim, as we know, that Hempel worked out the central ideas of his fundamental papers of the 1940s.

Hempel's thoroughgoing immersion in the practice of Carnapian "explication" during this period therefore makes perfectly good sense. Of all the leading members of the logical empiricist movement, Hempel had always been on closest terms, from a personal point of view, with Carnap. Now, he owed to Carnap his very presence in a new country and a new cultural and intellectual environment. Being himself particularly gifted in logical analysis, moreover, Hempel found himself surrounded by close friends and colleagues interested in this philosophical methodology above all. It is no wonder, then, that Hempel's major contributions of the time closely parallel Carnap's, and, more specifically, Carnap's developing work on the logical foundations of probability. Thus, for example, the argument for the "objective" and fundamentally logical character of the concept of confirmation cited above from Hempel's "Studies in the Logic of Confirmation," based, as it is, on Carnap's old, sharp distinction between psychological and logical questions, closely parallels the discussion in §§11–12 of Carnap's *Logical Foundations of Probability* (1950).[12]

The Carnap-Hempel ideal of "explication" remained the standard in philosophy of science until approximately 1960. At that time, as we know, winds of change began to blow over the discipline from a decidedly naturalistic direction. For Hempel personally, the work of Quine and Kuhn was particularly important. Quine's *Word and Object* (1960) propounded epistemological holism and attacked the sharp distinction, fundamental to Carnap's thought, between logical and empirical truth; and it used a motto from Neurath's "Protokollsätze" (1932b) as its emblem for such a holistic and naturalistic point of view. Kuhn's *Structure of Scientific Revolutions* (1962) developed a historical and sociological perspective on scientific change, which, especially in Hempel's own eyes, could be seen as applying such a fundamentally Neurathian perspective to the detailed study of science. The tension between a Carnapian and a Neurathian conception of philosophy of science, which had fundamentally shaped Hempel's earliest work but had long since lay dormant, was stimulated and came to life once again. With the work that finally resulted in "Provisoes"—which, as we have seen, recapitulates a defining moment from the earlier period—Hempel awoke from what he himself soon came to regard as a dogmatic slumber. This awakening is perhaps better described, therefore, as a reawakening.

I last saw Hempel in the spring of 1992, when I delivered some lectures at Princeton on Carnap's *Aufbau* and the development of logical empiricism. Hempel was extraordinarily gracious and kind, as always, but I still remember the gently bemused look on his face when I explained my revisionary reading of the *Aufbau* as owing less to traditional empiricism and more to contemporary strains of thought of a Kantian and neo-Kantian character. I remember one moment especially when Hempel irrepressibly exclaimed that I was now becoming more Carnapian than Carnap himself ever was. Soon after I returned home I received a paper in the mail. It was a duplicate of Hempel's own copy, signed to him by the author, of Neurath's 1928 review of Carnap's *Aufbau*.

Notes

Earlier versions of this paper were presented at a memorial for Carl G. Hempel at Princeton University and at the Logical Empiricism in North America Conference at Harvard University. I am indebted to Thomas Uebel and Thomas Ryckman for comments. A longer version appears in *Science, Explanation, and Rationality: Aspects of the Philosophy of Carl G. Hempel,* ed. James Fetzer (Oxford: Oxford University Press, 2000). I am grateful to James Fetzer and Oxford University Press for permission to publish the present version here.

    1. For Hempel's own recollections of these events, see Hempel 1973, 1981, 1991, and 1993.

    2. For an extremely careful and detailed analysis of the protocol-sentence debate, including the early hesitations about the *Aufbau* program expressed by Neurath and Schlick, see Uebel 1992. Neurath 1928 implicitly objects to "methodological solipsism" but also, and more explicitly, to Carnap's reliance on perfectly precise logical methods. Hempel's stay in Vienna comes at the end of what Thomas Uebel characterizes as "stage one" of the debate.

    3. See Hempel 1981, sec. 2, and also Hempel 1982: "In Fall 1929 I attended Schlick's seminar in Vienna and his lectures on natural philosophy with great interest, and I heard him and Neurath debate in the lively sessions of the Vienna Circle" (1982, 1).

    4. The more general theory of logical syntax, and the conception of philosophy as

Wissenschaftslogik, is developed at length in Carnap 1934. In §17, Carnap states "the *principle of tolerance*" as "we do not aim to set up prohibitions, but rather to stipulate conventions" ["wir wollen nicht Verbote aufstellen, sondern Festsetzungen treffen"].

5. Section 4 of Hempel 1993, entitled "Methodology of Science: Normative or Descriptive?" explains that "Neurath rejected Carnap's idea of a rational reconstruction, or explication, of science in terms of sentences all of which had precisely specified meanings" and further explains that Neurath held that *"[r]eliance on a universal slang with its fuzzy Ballungen is inevitable in the formulation of our ideas at any stage of scientific inquiry."* This should be compared with Hempel's prefatory remarks in Horwich (1993), which collects together the papers that were presented at the earlier occasion.

6. This characterization is given in an extremely polite letter to Schlick of May 27, 1935 (ASP, 102-46-25), written on receipt of Schlick 1935, which also carefully explains the invitation from Stebbing.

7. Carnap describes how he became acquainted with Tarski's semantic definition of truth and how he urged Tarski to present a paper at the Paris congress (published as Tarski 1936), in which the new conception was first made publicly accessible to the wider community of scientific philosophers; Carnap's and Tarski's papers at the congress provoked "vehement opposition" and "long and heated debates" (Carnap 1963, 61).

8. See Schlick: "It is one thing to ask how the system of science has been built up and why it is generally believed to be true, and another thing to ask why I myself (the individual observer) accept it as true. You may regard my article on 'Das Fundament der Erkenntnis' as an attempt to answer the last question. It is a psychological question" (1935, 69).

9. For the distinction between "pure" and "descriptive" syntax, see Carnap 1934: "*[Pure syntax]* is nothing but *combinatoric* or, if you prefer, *geometry* of finite discrete serial-structures of a certain kind. *Descriptive syntax* is to pure [syntax] as physical is to mathematical geometry; it treats of the syntactic properties and relations of empirically occurring expressions (e.g., of the sentences in a certain book)" (§2). See also Carnap 1963, 29, for the contrast between his and Neurath's ways of approaching the syntax of language.

10. The first endnote to Hempel 1988 explains that it "has grown out of a paper read in November 1980 at a workshop held under the auspices of the Center for Philosophy of Science at the University of Pittsburgh" (162–63). As I indicated at the very beginning, Hempel was already discussing the problem of provisoes in seminars in the early 1970s.

11. For a discussion of the cultural and political context of the Vienna Circle, including remarks, in particular, on the intellectual consequences of the American immigration, see Feigl 1969, Galison 1990, and Giere 1996. For Carnap's own recollections of his and Neurath's political attitudes, see Carnap 1963, 22–24, 51–52; for "language planning" and international languages, see Carnap 1963, 67–71.

12. Carnap's "On Inductive Logic" (1945a) and "The Two Concepts of Probability" (1945b) present his new conception of inductive logic for the first time. In the first, he explains that a "rational reconstruction" is "a theory . . . offered as a more exact formulation . . . of a body of generally accepted but more or less vague beliefs" (1945a, 95). And the second begins, "[W]e have here an instance of that kind of problem—often important in the development of science and mathematics—where a concept already in use is to be made more exact or, rather, is to be replaced by a more exact new concept. Let us call these problems (in an adaptation of the terminology of Kant and Husserl) problems of *explication*" (1945b, 513). To the best of my knowledge, this is the first time Carnap introduces the notion of "explication" in print. (Thus it is not true, as Giere 1996, 340, suggests, that Carnap first publicly introduced the notion in Carnap 1947.) Moreover, although the discussion is not

as developed as in Carnap 1950, §§11–12, Carnap 1945b, 524–25, also explicitly warns against the dangers of psychologism in both deductive and inductive logic.

References

Archives for Scientific Philosophy (ASP). University of Pittsburgh Libraries. References are to file folder number. All rights reserved.

Ayer, A. J., ed. 1959. *Logical Positivism.* New York: Free Press.

Carnap, R. 1928a. *Der logische Aufbau der Welt.* Berlin: Weltkreis. Translated as *The Logical Structure of the World.* Berkeley and Los Angeles: University of California Press, 1967.

———. 1928b. *Scheinprobleme in der Philosophie.* Berlin-Schlachtensee: Weltkreis-Verlag. Translated as *Pseudoproblems in Philosophy.* Berkeley and Los Angeles: University of California Press, 1967.

———. 1932a. "Die physikalische Sprache als Universalsprache der Wissenschaft." *Erkenntnis* 2:432–65. Translated as *The Unity of Science.* London: Kegan Paul, 1934.

———. 1932b. "Über Protokollsätze." *Erkenntnis* 3: 215–28. Translated as "On Protocol Sentences." *Nous* 21 (1987): 457–70.

———. 1934. *Logische Syntax der Sprache.* Wien: Springer. Translated as *The Logical Syntax of Language.* London: Kegan Paul, 1937.

———. 1936a. "Von der Erkenntnistheorie zur Wissenschaftslogik." In *Actes du Congrès international de philosophie scientifique,* vol. 1, 36–41. Paris: Hermann.

———. 1936b. "Wahrheit und Bewährung." In *Actes du Congrès international de philosophie scientifique,* vol. 4, 18–23. Paris: Hermann, 1936. Translated (in part, together with other material) as "Truth and Confirmation" in *Readings in Philosophical Analysis,* ed. H. Feigl and W. Sellars, 119–27. New York: Appleton-Century, 1949.

———. 1936–37. "Testability and Meaning." *Philosophy of Science* 3: 419–71; 4: 1–40.

———. 1945a. "On Inductive Logic." *Philosophy of Science* 12: 72–97.

———. 1945b. "The Two Concepts of Probability." *Philosophy and Phenomenological Research* 5: 513–32.

———. 1947. *Meaning and Necessity.* Chicago: University of Chicago Press.

———. 1950. *Logical Foundations of Probability.* Chicago: University of Chicago Press.

———. 1963. "Intellectual Autobiography." In *The Philosophy of Rudolf Carnap,* ed. P. A. Schilpp. La Salle, Ill.: Open Court.

Feigl, H. 1969. "The *Wiener Kreis* in America." In *The Intellectual Migration: Europe and America, 1930–1960,* ed. D. Fleming and B. Bailyn, 630–73. Cambridge, Mass.: Harvard University Press.

Galison, P. 1990. "Aufbau/Bauhaus: Logical Positivism and Architectural Modernism." *Critical Inquiry* 16: 709–52.

Giere, R. 1996. "From *Wissenschaftliche Philosophie* to Philosophy of Science." In *Origins of Logical Empiricism,* ed. Ronald Giere and Alan Richardson, 335–54. Minneapolis: University of Minnesota Press.

Hempel, C. G. 1935a. "On the Logical Positivists' Theory of Truth." *Analysis* 2: 49–59. Also in *Selected Philosophical Essays,* ed. R. Jeffrey, 9–20. Cambridge: Cambridge University Press, 1999.

———. 1935b. "Some Remarks on 'Facts' and Propositions." *Analysis* 2: 93–6. Also in *Selected Philosophical Essays,* ed. R. Jeffrey, 21–25. Cambridge: Cambridge University Press, 1999.

———. 1936. "Some Remarks on Empiricism." *Analysis* 3: 33–40. Also in *Selected Philosophical Essays,* ed. R. Jeffrey, 26–34. Cambridge: Cambridge University Press, 1999.
———. 1943. "A Purely Syntactical Definition of Confirmation." *Journal of Symbolic Logic* 8: 122–43.
———. 1945. "Studies in the Logic of Confirmation." *Mind* 54: 97–121. Also in *Aspects of Scientific Explanation,* 3–46. New York: Free Press, 1965.
———. 1965. *Aspects of Scientific Explanation.* New York: Free Press.
———. 1973. "Rudolf Carnap, Logical Empiricist." *Synthese* 25: 256–68. Also in *Selected Philosophical Essays,* ed. R. Jeffrey, 253–67. Cambridge: Cambridge University Press, 1999.
———. 1981. "Der Wiener Kreis und die Metamorphosen seines Empirismus." In *Das geistige Leben Wiens in der Zwischenkriegszeit,* ed. N. Leser, 205–15. Vienna: Österreichischer Budesverlag. Translated as "The Vienna Circle and the Metamorphoses of its Empiricism" in *Selected Philosophical Essays,* ed. R. Jeffrey, 268–87. Cambridge: Cambridge University Press, 1999.
———. 1982. "Schlick und Neurath: Fundierung vs. Kohärenz in der wissenschaftlichen Erkenntnis." *Grazer Philosophische Studien* 16/17: 1–18. Translated as "Schlick and Neurath: Foundations vs. Coherence in Scientific Knowledge" in *Selected Philosophical Essays,* ed. R. Jeffrey, 181–98. Cambridge: Cambridge University Press, 1999.
———. 1988. "Provisoes: A Problem Concerning the Inferential Function of Scientific Theories." *Erkenntnis* 28: 147–64. Also in *Selected Philosophical Essays,* ed. R. Jeffrey, 229–49. Cambridge: Cambridge University Press, 1999.
———. 1991. "Hans Reichenbach Remembered." *Erkenntnis* 35: 5–10. Also in *Selected Philosophical Essays,* ed. R. Jeffrey. Cambridge: Cambridge University Press, 1999.
———. 1993. "Empiricism in the Vienna Circle and in the Berlin Society for Scientific Philosophy." In *Scientific Philosophy: Origins and Developments,* ed. F. Stadler, 1–19. Dordrecht: Kluwer. Also in *Selected Philosophical Essays,* ed. R. Jeffrey, 295–304. Cambridge: Cambridge University Press, 1999.
———. 1999. *Selected Philosophical Essays.* Ed. R. Jeffrey. Cambridge: Cambridge University Press.
Hempel, C. G., and P. Oppenheim. 1945. "A Definition of 'Degree of Confirmation.'" *Philosophy of Science* 12: 98–115. Also in *Selected Philosophical Essays,* ed. R. Jeffrey, 135–61. Cambridge: Cambridge University Press, 1999.
———. 1948. "Studies in the Logic of Explanation." *Philosophy of Science* 15: 135–75. Also in *Aspects of Scientific Explanation,* 245–90. New York: Free Press, 1965.
Horwich, P., ed. 1993. *World Changes: Thomas Kuhn and the Nature of Science.* Cambridge: MIT Press.
Kuhn, T. 1962. *The Structure of Scientific Revolutions.* Chicago: University of Chicago Press.
Neurath, O. 1928. "Rezension: R. Carnap, *Der logische Aufbau der Welt* und *Scheinprobleme der Philosophie.*" *Der Kampf* 21: 624–26.
———. 1931. *Empirische Soziologie: Der wissenschaftliche Gehalt der Geschichte und Nationalökonomie.* Vienna: Springer.
———. 1932a. "Soziologie im Physikalismus." *Erkenntnis* 2: 393–431. Translated as "Sociology and Physicalism" in *Logical Positivism,* ed. A. J. Ayer, 282–317. New York: Free Press, 1959. And as "Sociology in the Framework of Physicalism" in *Philosophical Papers: 1913–1946,* ed. R. Cohen and M. Neurath, 58–90. Dordrecht: Reidel, 1983.
———. 1932b. "Protokollsätze." *Erkenntnis* 3: 204–14. Translated as "Protocol Sentences" in *Logical Positivism,* ed. A. J. Ayer, 199–208. New York: Free Press, 1959.

And as "Protocol Statements" in *Philosophical Papers: 1913–1946,* ed. R. Cohen and M. Neurath, 91–99. Dordrecht: Reidel, 1983.

———. 1934. "Radikaler Physikalismus und 'Wirklicher Welt.'" *Erkenntnis* 4: 346–62. Translated as "Radical Physicalism and the Real 'World'" in *Philosophical Papers: 1913–1946,* ed. R. Cohen and M. Neurath, 100–114. Dordrecht: Reidel, 1983.

———. 1983. *Philosophical Papers: 1913–1946.* Ed. R. Cohen and M. Neurath. Dordrecht: Reidel.

Quine, W. V. O. 1960. *Word and Object.* Cambridge: MIT Press.

Schlick, M. 1934. "Über das Fundament der Erkenntnis." *Erkenntnis* 4: 79–99. Translated as "The Foundation of Knowledge" in *Logical Positivism,* ed. A. J. Ayer, 209–27. New York: Free Press, 1959. And as "On the Foundation of Knowledge" in *Philosophical Papers,* ed. H. Mulder and B. van de Velde-Schlick, vol. 2, *1925–1936,* 370–87. Dordrecht: Reidel, 1979.

———. 1935. "Facts and Propositions." *Analysis* 2: 65–70. Also in *Philosophical Papers,* ed. H. Mulder and B. van de Velde-Schlick, vol. 2, *1925–1936,* 400–404. Dordrecht: Reidel, 1979.

———. 1979. *Philosophical Papers.* Ed. H. Mulder and B. van de Velde-Schlick. Vol. 2, *1925–1936.* Dordrecht: Reidel.

Tarski, A. 1936. "Grundlegung der wissenschaftlichen Semantik." In *Actes du Congrès international de philosophie scientifique,* vol. 3, 1–8. Paris: Hermann.

Uebel, T. E. 1992. *Overcoming Logical Positivism from Within: The Emergence of Neurath's Naturalism in the Vienna Circle's Protocol Sentence Debate.* Amsterdam and Atlanta, Ga.: Rodopi.

Rudolf Haller

4
On Herbert Feigl

In a retrospective on the philosophical movements of the last century, we cannot fail to notice in the early twenties the increasing importance in Vienna of logical empiricism, or "logical positivism,"[1] which rose to one of the leading movements of philosophy of the twentieth century. With the concentration of outstanding philosophers, logicians, mathematicians, and scientists forming this to some extent revolutionary philosophical movement, which consequently was given the name "Wiener Kreis" (Vienna Circle), Vienna, the former capital of an empire, in the meantime reduced to a very minor city, thereby regained the relative splendor of a fresh impetus, at least in the field of philosophy—and that not only in Europe but also in other parts of the world. Herbert Feigl was one of the advanced students who had the privilege of being accepted from the very beginning as a member of the Circle.

To add a personal note, I got to know Herbert Feigl in Vienna in 1964 when he was a visiting professor at the Institute for Advanced Studies (then called the Ford Institute), and especially in Salzburg that August, where Paul Weingartner of the Internationales Forschungszentrum für Grundfragen der Wissenschaften had invited him to take part in the fourth symposium of that center together with Harald Delius, Gerhard Frey, Paul Feyerabend, Jaakko Hintikka, Bela Juhos, Werner Leinfellner, and myself.

The topic of our symposium was analyticity or—as I formulated the title of the initial lecture—"Der Streit um die 'analytisch-synthetisch'-Dichotomie" ("The Quarrel Concerning the Analytic-Synthetic Dichotomy") (1964). Whereas I, following W. V. O. Quine's critique of the dichotomy, defended the porosity of this distinction, Feigl, who was the next speaker, argued strongly against the repudiation of the basic distinction.

The fact that a given string of signs (a sentence, a formula, and so on) may be the vehicle for either an analytic or a synthetic statement neither blurs the distinction nor introduces a neutral third category of statements (Feigl 1964). What Feigl observed was a distortion of the picture of scientific method. And according to his analysis, it had been the logicians who failed to recognize the "method of successive confirmation." This method presupposes that it is possible and even common to test successively logically independent postulates of scientific theories. Actually, Feigl finds, "it is curious that it should be the logicians who first undermine their own enterprise by a denial of the sharp distinction between the empirical and the logical" (1964, 180–81). Thus, he finds it "significant that Pierre Duhem's view of the impossibility of crucial experiments has been reiterated and re-emphasized primarily by the mathematicians and logicians (such as Poincaré and Quine)" (Feigl 1964, 186). Even if this sounds like an argument ad hominem, what it shows is that Feigl takes sides. Within the Vienna Circle, it seems that only Otto Neurath (who, by the way—except in his very early days when he studied Ernst Schröder's *Algebra and Logic*— never worked in the field of logic) based his understanding of scientific systems on Duhem's holistic point of view. Feigl, however, as we know and shall see, due to the influence of Moritz Schlick, throughout his life remained on the path marked by this admired teacher of his.

Therefore he could state that "the sharp distinction between analytic and synthetic statements is not only fruitful but indispensable . . . at least for the sake of mere clarity of thought" (Feigl 1964, 176). And Feigl adds some arguments for his point of view and against the revolutionaries who revolt against the analytic-synthetic dualism, a revolt that according to him rests "on a confusion of the logical analysis of (artificially fixed) languages with historical investigations of (growing, shifting natural) languages" (Feigl 1964, 177). Referring to Friedrich Waismann's series of articles on the analytic-synthetic in volumes 10–13 of the journal *Analysis* (1949–53), Feigl emphasizes that even granting the ambiguities, relativities, and vaguenesses of meanings in natural languages, "we cannot fail to notice the rule-governed aspects of linguistic behavior that establish synonymies of terms and consequently the analyticity of certain statements" (Feigl 1964, 177).

We see here that Feigl did not change his position in the lively quarrel about the synthetic-analytic dichotomy, whereas Neurath very early defended the Duhemian line both with regard to general philosophy of science and with regard to common, everyday language. As in many other questions, Feigl himself approved the mostly achieved solution similar to that of his teacher, Schlick. This holds especially true for one of Feigl's most discussed contributions to present-day philosophical thinking on the relation between mind and body, his identity theory, a topic that I shall deal with later. This topic has been widely discussed in English-speaking countries, especially in the 1950s, 1960s, and 1970s, and perhaps later, though, it seems to me, somewhat less intensively. It is true, as Dieter Sturma, a German philosopher, has recently observed, that in "German philosophy Feigl's works have been ignored almost entirely" (Sturma 1998). Some time ago, one of the best-known post–World War II Austrian philosophers, Wolfgang Stegmüller, who for decades ran a center for analytic philosophy in Munich, wrote in a short chapter on the mind-body problem, "The most accurate and most well thought-out investigations concerning this complex of questions, namely, the body-mind problem in the literature of present-day philosophers, stem from Herbert Feigl" (1969, 498). Unfortunately, this did not lead to further research in German-speaking countries.

At the time of our symposium in Salzburg, I think the major part of Feigl's work was well known to the scientific community, and since then it has maintained its place in the history of analytic philosophy, but very often only in the history. After all, only five papers out of the twenty-four in *Inquiries and Provocations: Selected Writings, 1929–1974* (edited by R. S. Cohen and published in 1991) were published after 1964.

My last talk with Feigl was on the phone in Minneapolis in 1987 when I thought I could meet him again. It was—I must say—the saddest conversation of my life. Feigl had fallen into such a deep depression that he asked me not to hope that he could talk at all.

Vienna

Here I am not going to try to give a general profile of Feigl's philosophical views or his biography. I should rather like to limit this sketch to the period of Feigl's time in Vienna, through the first period of the renewed relations between Austrian and American philosophers after the end of World War I, to the end of the 1930s.

Feigl himself was the first member of the Schlick circle to try to get a grant for visiting the United States. In 1927, he had finished his Ph.D. thesis "Zufall und Gesetz: Versuch einer naturerkenntnistheoretischen Klärung des Wahrscheinlichkeits- und Induktionsproblems" ("Chance and Law: An Attempt at an Epistemological Elucidation of the Problems of Probability and Induction in the Natural Sciences"). Much has been written about Feigl's work after World War II, but this cannot be said of his early work, except his 1929 book *Theorie und Erfahrung in der Physik (Theory and Experience in Physics)*. As I said, I think it worthwhile first to look back at his early studies in Vienna, which, after all, coincided with the formation of the Vienna Circle.

In 1922, Feigl went to Vienna to continue his studies in physics and mathematics, which he had started in Munich a year before. He was prevented from studying there by a new regulation that excluded all non-Bavarians from the University of Munich. While in Munich, he attended the courses in physics led by Leo Graetz, Arnold Sommerfeld, and Willy

Wien, and he extended his studies in Vienna to psychology and especially to philosophy. Psychology was taught there by the newly appointed Karl Bühler, and philosophy by the likewise newly appointed German philosopher Moritz Schlick. It had been Schlick's call to the University of Vienna that had attracted Feigl to the capital of Austria. Apart from philosophy and psychology, he continued his studies there in physics with Hans Thirring and in mathematics with Hans Hahn. Hahn—who, along with Philipp Frank and Otto Neurath, was a member of the "first Vienna Circle" from 1907 to 1912—was offered the vacant chair of mathematics after Gustav von Escherich, and after holding positions in Czernowitz and Bonn he returned to Vienna in 1921. It was mainly Hahn who initiated and organized the faculty's vote to invite Schlick to accept the chair formerly held by Ernst Mach (Haller 1995; and see Schmetterer and Sigmund 1995). Schlick arrived in Vienna in the autumn of 1922 and started lecturing there in the winter term, from November to the end of January. Feigl had arrived a few months earlier. He knew Schlick's main work, *Allgemeine Erkenntnislehre (General Theory of Knowledge),* which he regarded as a masterpiece, but he did not know at that time that in a few years he would have to correct the proofs of the extended second edition. From the first day, he was among those who could not have been overlooked. And I think it was even he himself who lured Schlick to produce a second edition.[2]

Feigl was then twenty-one years old. In a competition in his first year in Vienna, he was awarded a prize for his extended paper "The Philosophical Significance of Einstein's Theory of Relativity." The referees were Ernst von Aster, Max von Laue, and Schlick. In 1915, Schlick had published "Die philosophische bedeutung des Relativitätsprinzips," an important paper that presumably was the impetus for Feigl's trip from Munich to Vienna. In this paper, Schlick argued that of the philosophical movements then available, only two remained in close association with exact science: Kantianism and positivism. The interesting and important step in Schlick's essay consisted in his critique of both positions. Against the Kantians, Schlick emphasized that the Newtonian conception of time could from then on not be accepted as "the only possible one," a step that puts an end to the concept of absolute time. Against the positivists, Schlick had to argue the other way round. If we claim, in accordance with classical positivism, that only what can be perceived is real and that the world is to be constructed solely from immediate given "elements," then only relative motions are perceivable and therefore only they are real. However, Schlick refuses this line of reasoning and is eager to show that the relativity principle is "not a mere consequence of the general relativistic position" (Schlick 1979, 179). Schlick's attempt to clarify the revolutionary step in physics, especially in regard to its philosophical consequences, aimed at using the newly discovered principle as "a sort of criterion for the soundness of philosophy" (154). With Einstein's discovery, the best-confirmed part of scientific knowledge—Newtonian mechanics—lost its cornerstones, the concepts of absolute time and space.

The problem in the way we philosophers analyze and judge theories or parts of theories was one of the earliest ones Feigl raised in his letters and discussions with Schlick. In writing a critical review of a book by Alfred C. Elsbach, *Kant and Einstein* (1924), Feigl criticizes the "undecided formalizing way of consideration." It is, Feigl says, a mere "*morphological characterization* of the course of science, but not an *exact analysis* of knowledge with regard to its origins, presuppositions and methods."[3] From this characterization, one can easily grasp the demands for a new way of philosophical investigation: the change from a purely descriptive or phenomenological to an *analytic* method and style.

In his first publication, *Theory and Experience in Physics* (1929a), Feigl had already developed a general picture of the nature of theories. According to this early view, theories are first of all constructions that—in principle—contain a certain arbitrariness when compared to facts of experience. But on the other hand, this arbitrariness is limited and restricted by the idea and norm of *simplicity*. As sources of his anatomy of theories, he mentions "Poincaré, Mach, Duhem, Enriques, Schlick, Reichenbach, Carnap, and others" (Feigl 1929b, 126).

Knowledge, the young Feigl argues, is not pure representation, and theories are not just representations of reality. "Knowing means discovering relationships and the formulation of knowledge will always consider it its highest aim to strive toward representing these relationships in the most exact form, i.e., a strictly logical and mathematical form" (Feigl 1929b, 117, 127). The proper way of construction—or perhaps the ideal form of theories, as we have heard—is to continue them as hypothetico-deductive systems, put into an axiomatic form, where the axioms are independent from one another and complete; all "propositions of the field in question" should be deducible.

I am not going further into this program. Feigl himself refers only in general to a few examples, where the individual physical theories have been axiomatized. Apart from the formal requirements, there remains the most important one, which is called *the empirical requirement* for *truth*, which is, as he writes, "agreement with experience." Surely, that is what we need when we want to know if our thoughts or guesses fit the facts. The test of the validity of a theory has to be carried through by induction, because only thus can the ability to encompass certain factual material be observed. Therefore, Feigl states, "what we might call factually justified trust in a theory . . . is its objective probability" (1929b, 130).

This optimistic statement had to be relativized by another discovery, namely, "that verification can never be ultimately decisive. The leeway which is always left by inadequate establishment of facts can be used to construct several theories which embrace the facts just as precisely" (Feigl 1929b, 131).

If, however, one can construct different theories with regard to one kind of fact, it is questionable how far we can state the identity of the fact in question. If we accept that what is a fact could be identified without referring to one of the different theories, then the question remains of what other theory functions as its conceptual frame. In this case, we would have to decide this question on the basis of some theoretical construction used as a frame. Presumably the decisive step, then—disregarding borderline cases—will be to use the criterion of simplicity in choosing between competing theoretical constructions.

In the winter of 1924–25, there existed something that was later called "the Circle" but that started as a "Colloquium" or "Evening Colloquium" on the philosophy of mathematics. A year before, the young Feigl had had the chance to visit Albert Einstein in Berlin, a meeting that impressed him deeply and that he also related to Schlick, who had been in contact with Einstein since his essay on relativity.[4] From the time of his visit to Berlin, where he attended lectures, not only by Max Planck and other physicists and chemists but also by Wolfgang Koehler, Max Wertheimer, and Max Dessoir, Feigl's interests changed to biology, mathematics, set theory, and psychology, especially the psycho-physical problem, which he characterized as the most burning question.[5]

The following year, it was again the theory of probabilities that "immensely fascinated" him, and he started to write his Ph.D. thesis, submitting it to the faculty in July 1927.[6] The general philosophical position that Feigl had then adopted was a kind of critical realism, for which Feigl himself named Alois Riehl, Oswald Külpe, and, clearly first of all, Schlick as the main sources. He also mentions Riehl, Schlick, and the Bertrand Russell of *Human Knowledge* as main sources, and the last two as the main supporters of his mind-body monism (Feigl 1974, 2–3). From the early letters to Schlick, we immediately see him on the way to his own positions.

I should like to say a few words about Feigl's earliest work and why it has not yet seen the light.

"Chance and Law" was already advertised for *Schriften zur wissenschaftlichen Weltauffassung,* like Waismann's *Kritik der Philosophie durch die Logik* or Neurath's *Der wissenschaftliche Gehalt der Gesellschafts- und Wirtschaftslehre,* but like these two books it never appeared at all. The topic was hinted at by its subtitle: an analysis of the place of chance in a world whose laws of nature seem to exclude chance. The question was, might chance be compatible with the idea of a strict causal order in the world? And Feigl points to two problems that divide the sciences like *"Abgründe"* ["chasms"]. The one concerns the regularity of life, the basic problem of biology; the other concerns the basic problem of modern physics, the question of the character of the elementary physical laws. All these are questions concerning our world, concerning the structure of reality. The rapid development of quantum theory has changed the picture of nature that science has built up so far, perhaps even more than relativity, Feigl says. The task of the philosopher confronted with these new, changing developments is first of all elucidation *(Klärung)*. And in the process of these clarifications, one cannot stop before reaching the last stratum of problems, what has been called the foundations of our knowledge.

Feigl states that if we reach the basic problems of chance and law in nature, we also reach the foundations of philosophy. Insofar as the problems of the laws of nature are linked with the problem of induction, we are posing the question of the validity of our methods of knowing reality. David Hume's problem is nearby. Thus, Feigl hints at the horizon of his investigation, which should lead "to the last principles of the knowledge of nature" ["die letzten Prinzipien des Naturerkennens"] (Feigl 1927, 12).

From this point on, he thinks we may return to the finite character of our knowledge of nature, understanding the importance of the concept of simplicity.

Feigl's "Chance and Law" claims not to deal with general epistemological problems but to investigate the logical-philosophical problems of special induction, as they present themselves in the question of the *Anwendungsproblem,* that is, the applicability of probability theory to reality in questions concerning the knowledge of natural laws by means of the exclusion of chance. In his thesis, Feigl's aim is ambitious: to clarify the nature of the problems of probability and induction. This is, as everyone will agree, a broad field of scientific and philosophical problems. As Feigl himself says, it is not the general epistemological problems that are at the center of his interest but the logical problems of that special induction that concern the applicability—*die Anwendbarkeit*—of the calculus of probability to reality, the nature of causal and statistical laws and their relations.

Feigl distinguishes between the formal and the material conditions for the applicability of theory of probability. Concerning the first, he (as usual) distinguishes subjective theories from objective ones, the first as grades of subjective certainties, the latter as grades of objective justified surmises *(Vermutungen)*. As far as I can see—but I am not at all an expert in this field—I find Feigl's position not particularly unusual. Thus, it was only left to decide which of the two general projects should be followed. Feigl decides against subjectivist positions, and his point of view does have connections *(Berührungspunkte)* with the objective theory of probabilities. The main purpose of his thesis, it seems to me, is to corroborate the proposition of the uniqueness of physical causality as the strong lawfulness of all events. That should mean that the physical laws of nature represent the only existing kind of law ("Der Typus der physikalischen Naturgesetze soll die einzige überhaupt vorhandene Art von Geschehensgesetzmäßigkeit repräsentieren"; 1927, 146).

Feigl in the United States

I will now turn to the decisive change in Feigl's life and career. When in 1930 his application for an International Rockefeller Research Fellowship was granted, he almost immediately (after the famous conference of Königsberg) crossed the ocean and arrived at Harvard. Feigl's application for this fellowship was supported by Einstein, Schlick, Carnap, and several Viennese physicists.

In spite of the fact that in the bad times after World War I and after the end of the Austrian empire a lot of people emigrated, it was rather exceptional that a philosopher was among them. There were friends involved, Americans whom Feigl had gotten to know during his last four years in Europe. One of them was Dickinson Miller, who, Feigl reports, "was a visiting member of the Circle in 1926" (Feigl 1974, 7). Since in no other document of the Vienna Circle is Miller's name mentioned, he may have been invited by Schlick himself. Anyway, it was Miller, together with a professor emeritus from Columbia University, C. A. Strong, who invited Feigl to stay with them in Florence, Italy, where they discussed matters of epistemology.[7] A year later, in 1928, Feigl accepted another invitation to Fiesole. But according to his own report, it was not these two generous friends but Albert Blumberg who turned Feigl's interest to American philosophy and to the idea of visiting the United States. Blumberg, then a young student, also attended the meetings of the Schlick circle in 1929–30, and Feigl, always friendly, helped him "with the German language as well as philosophically" (Feigl 1974, 7).

In October 1930, Feigl arrived at Harvard, and he stayed there for "about eight months," as he remembered. In his own words,

> My main first contact there was the physicist-philosopher Percy Williams Bridgman (1882–1961), the originator of the "operational" approach in the logic of science. I also became acquainted with Alfred North Whitehead,

Henry Sheffer, C. I. Lewis, R. B. Perry, Susan Langer; and among the younger scholars or graduate students, Paul Weiss and Willard V. O. Quine. (Feigl 1974, 10)

Naturally, the contact with Bridgman, whose general approach was known in Vienna, meant a lot to Feigl at the beginning of this new phase in his life and gave him a strong feeling of being in the right place. Feigl himself—at Blumberg's suggestion—had read Bridgman's *Logic of Modern Physics* (1927) the year before, and he saw Bridgman's position as being close to the views of Carnap, Frank, and Richard von Mises. This concerned first of all the operational analysis of the meaning of physical concepts (Feigl 1974b, 69).

In his "Der Wiener Kreis in America" (1969), Feigl mentions how two other people helped him in the first year: One was Paul Weiss, then an instructor at Harvard and at that period—contrary to his later work—interested in logic and epistemology. The other was George Morgan, a secretary to Dickinson Miller, whom Feigl had already met in Fiesole. In his reminiscences of the Vienna Circle, Karl Menger also mentions the period he spent with Feigl at Harvard in the winter of 1930–31. He refers to the many conversations they had and underlines Feigl's "limitless admiration for Carnap" (Menger 1994, 66). Actually, Menger took that as an explanation of why Ludwig Wittgenstein had ended his relationship with Feigl immediately when he, Wittgenstein, had excluded Carnap from the circle around Schlick, which finally included only Schlick, Wittgenstein, and Waismann. Feigl wrote, in a letter to Schlick of December 6, 1930,

> Cambridge, Mass., 21, Forest Street
> Unfortunately there does not exist here an analogue to the Vienna Circle—only in the last days I have heard from the idea to establish—unfortunately only next year—a discussion-Circle for logical problems, a circle for which Whitehead, Lewis, Sheffer and Huntington should serve as presidents. Lewis' book *(Mind and World Order)* is certainly the best theory of knowledge within English literature. He surely is the one here who in his thinking is next to ours. I took part in his seminar (theory of Truth) where there were often interesting discussions. Unfortunately personally he is reserved, in spite of being kind and friendly. What Lewis means by pragmatism is hardly distinguishable from our positivism.[8]

In the same letter, Feigl reports that Russell had been at Harvard the year before and had reportedly said that if he were still young, he would invest all his energy in the probability problem, for it was the most interesting and the most difficult problem he knew of.

As I have already mentioned, the article that Blumberg and Feigl wrote at Christmastime in 1930, "Logical Positivism: A New Movement in European Philosophy," which was published in spring 1931 in the *Journal of Philosophy* and which introduced the term "logical positivism," was not only a piece of propaganda but the first historical outline of logical empiricism since the publication in 1929 of *Wissenschaftliche Weltauffassung: Der Wiener Kreis* (written by Neurath and Rudolf Carnap with the help of Feigl and Waismann and dedicated to Schlick). However, even more important in this period was the fact that in Germany, *Die Annalen der Philosophie und philosophischen Kritik,* which ended its publication in 1929, was succeeded by a new journal, *Erkenntnis,* which on behalf of the Berlin Society for Empirical Philosophy and the Ernst Mach Association started with its first volume in 1930–31 under the editorship of Carnap and Hans Reichenbach. The first part of the new journal started with Schlick's programmatic article "Die Wende der Philosophie," followed by Carnap's "On the Old and New Logic" and Reichenbach's "On the Philosophical Significance of Modern Physics." The comprehensive second part provided the proceedings of the first conference for the Epistemology of the Exact Sciences, held in Prague in 1929. It is in this volume that Feigl provided a kind of short summary of his "Zufall und Gesetz," announcing that it would be published in *Schriften zur wissenschaftlichen Weltauffassung,* to be edited by Frank and Schlick.

The main reason Feigl hesitated to publish his thesis may have been the fact—a fact that impressed him very much—that with the progress of quantum mechanics, causality had broken down, and he may have feared that at the moment his clearly written, almost hymnal, appraisal of the causal laws was published, he could or would be harshly criticized. Still, it seems strange that as a young and ambitious philosopher, he dispensed with a chance rarely available to a young philosopher.

In this last part of my paper, I will add some remarks on that part of Feigl's work that was most dear to him: the venerable mind-body problem.

In 1931 Feigl started his academic career at the University of Iowa, where he stayed as lecturer, assistant professor, and associate professor until 1940. There he did not have much time to write articles or books. But there was always time to write letters, for instance, a dozen to his beloved teacher Schlick.

It may be, however, that at that time Feigl was already concerned about his future, because shortly after the publication of the article he coauthored with Blumberg, there began a new and constant period which he later associated with the "expansive spirit of Neurath and Reichenbach, and to some extent also of Carnap" (Feigl 1969, 71). He wanted to encourage some kind of mission in America, which was the reason he later gave for his participation in the meetings of the American Philosophical Association in Ann Arbor in 1932, in Chicago in 1933, and in St. Louis in 1934.

In Chicago, he gave a talk, "A Logical Analysis of the Psychophysical Problem," that met with "enthusiastic approval and devastating criticism" (Feigl 1969, 71) as he remembers. It was also in Chicago in 1933 that he met Charles W. Morris for the first time. Morris invited Feigl to give a colloquium paper there the following year, in 1934. Feigl convinced Morris to visit Carnap, who at that time was a professor in Prague. Morris went to Prague in 1934, and two years later Carnap, through Morris, received his position at the University of Chicago, followed by his young friends Carl G. Hempel and Olaf Helmer. Morris also remained in close contact with Neurath, and because of the great project of the *International Encyclopedia of Unified Science,* Chicago became a special place on the philosophical map. But that is part of another story (see Reisch's chapter 8, in this volume).

The part of Feigl's work that played the most important role in later periods of his life was his theory concerning the identity of mind and body. The problem invites the typical metaphysical thesis reducing different kinds to one, and that seems at first to describe Feigl's view about the identity of mind and body. But if we look closer at how Feigl develops his general theory, we may see it differently.

The basic picture from which Feigl starts is the following: In order to show that our subjective experiences of real things are identical with the physical processes, Feigl must identify the extensions of the terms of the supposed two languages we are using: The private language of our subjective experience (sense impressions, feelings, and so on) must be identified with the objective facts of our brain states, the cortical occurrences that the anatomy and physiology of our brains take as objects of their observations. What is taken for granted and is presupposed is first, the possibility and the factuality of our *direct knowledge* of our own feelings and impressions and second, the possibility that knowledge *is* expressible in a private language, which by definition can only be understood by the person using it. These private data are, so to speak, the "deepest level of evidence" (Feigl 1958, 392). It would be hard to remain on this level if it were as private as the language with which we describe it. Fortunately, this is not the case, because, according to Feigl, our sense impressions are definable in the public language as a state (or states) of a person. This public language says that these states of a person dispose him or her to behave in a certain way.

However, we reach the level of the public, that is, *inter*subjective, language via theoretical states of a person. And this enables Feigl—who calls these states that are describable in the general public language "central states"—to postulate the identity theses on the basis of the identity of reference of terms of two radically different languages. What really has been done so far to foster the empirical claim is to find correlations between the different levels. But Feigl thought that this was enough to accept the identity thesis. According to my understanding, this rescues the thesis only if we accept the semantic theory supposed to explain it, namely, that the behavior of a person defines the meaning of a word. And I doubt that such an argument succeeds.

These remarks are not meant neither as an analysis of Feigl's main ideas nor as a sufficient exegesis of the two topics that stood at the center of his interest and therefore of his work in his philosophical career. They are meant to recall Feigl's work as something that naturalistic philosophy should not neglect. We should not overlook the impact of the movement of the Vienna Circle in general: Most of the problems of present-day analytic philosophy, and especially of present-day philosophy of science, have their roots in this movement if we take it in its broad frame from logic, language, and truth to the logic and history of scientific research. Too often we forget what we have learned from those we like to criticize or neglect today. It is in this sense that historical studies should interrupt our daily work.

Notes

1. As it was called by Albert E. Blumberg and Herbert Feigl in an early paper (Blumberg and Feigl 1931).
2. Feigl to Schlick, letter of September 4, 1924. Schlick-Nachlass, Rijksarchief in Noord Holland/Haarlem. Feigl goes on to ask, "What about the second edition?" He is sorry for reminding Schlick, but says Schlick owes this work to himself and to the world! ("Wie steht es mit der zweiten Auflage? Verzeihen Sie, hochverehrter Herr Professor, daß ich Sie daran erinnere, aber Sie wissen, daß Sie diese Arbeit sich und der Welt schuldig sind.")
3. Feigl to Schlick, letter of September 4, 1924.
4. See Feigl to Schlick, letter of July 26, 1923. Schlick-Nachlass, Rijksarchief in Noord Holland/Haarlem.
5. Feigl to Schlick, letter of September 4, 1924.
6. The thesis, comprising 186 pages, has been published along with the theses of Marcel Nakin, a close friend of Feigl and Kurt Gödel, and of Tscha Hung, the Chinese student of Schlick (Haller 1999).
7. Strong also dedicated a copy of his book *Theory of Knowledge* to Feigl.
8. Feigl to Schlick, letter of December 6, 1930. Schlick-Nachlass, Rijksarchief in Noord Holland/Haarlem.

References

Blumberg, A. E., and H. Feigl. 1931. "Logical Positivism: A New Movement in European Philosophy." *Journal of Philosophy* 28: 281–96.

Elsbach, A. E. 1924. *Kant und Einstein: Untersuchungen über das Verhältnis der modernen Erkenntnistheorie zur Relativitätstheorie.* Berlin: Walter de Gruyter.

Feigl, H. 1927. "Zufall und Gesetz: Versuch einer naturerkenntnistheoretischen Klärung des Wahrscheinlichkeits- und Induktionsproblems." Ph.D. diss., University of Vienna. Reprinted in *Zufall und Gesetz: Drei Disserationen unter Schlick,* ed. R. Haller, 1–187. Studien zur Österreichischen Philosophie, vol. 25. Amsterdam: Rodopi, 1999.

———. 1929a. *Theorie und Erfahrung in der Physik (Theory and Experience in Physics).* Karlsruhe: G. Braun.

———. 1929b. "Meaning and Validity of Physical Theories." Chap. 3 of *Theorie und Erfahrung in der Physik (Theory and Experience in Physics).* Reprinted in H. Feigl, *Inquiries and Provocations: Selected Writings, 1929–1974,* ed. R. S. Cohen. Vienna Circle Collection, no. 14. Dordrecht: Reidel, 1981.

———. 1958. "The Mental and the Physical." In *Concepts, Theories, and the Mind-Body Problem,* ed. H. Feigl, M. Scriven, and G. Maxwell. Minnesota Studies in the Philosophy of Science, no. 2. Minneapolis: University of Minnesota Press.

———. 1964. "The Distinction between Analytic and Synthetic Arguments as a Cornerstone of Logical Empiricism." In Die analytischen Sätze und die Grundlagen der Wissenschaften. Viertes Forschungsgespräch, 1964. Also in *Analytizität, und Existenz,* from which the quoted page numbers are taken. Drittes und viertes Forschungsgespräch. Ed. P. Weingartner. Salzburg and Munich: A. Pustet, 1966. Originally published in *The Foundations of Science and the Concept of Psychology and Psychoanalysis,* ed. H. Feigl and M. Scriven. Minnesota Studies in the Philosophy of Science, no. 1. Minneapolis: University of Minnesota Press, 1956.

———.1969. "Der Weiner Kreis in America." In H. Feigl, *Inquiries and Provocations: Selected Writings, 1929–1974,* ed. R. S. Cohen. Vienna Circle Collection, no. 14. Dordrecht: Reidel, 1981.

———. 1974. "No Pot of Message." In H. Feigl, *Inquiries and Provocations: Selected Writings, 1929–1974,* ed. R. S. Cohen. Vienna Circle Collection, no. 14. Dordrecht: Reidel, 1981.

Haller, R. 1964. "Der Streit um die 'analytisch-synthetisch'-Dichotomie." In Die analytischen Sätze und die Grundlagen der Wissenschaften. Viertes Forschungsgespräch, 1964. In *Deskription, Analytizität, und Existenz.* Drittes und viertes Forschungsgespräch. Ed. P. Weingartner. Salzburg and Munich: A. Pustet, 1966.

———. 1995. "Hans Hahn: Wien—Czernowitz—Bonn—Wien (1879–1934)." In *Die Bukowina: Vergangenheit und Gegenwart,* ed. I. Slawinski and J. Strelka, 179–90. Bern: Peter Lang.

———, ed. 1999. *Zufall und Gesetz: Drei Disserationen unter Schlick.* Studien zur Österreichischen Philosophie, vol. 25. Amsterdam: Rodopi.

Menger, K. 1994. *Reminiscences of the Vienna Circle and the Mathematical Colloquium.* Ed. L. Golland, B. M. C. Guinness, and A. Sklar. Vienna Circle Collection, no. 20. Dordrecht, Boston, and London: Kluwer.

Reisch, G. 2003. "Disunity in the *International Encyclopedia of Unified Science.*" In *Logical Empiricism in North America,* ed. G. L. Hardcastle and A. W. Richardson, 197–215. Minneapolis: University of Minnesota Press, 2003.

Schlick, M. 1915. "Die philosophische Bedeutung des Relativitätsprinzips." *Zeitschrift für Philosophie und philosophische Kritik* 159, no. 2: 129–75. Leipzig: Johann Ambrosius Barth.

———. 1979. "The Philosophical Significance of the Principle of Relativity." In M. Schlick, *Philosophical Papers,* ed. H. L. Mulder and B. F. B. van de Velde-Schlick, vol. 1. Dordrecht, Boston, and London: Reidel.

Schmetterer, L., and K. Sigmund. 1995. "Hans Hahn: A Short Biography." In *Gesammelte Abhandlungen/Collected Works I,* ed. L. Schmetterer and K. Sigmund, 29–35. Vienna and New York: Springer.

Stegmüller, Wolfgang. 1969. *Hauptströmungen der Gegenwartsphilosophie.* 4. erweiterte Auflage. Stuttgart: Alfred Kröner.

Sturma, D. 1998. "Reductionism in Exile? Herbert Feigl's Identity Theory and the Mind-Body Problem." *Grazer Philosophische Studien* 54: 45–70.

Weingartner, P., ed. 1966. *Deskription, Analytizität, und Existenz.* Viertes Forschungsgespräch 1964: Die analytischen Sätze und die Grundlagen der Wissenschaften. Salzburg and Munich: A. Pustet.

Diederick Raven

5
Edgar Zilsel in America

> Zilsel [lived] an unhappy and isolated existence in this country.
> —**Herbert Feigl, "The *Wiener Kreis* in America"**

> But my poor, rational, impractical father never had a chance. With his great powers of intellect and insight, he tried to function and cope with the world: with the unfamiliar language and the . . . labyrinth that the city of New York presents to strangers. . . . I remember him during those New York summers, how he lived in a series of increasingly smaller and shabbier rooms on Manhattan's Upper West Side. . . . He was scarcely fifty years old, but, for all the world, he looked like a piece of scrap . . . and was treated as such.
> **Paul Zilsel, "Portrait of My Father"**

The leading figures of the Vienna Circle were in general very successful in building new academic careers and exerting tremendous intellectual influence in their newfound homeland, America. The one exception was Edgar Zilsel (1891–1944). His tragic, self-chosen death on March 11, 1944, the sixth anniversary of the Austrian *Anschluss*, makes his emigration story exceptional. Michael Stöltzner has remarked that when "[c]omparing Zilsel's fate with the success of the other members of the Vienna Circle an explanation is wanted" (1995, 335). I do not think that given the huge gaps that still exist in our knowledge about Zilsel's life and work in general and about his period in America in particular, we are even close to providing an explanation. Still, I think Stöltzner's point is a valid one that should be taken seriously. Given the current state of research, the best that can be done is to bring together a number of issues that I see as contributing factors, without claiming that they are exhaustive. In my view, any understanding of Zilsel's fate in America must encompass an answer to the question of why Zilsel remained marginal in Vienna, too. In other words, I see Zilsel's ill-fated stay in America as an extension of his ill-fated career in Vienna. As my narrative framework, I will use Friedrich Stadler's proposed three-stage scheme of transfer, transformation, and influence (see his chapter 9, in this volume) to analyze the exodus of scholars to the United States. To make this scheme applicable to an individual story, I will discuss the biographical/individual aspects and the structural features of Zilsel's emigration story separately before bringing the two together in the third part of the chapter.

Biographical Aspects

Zilsel in Austria

At the time of the *Anschluss* (March 11, 1938), Zilsel worked as a physics and mathematics teacher at a *Gymnasium* in Vienna. It is unclear when the Zilsel family left Vienna for London. In a letter to Otto Neurath, dated January 17, 1939, Zilsel wrote, "We indeed did manage to escape the Nazi jailhouse and have been living safely in England for some time now."[1] Very little is known about his time in England. He left England via Southampton on March 26, 1939, to arrive in New York on April 4. His son Paul—sixteen at the time—stayed in England until the summer in order to prepare for his final exams, and he arrived a few months later.

Why did Zilsel leave Austria only after the *Anschluss*? After all, in 1933 he had written two essays, under the pseudonym of Rudolf Richter, in which he commented on Third Reich policies (Zilsel 1993a, 1933b). The fact that Zilsel used a pseudonym suggests that he was keenly aware of the dangers facing him and his family. With Germany coming more and more under the influence of the Nazis, one almost automatically assumes that the position of Jews, as Zilsel and his family were, was rapidly becoming more and more desperate. The historian Herbert Strauss (1983) has persuaded me that this line of thought is based too much on knowledge of things to come, knowledge Zilsel, of course, could not have. Before *Kristallnacht* (November 9–10, 1938), things may not have looked very promising to the Jews in Germany, but they also did not look very bleak either. As Strauss points out, the reason for this misperception is that the Nazi policies against the Jews were not perceived as a uniform threat: "They rather appeared, and in fact were, polymorphous and discontinuous. For the first few years, they also were polycentric: more than one power center was competitively involved in decision making about Jews." He continues:

> We cannot understand why so many intellectual émigrés did not emigrate from Germany following the Law of April 7, 1933, or the Nuremberg Laws of September 15, 1935, unless we remember that they were faced with a confusing and, in fact, confused reality that they could legitimately interpret in several selective frameworks. (1983, 49)

Before the *Anschluss,* Austria was an independent country, and this obviously is a significant factor as well. Surely it had its own anti-Semitic movement, but before the *Anschluss* about 150,000 Jews left Germany for Austria, indicating that they considered it a safe country. At the individual level, things can sometimes look even more complicated. Zilsel's wife was suffering from manic depression. Her illness started to manifest itself in the early 1930s. She very much loved her country, and Zilsel was aware that emigration might seriously upset her already precarious state of mind. He left the country only because he felt his son Paul would have no future there.[2]

But what kind of future career was open to Zilsel? What kind of professional life had he lived in Austria? Nearly all his working life, Zilsel had earned his living as a schoolteacher, first at a *Gymnasium* and later at an institution for adult education *(Volkshochschule)*. The latter job he held from 1922 to 1934. He then again became a teacher at a *Gymnasium*. His ten hours a week of evening classes at the *Volkshochschule* allowed him ample time for research. To pursue a career at the university required him to write a *Habilitationsschrift*. He handed in his habilitation in the summer of 1923. Because of strong opposition, he withdrew his application a year later. Any hopes of pursuing an academic career were dashed. But his philosophical interests were not the least diminished by this episode. He used the habilitation as the basis of his best-known book, *The Development of the Concept of Genius* (1926). Shortly afterward, he started to work on two manuscripts in which he planned to solve what he took to be greatest philosophical challenge of his time: "the problem of unifying science and the humanistic studies" (Zilsel 2000, 211).

Zilsel in America

When Zilsel arrived in the United States, the search for a job was on right from the start. Within days he was able to establish contact with Max Horkheimer, the director of the International Institute of Social Research (IISR), the emigrated Frankfurt school. Although they did not have the necessary means to support Zilsel, they did actively assist his efforts to find funding.[3] These first few months were spent in the time-consuming work of writing project proposals and obtaining recommendations. Zilsel was able to make a little money (four dollars a week) by privately tutoring classical Greek to an émigré he met on the boat.

In June 1939, Zilsel received a Rockefeller Fellowship to work on the origins of early modern science. He was tremendously excited about the prospect of working full-time on the project. To Neurath he wrote, "Good news: I obtained the Rockefeller fellowship, and am thus taken care of for a year. I've never in my whole life had the opportunity to scientifically work without a secondary job. This is indeed a beautiful country."[4] From the letters written in the first few month of his exile, it is clear that he was very excited and that he was thrilled about the opportunities he was given in the new world. The swiftness with which Zilsel was able to get a research grant suggests that the relief agencies thought Zilsel had good prospects for getting a job at a college or university. (The level of teaching in the final years of a *Gymnasium* overlaps with that of the lower years of the average American college.)

Zilsel was initially given money for one year and was able to get a renewal for another year. But a continuation beyond a two-year period was out of the question. New money was only available if Zilsel could affiliate himself with another institution. There is very little information on this crucial period, but their correspondence shows that Zilsel tried to get George Sarton, who was at Harvard, interested. The IISR acted as an intermediary between the relief agencies and Zilsel. He had to do all the preparatory work and negotiating with the agencies, but the IISR would sign all the official documents, so that it looked as if Zilsel was affiliated with them and as if they were taking care of him. Zilsel inquired with Sarton if he could work at his institute. Sarton indicated he could, although he pointed out that the label "institute" was a little much for what was basically one room with his desk. It is clear from Zilsel's letters that he was taking the way the IISR handled the situation as the norm, that is, that he had to arrange the funds and have Sarton sign all the documents. Sarton, on the other hand, assumed that Zilsel already had funds, and he started to feel very uncomfortable with the prospect of having to sign documents and began to give Zilsel the cold shoulder in no uncertain terms. Not only did he indicate that Zilsel's presence was not needed, but he even wrote, "Your presence would be somewhat annoying, as the desk which you would occupy is sometimes needed . . . for the occasional visitors."[5] This argument cut no ice: Zilsel was asking not whether a desk was available close to Sarton—Zilsel could work in the library, as he had done in his New York period—but whether Sarton was willing to sign papers. But the egocentric Sarton was not able to see things that way.[6]

Zilsel and Mills College

Zilsel would have liked to see the whole episode as a "comedy of errors," if not for "the rather serious consequences for my work and my family" (Zilsel to Sarton, May 2, 1941). It is likely that Zilsel tried other avenues, but nothing came from them. It becomes almost impossible to keep track of Zilsel's whereabouts after the summer of 1941. It is clear that somehow he was able to secure a part-time job at Hunter College, New York, teaching an introductory course in elementary physics, and his scholarly output stopped, as did the exchange of letters between him and his former friends and colleagues. The gap in archival material makes it impossible to tell what happened to Zilsel until, in the spring of 1943, he secured a grant of $1,000 from the American Philosophical Society (APS) to finish his work "on the social origins of modern science." Shortly thereafter, on what appears to be Lynn White Jr.'s initiative, Zilsel was offered a job at Mills College in Oakland, California. Like Hunter College, Mills College was an all-female institution. White had just moved from Stanford to Mills College to become its new president. He was familiar with Zilsel's recently published historical work and work on similar subjects. Mills was at that time a "small and impoverished institution"[7] that could not afford a specialist solely assigned to teach the history of science. White nevertheless indicated to Zilsel that he would be given the opportunity to teach a course on the history of science. The initial offer of a half-time job was, as White put it, "to teach physics."[8] Zilsel also initially spent a lot of time "cleaning up my laboratory and reassembling and repairing apparatus."[9] In February of the following year, however, Zilsel wrote to his son Paul that "having escaped the worst of difficulties, I no longer take the lab so seriously."[10]

Zilsel was apparently not very happy at Mills College. Due to White's time-consuming administrative duties as president of the college, Zilsel hardly ever had a chance to exchange ideas with him. Zilsel's wife gave him serious cause for concern, not only because of her unstable mental condition but also because she did not want to follow him to California. Nor did he see much of his son, Paul, who was studying physics at the University of Wisconsin. All in all, Zilsel found Mills College a very lonely place. All his fellow teachers were married and went home for lunch or dinner, whereas he had to remain on campus. He found having to eat with the "girls in the hall" embarrassing. He never managed to adjust to Mills very well, and he regarded it as "quite provincial." Besides, he hardly had "any possibility of scientific stimulation or discussion."[11] It almost seems as if Zilsel felt he was imprisoned at Mills College. Transportation was getting very bad at the time, and he was hardly able to leave the campus. Also, the library of Mills College was far too impoverished to assist him in any serious way in his research. The research library of the University of California, Berkeley, was for him—not having a car—too far away. The one thing that really excited him was the monthly meeting of the History of Science Dinner Club, an unofficial Berkeley amateur club, founded by the biologist Herbert Evans, devoted to the history of science (Hahn 1999).

Still, Zilsel must have become deeply depressed, for on the night of Friday, March 11, 1944, he failed to return to his place of residence. Instead, he stayed in his office and wrote three letters, one addressed to White, one addressed to his son Paul, and a third that he put on his desk. Having done that, he "fashioned a pillow from excelsior and his jacket, took poison and then reclined on the floor awaiting death—his hands in his pockets."[12] The note on his desk nearby read,

> No fuss, please!
> Just inform Dr. French, don't tell anybody else, please. Keep silent, please!
> Nobody must know of the suicide, everybody must be told that I died through a traffic accident.
> No students must see the body.
> Please, please, don't try to wake me up again.
> I am sorry to have inconvenienced you.
> Thank you.

There was a postscript to this note:

> If the janitor finds me, he may keep the $10 bill as compensation for the shock.

The money was on his desk.

Zilsel was found by David French, the dean of the faculty, who had been alerted that Zilsel had not turned up at home.

Structural Features

An Émigré in America

In order to understand the options that were open to an émigré scholar like Zilsel, I will focus my attention in this second part of my chapter on the structural aspects of life in the United States for such scholars and on how Zilsel negotiated them.

The "Herculean efforts" (Beyerchen 1983, 30) of the private relief agencies were of course in stark contrast to the "passivity and indifference" (Krohn 1993, 26) at the governmental level. The private relief agencies, however, never expected to be in business for as long as they eventually had to be. The Emergency Committee in Aid of Displaced German Scholars, for example, had hoped to finish its work in two years (they started in 1933). This proved to be an illusion and created all sorts of problems, financial constraints being one of them and the most pressing. For all the relief agencies, the extended period of need meant that financial aid to scholars was reduced from two years to one and also that the salary paid had to be drastically reduced. In 1933, relief agencies could pay $2,000 and counted on other sources to pay the second half of a scholar's salary, which would bring the total amount a scholar would receive to $4,000. This was not much compared to what an American scholar at the top of his field could expect: $12,000 to $15,000 (see Krohn 1993, 29). But by 1937, an émigré scholar could only expect to receive $1,400 total annually, and a year later, even this was reduced to $1,000. After the defeat of France, it sank to less than $650 (Krohn 1993, 28–9). Paul Zilsel, not unfairly, described these grants as "starvation grants" (P. Zilsel 1982, 12).

The aim of the various relief agencies was "to save learning, not to provide personal help for individual scholars."[13] Their grants were intended to provide the émigré scholar with an opportunity to present himself to American scholars, through publications or otherwise. Hence, all the scholars needed to get their essays or books published as quickly as possible. Zilsel felt very frustrated at the lack of "tempo" of the American publishers.[14] Still, Zilsel managed to publish a large number of smaller and larger essays, and they are not all on the topic of "the social origins of modern science," for which he was to be become best known. (All his wartime essays are collected in Zilsel 2000). He also published a number of smaller essays in which he took issue with the antinaturalist philosophy of social science of such people as Heinrich Rickert, Wilhelm Windelband, Wilhelm Dilthey, and Max Weber. This outburst of creativity and the wide range of topics not only show that Zilsel had been working on these two topics for a very long time—from the archival material I have collected, it is clear that he started to work on these two topics somewhere in 1928 or 1929—but also reflect the real need to be noted by others. Anything he could get into print enhanced his chances to get an academic position somewhere.

Publishing essays is social networking at a distance. How about the local networking? How about getting into contact on a face-to-face basis? There were two networks of people that were important to Zilsel's intellectual activity in America. The first was the members of the Vienna Circle, and the other was the IISR. To both groups he was, to a large extent, marginal. His old friends from Vienna were, of course, more than willing to help him, but many of them were themselves in marginal positions.[15] Furthermore, it appeared that Zilsel had drifted away from scientific philosophy to history of science. But through the yearly meetings of the Unity of Science movement, Zilsel was given an opportunity to meet other people and air his views. He was also willing to try less conventional ideas, like asking to have his name included on the stationery of the *International Encyclopedia of Unified Science;* this request was refused.

Was the IISR of any help to Zilsel in getting Americans to take notice of him? I do not know why Zilsel contacted the IISR nor why the institute was interested in helping him. I do not believe that much depended upon it in the early stages of Zilsel's stay in New York. It would have mattered very little who would help Zilsel out. What really mattered was that he got help in the first place. As it happened, the IISR was first. That leaves open the question of why they were willing to do so. Both Franz Borkenau and Hentyk Grossmann, when they were attached to the IISR in the early 1930s, had worked on the same problem as Zilsel. Grossmann, during his time at the IISR, was working on a book about Descartes and would have been thematically close to Zilsel. One is bound to infer that this played a part in the IISR's decision to help Zilsel. This, however, seems very doubtful. In a letter to Sarton, Zilsel says this about his relation to the IISR:

> Two years ago, on occasion of a visit, its head, Dr. Horkheimer, invited me to work at his institute. He told me he had no money, no room, and no desk for me but offered me to apply for grants in my favor, if I could find out myself institutions willing to make them. At this time I was a newcomer and my wife already was ill. I accepted, therefore, the offer very gratefully, got several scientific recommendations on account of my previous publications, and eventually the grants. I had to do the negotiations myself. The Rockefeller grant was given by the Natural Science Department; Dr. Horkheimer is a sociologist and a Hegelian. As far as I know the Institute never has maintained to anybody that they need me. Moreover, they never showed the slightest interest in my work. I never had an opportunity to discuss one line of it. Up to now I could not find out why, apart from humanitarian reasons, Dr. Horkheimer has invited me. By chance I learned a few weeks ago he has left New York and settled down in California. The Institute still is in New York.[16]

In an earlier letter, again to Sarton, he wrote, "At present I am a 'research associate' on the staff of the IISR, New York City. Actually, however, I am working in the New York Libraries and drop in at the Institute about once a month only, since neither the library nor the members of the Institute, as friendly as they are, have any connections with the subject of my study."[17] This brings up the next question: Was the IISR the best place for Zilsel to start an academic career in the United States? This, of course, is a difficult question to answer, but I am inclined to answer it negatively, because the IISR was, in Zilsel's words, "run by German immigrants and scarcely gives opportunity to make contacts with American scientists."[18] He also points out that the "greatest difficulty scientists expelled from Germany are faced" with was how "to induce American colleagues to become acquainted with their work." The IISR was clearly a hindrance in this respect, since most of its staff were paid by the institute and therefore were in no rush to make outside contacts. Hence, they were also not in a position to be of much assistance in this respect. At this point, the fact that Zilsel was working on a topic far removed from the main concerns of the IISR could be of some significance.

Still, I believe a case can be made that Zilsel was able to get himself noticed by his American colleagues. On December 29, 1941, the American Historical Association met in Chicago, with Dana B. Durand as one of the keynote speakers. In her talk "Tradition and Innovation in Fifteenth Century Italy," she restricted her analysis to science proper, "leaving aside the fruitful work of artists and craftsman who bridge the gap between the theoretical and the productive or practical activities. The happy union of the hand, the eye, and the mind . . . was undeniably a chief glory of Quattrocento Italy. . . . But the task of assessing the relative importance of the operational and the theoretical contributions to the whole of Quattrocento science must wait for further monographic research" (Durand 1943, 1). She quoted both Leonardo Olschki and Zilsel as her references. Paul Kristeller, who was one of the commentators on Durand's talk, took issue with the point Durand raised: He rejected the idea that humanists were the only representatives of science and learning in the fifteenth century and added that "there were the artisans and engineers who through their practical work came face to face with mathematical and scientific problems and sometimes made important contributions, as has been recently emphasized" (1943, 61). A little later he directly challenged Zilsel's view on the role of the humanist: "The number of artists and engineers who made active contributions to science was still comparatively small in the fifteenth century as compared with the sixteenth. But the case of Leon Battista Alberti shows that this scientific activity of the artists cannot be separated from, or opposed to, contemporary humanism" (1943, 61–62).[19] From the printed essays of this meeting, it is clear that Zilsel's ideas were having an impact.

Another example Zilsel's influence is Alexander Koyré, who started to publish a number of influential papers on the nature of the scientific revolution about this time. Clearly Koyré did not agree with Zilsel, but that is not the point. What is significant is that he felt obliged to say of Galileo and Descartes, in what his references clearly show is a direct challenge to Zilsel's position, "Their science is made not by engineers or craftsmen, but by men who seldom built anything more real than theory" (1943, 33). One final example of Zilsel's influence is to be found in Cyril Smith's introduction to the translation of Vannoccio Biringuccio's *Pirotechnia*. Smith suggests that the development of metallurgy up to the sixteenth century was seriously hampered because "there was little connection between the worlds of those who wrote and those who worked in metals," and later he wrote, "It was men like Biringuccio, the practical metalworkers, dyers, pottery makers, alum boilers, and kindred artisans, who accumulated the basic facts for the chemical science during the period [sixteenth century], in which learned men of church and university were engaged in lengthy but barren theological disputation." Smith even goes so far as to proclaim, "The artisans were the true scientists of this period" (1943, xi, xv).

Zilsel's essays were noted and were taken seriously, but not much followed from this. Why? This basically is the question: What kind of position could Zilsel expect to be offered?[20] Zilsel's analyses of the origin of early modern science may be sociological in nature, but they did not qualify him as a sociologist. His knowledge and familiarity of that subject were too rudimentary. A position as a philosopher or as a historian of science does seem most appropriate. But this, of course, also made his situation difficult. His main publications in the United States were on the history of science. But as Arnold Thackray points out in his study on the prehistory of the history of science in the United States, "[I]n 1941 the history of science seemed doomed to live in intellectual twilight" (1984, 416). In other words, if Zilsel was looking for a position as a historian of science, he was doing so at a very bad time. I do not think that he could have hoped for a position as a philosopher. Surely he was gifted, but he had no really significant publications of recent date to his name. Besides, there were too many people better qualified than he and also looking for a position.

Integration

All Coherence Gone

When White offered Zilsel a job at Mills College, that was, given the difficult circumstances for everybody, about the best he could expect to be offered. From here things could only get better. Zilsel had a position, and Princeton University Press had indicated that they were excited about his book. But things went wrong. Why? This, of course, is an impossible question to answer. Besides, I am under no illusion that one is likely to get any agreement on what establishes a satisfactory explanation in a case like this, for the simple reason that it is unclear what counts as a satisfactory explanation of a self-chosen death.[21] Describing Zilsel's death as an example of what Emile Durkheim calls anomic suicide—and this may well be a correct description—is nothing more than labeling it and clearly falls short of explaining it. Furthermore, I am not sure that the self-chosen death is what should be the main issue in understanding Zilsel's life and work. For me, the greatest puzzle has always been his self-chosen marginality.

By far the best characterization of Zilsel's life I can think of is this: continuously many-sided marginal man. Whether at the University of Vienna, in the Austrian Socialist Party, or in the Vienna Circle while it was still functioning in Vienna, Zilsel always found himself in a marginal position. This position should not be understood as one in which his contributions are not valued—quite the contrary. But Zilsel somehow was not the man to feel comfortable near the center of a movement or society, or group. He always ended up at the margins. Clearly, this is a sign of having an independent mind, but it also signals a single-minded hardheadedness that makes compromises difficult.

There are many ways to illustrate this. I will give two brief examples: his stance on Marxism and his reaction to the suggestions that he should hand in a new *Habilitationsschrift*. In 1929 he wrote about Marxism, "It represents . . . a melting together of natural, scientific, and historical-sociological ideas . . . which is indispensable for any comprehensive philosophical theory" (Zilsel 1929, 186). Nonetheless, he opposed a party dogmatism that attempted to bind party membership and political activity to an unconfirmed theory of historical materialism:

> Precisely because I consider Marx's theory of history, in its most radical form, to be correct, I resist its misuse . . . . Turning it into a party dogma does equal damage to party and theory. Every scientist is aware of the enormous difficulties that necessarily attach themselves to any far-stretched theory. As a scientist, I in no way want to be a member of a party that really expects 400,000 members to agree with an incredibly demanding theory—one which can only be confirmed through the most careful studies. Such a thoughtless party must sooner or later founder politically. (Zilsel 1931, 214)

In the sad and tragic episode of his rejected *Habilitationsschrift,* this uncompromising stance again comes to the fore. Of course, political issues were involved in the rejection of Zilsel's habilitation (see Stadler 1991), but even today a manuscript like the one Zilsel presented in 1923 would not easily qualify as a philosophy thesis. The manuscript Zilsel presented to the Faculty of Philosophy was to a large extent a philological and historical analysis of the genealogy of the concept of genius; added to it were a large number of lawlike statements specifying under what socioeconomic and institutional conditions the concept of genius could develop. The opposition to this manuscript by Robert Reininger and Richard Meister should to a large extent also be construed as a dispute about what constitutes the subject matter of philosophy proper. Precisely on this point, Zilsel had very outspoken ideas, and he was by no means willing to compromise. When Moritz Schlick asked Zilsel to withdraw the manuscript and submit a new one that would be more strictly philosophical, Zilsel refused. In his reply to Schlick, he shows his strong commitment to stick to what from now on one can identify as his research program:

> In continuing my work in philosophy and physics on phenomena of chance and large numbers in inanimate nature, my interests, in recent years, have turned toward the application of natural scientific methods to the humanities as well as toward the disclosure of fairly exact laws concerning the events in these fields. This research area shall occupy my mind for a longer period of time. I have already collected rather extensive material, especially regarding the history of the concept of genius. The results of the finished parts are formulated with respect to this material. How long the final completion of the other parts will occupy me, I, of course, cannot say today. I could not, however, justify to myself that the direction of my scientific work be influenced by any considerations other than by the problems themselves, my interest in them, and my previous work. (Quoted in Dvorak 1981, 130–31)

In a later letter to the dean, Zilsel indicates that it would be highly improbable that he would " change his method of work fundamentally" and that it would therefore be unlikely that he would submit another manuscript "which belonged to the most narrow definition of philosophy," as had been requested. He attempted one last justification of his endeavor. His aim was "to develop contributions to the philosophy of nature and of history with respect to physical facts and the details of the history of the humanities . . . in the opinion that this will serve philosophy better than if it remains separate from the fruitful grounds of the disciplinary sciences."[22]

These two episodes indicate what I believe is true for Zilsel's life as a whole: He was a man of strong and radical views and unlikely to compromise on matters dear to his heart. If it meant he had to stay on the sidelines, then for him that was the end of the matter. He did so without ill feelings. The problem, of course, with taking such a stance is that although it may be intellectually satisfactory to remain faithful to one's own principles, in the end it makes life hard. Social interaction requires compromises, and Zilsel was not the easiest person around when it came to compromise. His pseudonym, Richter, is also telling in this respect: *"Richter"* means "judge."

Financially, life was difficult for all late-arriving émigré scholars, regardless of their former academic status. Zilsel's income started at $1,800 in the first year and slowly declined to less than $1,000. All available information indicates that $1,000 a year was just not enough to meet the barest minimal costs of living. Kristeller, who, although he had a position at Columbia, was on a salary similar to Zilsel's, told me he could survive only because his wife was able to secure a small job. This story is corroborated by Otto Kirchhemer, a part-time member of the IISR who received $120 a month. He indicated that this was "a minimal income, which thanks to my wife's job and some occasional supplementary earning makes it possible to get through on an extremely modest basis" (quoted in Wiggershaus 1994, 264). To make matters worse, Zilsel had to pay himself the cost of the periods his wife had to be taken to a "sanatorium." These costs could be as high as four-fifths of his salary.[23]

If one ignores for the moment the amount of money Zilsel received from the various agencies, he did very well in getting funding. Not many people were as successful as Zilsel was. In a sense his story was a success story. He was able to work full-time on his big idea for the first two years of his emigration. The rules of the game changed drastically after he failed to find another institution that would look after him. It is clear that from the spring of 1942 the Emergency Committee was actively trying to arrange a teaching position for him. They contacted F. Drake University (Des Moines) in March 1942, Goddard College (Plainfield, Vermont) in July 1942, Queens College in August 1942, Queen's University (mislocated in the documents of the committee in Toronto, Canada) in July 1942, and Mills College (Oakland, California) in May 1943. In November 1943 they closed the file on Zilsel with the words "Zilsel is taken care of during this academic year at least, and perhaps permanently. Close."[24] From the files, it is unclear what exactly happened with these various contacts and why Zilsel decided to accept the offer from Mills College.

What did Zilsel hope to get out of the situation he was in? He indicated to his son that he did not want to go back to Austria, so his mind was focused on settling down in America. Whether or not he saw his forced emigration as the opportunity to get a job in academia is totally unclear to me, but I believe one has to seriously consider this possibility. Independently of how one answers this question, I believe that he was very serious about finishing at least one of the two books he had been working on since the late 1920s. One of them was his book on the origins of early modern science; the other was on the idea that the concept of law is applicable in the sociohistorical sciences. Being able to publish one book—and he had made the most progress on the history of science book—would have improved his chances of getting an academic job. Given the very grim job situation for everyone, my own feeling is that a book would not have harmed Zilsel's chances, but I doubt whether they would greatly have improved them. The going was tough for everyone.

When Zilsel applied for financial assistance to the APS in 1943, he explained his plea thus: "The grant is desired to enable me to teach not more than six hours a week at Hunter College." About the duration of the investigation he wrote, "I underestimated the duration of the investigation in my application of 1941. In the mean time so much further material was collected that *completion by the fall 1944 can be expected practically with certainty.*"[25] As the APS made clear, their grant to Zilsel was a final one, and they pointed out that they were "not making this grant so you may have only six hours of teaching but because of the importance of the project which you are dealing with." They strongly advised him to establish himself as a teacher as soon as possible. Zilsel felt the need to reply to this point as follows: "When I stated in the application that I wanted to restrict my teaching to six hours a week I was naturally not motivated by an unwillingness to work. My book on which I have been working for many years means very much to me. I have to complete it and realize that this simply can not be done if I can not concentrate upon my research. As soon as the book is finished I shall gladly devote all my time to teaching."[26]

Elsewhere have I argued that the two books he was working on were part of one big, overall project: to single out the unique features of modern science and to investigate the sociological conditions of its origins (Raven and Krohn 2000). Based on a reconstruction of what these two books should have contained and a comparison of that reconstruction with what is published in *The Social Origins of Modern Science* (Zilsel 2000), it is quite clear that Zilsel never was close to finishing even one of these books. Even for the history book, significant parts are missing. What makes the situation difficult to assess is that it looks as if Zilsel had boxed himself into a very difficult corner. His history book was "merely" a case study for the kind of argument he wanted to make in the other philosophical book: that in the sociohistorical sciences, lawlike statements are possible. The problem he faced was that he needed a conceptual clarification of the notion of law in order to present his historical material so that it could function as a case study, while at the same time he hoped that his historical analysis of the emergence of modern science would provide the empirical data to justify the conceptual clarification he was proposing. A fatal circularity is looming large here. The currently available archival evidence does not provide grounds for further conclusions. However, my own feeling is that he was intellectually in trouble and that these troubles partly explain why his publications ceased after 1942.

More significant than these speculations is the fact of his changing intellectual milieu: When Zilsel started to work on his two books in the late 1920s, he may have chosen a topic that was marginal to the interests of the Vienna Circle at the time, but at least he was working on something that made sense given the intellectual tradition he was part of. Continuing these interests after he arrived in the United States was quite another matter. Not only did these interests cause him to miss out on the changes that were happening to the Vienna Circle (see Galison 1998; Holton 1993,

1995), but they most certainly did not enhance his chances to be offered a job at a college or university. The job at Mills College was probably the best he could expect to be offered.[27]

But Zilsel was not only intellectually marginal and isolated. As indicated above, when Zilsel left Austria his wife was mentally ill. From all the information I have, it is clear that Zilsel's wife could not cope with the United States. Except for a number of short periods, she stayed in mental institutions while World War II lasted, suffering from what was most likely manic depression. When she was in the hospital, her husband went to see her every week. However, when Zilsel left for California, she did not join him. I have reason to assume that the separation was not just because of his wife's mental problems but because she had effectively terminated their relationship, although legally they were still married. Later in life, she would tell her son that she felt she had hurt her husband deeply. Zilsel told his son that the one thing a man needed most was an enduring and reliable relationship with a woman. His wife later in life converted to Roman Catholicism and liked to be called Maria Magdalene. There is clearly some deeply metaphorical meaning to this change of name.

I believe that it is not unreasonable to assume that Zilsel's political vision was falling apart as well. The argument for this is as follows: His Marxism and his adherence to logical positivism were both expressions of his optimism for the future and were united in the sense that both movements were struggling to change the world into a better place to live, more democratic and with less-exploitative structures in society. The rising anti-Semitism in Austria in the 1930s and the growing influence of the conservative Catholics at the time made it clear that his political hopes were no longer carrying the day. As Stadler (1991) has shown, the scientific worldview of logical positivism did not have an easy time either. It always remained in a marginal position at the University of Vienna. Both the political radicalism of his logical positivism and of his Marxist leanings were losing credibility, albeit from different directions. Whereas the radical democratic aspirations of logical positivism were crushed by the rising tide of Nazi sympathizers, the radical Marxist hopes were undermined by the show trials Stalin was setting up in Russia in the late 1930s, indicating that somehow that political system was as corrupt as capitalism. What the effect of this was on Zilsel is difficult to tell, but it probably dawned on him that his political hopes for the future were slowly going up in smoke. Shortly before he left New York to go to the West Coast, he broke all ties with the Austrian Labor Committee, with whom he had been actively involved ever since his arrival in the United States. In the letter in which he motivated his break he wrote,

I joined the Austrian Social-Democrats in October 1918 and, in the party, I have become another man, no matter what I may still experience. I had the good fortune to live with the Viennese proletariat for twelve years as a folks–high school teacher; I know very well that the wonderful idealism of the Viennese workers originate in the spirit of socialism as this was brought forward by the Austrian party. I am proud to have belonged to a party that, in the February struggle in 1934, while all the others capitulated without a struggle, saved the dignity of socialism for the future. I am fortunate to have been able to be for years in friendly contact with Otto Bauer. And I will always remember the manifold examples of courage, decency, and efficiency that I have met with in the Austrian party.

I cannot attest to how earnest my step is to me other than through the, at least sloganistic, representation of the political conception that seems right to me in the enclosed letter. You may show the outlines to everyone if you wish. If I have the opportunity to publish my theoretical conceptions in the future, I hope that our (?) contratemps can proceed without personal bitterness. I will certainly take pains to avoid personal bitterness and personal criticism.[28]

I have not seen the captionlike phrases, but my understanding of the letter is this: It shows someone who is trying to think through a political situation that had radically changed since he first became politically active in his student days, way back in Vienna in the 1910s. It seems fair to assume that this letter was written partly because he no longer believed, as apparently the leadership of committee did at the time, that they—the old guard—could again become the leading force in politics in Austria after the war. In this, Zilsel turned out to be right.

This concludes my analyses of Zilsel's American tribulations. I hope to have made clear that in terms of academic prospects, his exile should not directly be seen as a failure. His succeeded in getting a position at a well-respected college, he was able to attract the attention of people with the essays he wrote in a relatively short period of time while in the United States, and Princeton University Press had more or less indicated that they would publish his book. On a personal level, things were clearly less rosy. His personal biography seems to make less and less sense. Zilsel, the committed Marxist radical, who enjoyed teaching evening classes for about fifteen years to the socialist workers in Vienna, found himself assigned to teach at a middle-class to upper-middle-class girls school in California. He was always very much in favor of technological progress, but he did not like what he saw. He certainly could not cope with it himself. In a letter to his son a month before he died, Zilsel wrote, "Believe me, I have learned above all else in the last few years that one can only be happy if one binds oneself to another: the man who is alone, without friend or wife who he truly values and likes, is not happy and will never be so."[29] This is written by a man who was, after having been in the United States for almost four years, intellectually isolated; whose main intellectual project was in dire straits; who did not have anyone around at Mills College whom he could call a friend; and whose wife had just left him.

Notes

Part of the research leading to this essay was made possible by a grant I received from the American Philosophical Society.

In this essay I have quoted or referred to archival material. The acronyms used are APS/Z: the Zilsel file held by the American Philosophical Society in Philadelphia; EC/Z: the Emergency Committee in Aid of Displaced Scholars, file Zilsel, which is in the New York Public Library; Astor, Lenox, and Tilden Foundations, Manuscripts and Archives Division; and NP/Z: the Neurath papers, file Zilsel, Vienna Circle Archive in the Rijksarchief of Noord-Holland.

    1. Zilsel to Neurath, quote from the Neurath Papers (NP), Wiener Kreis Archief, Haarlem, The Netherlands, file Zilsel (Z). If necessary, the abbreviation NP/Z will be followed by a date.

    2. Paul Zilsel has claimed (in a personal communication) that his father believed he and his wife would somehow have been able to survive the Nazi troubles. Seen against the remarks I just quoted from Strauss, this may not have been as strange an assessment of the situation as it looks at first sight.

    3. In all the institute's official correspondence with outside funding agencies, its main line was, "The present financial situation of this Institute does not permit us as yet to offer Dr. Zilsel a permanent position with us." As background, the following should be mentioned: The institute had been founded with funds from the Weil Foundation, and it had just done badly with investments on the stock exchange. Horkheimer, in his function as director, ordered a cutback in spending. Understandable or not, this cutback did not prohibit Horkheimer from setting aside—in a move that was most likely not according to the rules—fifty dollars for himself (cf. Wiggershaus 1994, 261). Remarks by members of the IISR critical of Horkheimer's management of the funds of the Weil Foundation are quoted in Wiggershaus 1994, 261.

    4. NP/Z, July 27, 1939.

    5. Sarton to Zilsel, April 30, 1941, quote from the Sarton Papers (SP), file Zilsel (SP/Z), which are in the Houghton Library of Harvard University. If necessary, the abbreviation SP/Z will be followed by a date.

    6. At exactly the time Zilsel was trying to persuade Sarton to help him, Sarton himself was in dire straits. His main financial supporter—the Carnegie Foundation—was indicating that they were not going to continue his work after he retired. (Harvard at best paid only 40 percent of Sarton's salary.) Furthermore, they were no longer willing to finance his assistant, I. B. Cohen. The latter was especially upsetting Sarton, for he feared that Cohen would not write his Ph.D. not on the history of science but on something else instead. If the latter happened, Sarton believed he would never again be able to find an able person to become his successor. (He described Cohen to James Conant as likely to be "the leading historian of science twenty five years hence" [SP, Sarton to Conant, June 23, 1941].)

Sarton was so occupied with his own troubles that he even interpreted Zilsel's request as an attempt to occupy the position he was trying to find for Cohen.

7. White, in private correspondence to Wolfgang Krohn (private file with the author). For material that is not in any manner publicly available, I use the abbreviation PF followed by the date of the letter.

8. White to Evans (August 28, 1943), quote from the Evans Papers (EP), file White, which are in the Bancroft Library, University of California, Berkeley. Later, in the already mentioned letter to Krohn, White was to recollect—mistakenly I would argue in light of his letter to Evans—that he had invited Zilsel to Mills College to teach mathematics.

9. Zilsel to Evans (EP), October 21, 1943.

10. Edgar Zilsel to Paul Zilsel, PF, February 3, 1944.

11. Ibid.

12. *Oakland Tribune,* Sunday, March 12, 1944, A7.

13. This was the stated aim of the Emergency Committee in Aid of Displaced Foreign Scholars and the Rockefeller Foundation. A useful source of information on the Emergency Committee is Duggon and Drury 1948; the quote is from page 19.

14. Zilsel to Neurath, NP/Z, November 11, 1940.

15. The position Heinrich Gomperz had is not untypical. Gomperz was a visiting professor at the University of Southern California when the outbreak of the war prohibited his return to Vienna. He was allowed to keep this "visiting chair" as a personal gesture. When he died in February 1943, a number of people, among them Hans Reichenbach, suggested using the "vacancy" to help another refugee, only to discover that Gomperz had had a personal chair. Besides, they were told that the dwindling number of students suggested a downsizing of the philosophy department anyway. Reichenbach in a letter to Zilsel, February 16, 1943. The Reichenbach Papers are located at the Archives of Scientific Philosophy, University of Pittsburgh. The inventory control number of this letter is HR 038-17-08.

16. Zilsel to Sarton, SP/Z, May 2, 1941.

17. Zilsel to Sarton, SP/Z, March 31, 1941.

18. Ibid.

19. Kristeller and Zilsel were at this time both in New York and in contact with each other. Zilsel had a rather negative view of the role and function of the humanists in bridging the gap between brain work and hand work; see his essay "The Method of Humanism," which was published in Zilsel 2000 for the first time but was written in 1942–43. Kristeller knew this essay and wrote extensive comments on it, which are also contained in Zilsel 2000. Kristeller clearly did not share Zilsel's negative judgement of the humanist and used his comments on Durand's paper not only to confirm Zilsel's general view but also to express his own forcefully where he disagreed with Zilsel.

20. Thanks to I. B. Cohen, who, in private correspondence, made some useful suggestions on this question.

21. There are two additional reasons why venturing into the reasons for Zilsel's suicide is a hazardous enterprise. For all who were close to him, it not only proved to be a great emotional shock but was also an unexpected event: No one had seen it coming. A completely different issue is that, as Dahms 1987 shows, there is no pattern whatsoever to be found in all the many emigration stories. Hence, it is impossible to say if and in what way Zilsel's story—apart from his suicide—conforms to or deviates from a typical emigration story.

22. Quotes are from the personal file, Zilsel, University Archive, Vienna.

23. Zilsel to Neurath, NP/Z, November 11, 1940.

24. Emergency Committee in Aid of Displaced Scholars, New York Public Library,

Astor, Lenox, and Tilden Foundations, Manuscripts and Archives Division, file Zilsel (EC/Z), November 1943.

25. Italics added. Quoted from the Zilsel file held by the American Philosophical Society in Philadelphia. I will use the abbreviation APS/Z followed by a date in my references to this collection. The quote is from Zilsel's grant application of February 28, 1943.

26. Zilsel to APS, APS/Z, May 17, 1943.

27. I have no idea what Zilsel expected to find at Mills College, but Lynn White, in an autobiographical essay published in 1970, has this to say on the situation he found Mills College to be in when he was offered its presidency late in 1942: "Mills was underendowed, deep in debt, and was saddled with an inefficient campus. Its chief merit was a faculty of a quality far higher than the salaries they received. Mills was not then attractive" (White 1970, 56).

28. Quote from the obituary in *Austrian Labor Information* 26 (March/April 1944): 9.

29. Edgar Zilsel to Paul Zilsel, PF, February 3, 1944. The quote is out of context, for Zilsel is encouraging his son to make friends in the United States; in the background is a remark that he would like to see his son stay in the United States instead of going back with his mother to Austria after the war. I take it that Zilsel is talking as much to his son as to himself. In the same letter, he writes that one has to accept one's fate: "Aber man muss sich eben fügen."

References

Archives of Scientific Philosophy (ASP). Hans Reichenbach Papers. University of Pittsburgh.

Beyerchen, A. 1983. "Anti-Intellectualism and the Cultural Decapitation of Germany under the Nazis." In *The Muses Flee Hitler: Cultural Transfer and Adaptation, 1930–1945,* ed. J. C. Jackman and C. M. Borden, 29–44. Washington, D.C.: Smithsonian Institution Press.

Dahms, H.-J. 1987. "Die Emigration des Wiener Kreises." In *Vertriebe Vernunft: Emigration und Exil Österreichischer Wissenschaft, 1930–1940,* ed. F. Stadler, 66–122. Munich: Jugend und Volk.

Duggan, S. P. H., and B. Drury. 1948. *The Rescue of Science and Learning: The Story of the Emergency Committee in Aid of Displaced Foreign Scholars.* New York: Macmillan.

Durand, D. B. 1943. "Tradition and Innovation in Fifteenth Century Italy." *Journal of the History of Ideas* 4, no. 1: 1–20.

Dvorak, J. 1981. *Edgar Zilsel und die Einheit der Erkenntnis.* Vienna: Löcker Verlag.

Evans Papers. Bancroft Library. University of California, Berkeley.

Feigl, H. 1969. "The *Wiener Kreis* in America." In *The Intellectual Migration: Europe and America, 1930–1960,* ed. by D. Fleming and B. Bailyn, 630–73. Cambridge, Mass: Harvard University Press.

Galison, P. 1998. "The Americanization of Unity." *Daedalus* 127, no. 1: 45–71.

Hahn, R. 1999. "Berkeley's History of Science Dinner Club: A Chronicle of Fifty Years of Activity." *Isis* 90: 182–91.

Holton, G. 1993. "From the Vienna Circle to Harvard Square: The Americanization of a European World Conception." In *Scientific Philosophy: Origins and Development (Vienna Circle Institute Yearbook [1993]),* ed. F. Stadler, 47–73. Dordrecht: Kluwer.

———. 1995. "On the Vienna Circle in Exile: An Eyewitness Report." In *The Foundational Debate: Complexity and Constructivity in Mathematics and Physics,* ed. W. DePauli-Schimanovich, E. Köhler, and F. Stadler, 269–92. Dordrecht: Kluwer.

Koyré, A. 1943. "Galileo and Plato." *Philosophical Review* 52: 333–46.
Kristeller, P. O. 1943. "The Place of Classical Humanism in Renaissance Thought." *Journal of the History of Ideas* 4: 59–63.
Krohn, C.-D. 1993. *Intellectuals in Exile: Refugee Scholars and the New School for Social Research*. Amherst: University of Massachusetts Press.
Raven, D., and W. Krohn. 2000. "Edgar Zilsel: His Life and Work (1891–1944)." Introduction to E. Zilsel, *The Social Origins of Modern Science*, ed. D. Raven, W. Krohn, and R. S. Cohen, xix–lix. Dordrecht: Kluwer.
Sarton Papers. File Zilsel. Houghton Library of Harvard University.
Smith, C. 1943. Introduction to *Pirotechnia*, by V. Biringuccio. Trans. and ed. C. Smith and M. T. Gnudi. New York: Dover, 1990.
Stadler, F. 1991. "Aspects of the Social Background and Position of the Vienna Circle at the University of Vienna." In *Rediscovering the Forgotten Vienna Circle: Austrian Studies on Otto Neurath and the Vienna Circle*, ed. T. E. Uebel, 51–77. Dordrecht: Kluwer.
———. 1997. *Studien zum Wiener Kreis: Ursprung, Entwicklung, und Wirkung des Logischen Empirismus im Kontext*. Frankfurt am Main: Suhrkamp. Translated as *The Vienna Circle: Studies in the Origins, Development, and Influence of Logical Empiricism*. Vienna: Springer, 2001.
———. 2003. "Transfer and Transformation of Logical Empiricism: Quantitative and Qualitative Aspects." In *Logical Empiricism in North America*, ed. G. L. Hardcastle and A. W. Richardson, 216–233. Minneapolis: University of Minnesota Press, 2003.
Stöltzner, M. 1995. "Vienna-Berlin-Prague: Centenaries Carnap, Reichenbach, Zilsel." In *The Foundational Debate: Complexity and Constructivity in Mathematics and Physics*, ed. W. DePauli-Schimanovich, E. Köhler, and F. Stadler, 317–42. Dordrecht: Kluwer.
Strauss, H. A. 1983. "The Movement of People in a Time of Crisis." In *The Muses Flee Hitler: Cultural Transfer and Adaptation, 1930–1945*, ed. J. C. Jackman and C. M. Borden, 45–60. Washington, D.C.: Smithsonian Institution Press.
Thackray, A. 1984. "The Pre-history of an Academic Discipline: The Study of the History of Science in the United States, 1891–1941." In *Transformation and Tradition in the Sciences: Essays in Honor of I. B. Cohen*, ed. E. Mendelsohn, 395–420. Cambridge: Cambridge University Press.
Weiner Kreis Archief, Haarlem, The Netherlands.
White, L. 1970. "History and Horseshoe Nails." In *The Historian's Workshop: Original Essays by Sixteen Historians*, ed. L. P. Curtis, 49–63. New York: A. A. Knopf.
Wiggershaus, R. 1994. *The Frankfurt School: Its History, Theories, and Political Significance*. Trans. M. Robertson. Cambridge, Mass.: Polity Press.
Zilsel, E. 1926. *Die Entstehung des Geniebegriffes: Ein Beitrag zur Ideengeschichte der Antike und des Frühkapilismus*. Tübingen: Mohr. Reprinted, with a preface by H. Maus. Hildesheim and New York: Olms Verlag, 1972.
———. 1929. "Philosophische Bemerkungen." *Der Kampf* 22: 178–86.
———. 1931. "Partei, Marxismus, Materialismus, Neukantianismus." *Der Kampf* 24: 213–20.
——— [pseud. R. Richter]. 1933a. "Das Dritte Reich und die Wissenschaft." *Der Kampf* 26: 486–93.
——— [pseud. R. Richter]. 1933b. "SA Philosophiert." *Der Kampf* 26: 393–402.
———. 2000. *The Social Origins of Modern Science*. Foreword by J. Needham. Introduction by D. Raven and W. Krohn. Ed. D. Raven, W. Krohn, and R. S. Cohen. Dordrecht: Kluwer.
Zilsel, P. R. 1982. "Portrait of My Father." *Shmate: A Journal of Progressive Jewish Thought* 1, no. 1: 12–13.

Thomas E. Uebel

6
Philipp Frank's History of the Vienna Circle: A Programmatic Retrospective

It is no secret anymore that despite much effort to retain a common public front against "school philosophy," the Vienna Circle was riven by internal divisions. The best known is that between the Moritz Schlick–Friedrich Waismann wing of Wittgensteinian fellowship and the so-called left wing of the Circle, with Rudolf Carnap, Philipp Frank, Hans Hahn, and Otto Neurath, especially the rift between Schlick and Neurath; a more subtle one is the division within the left wing between Carnap as a rational reconstructionist and Neurath as a naturalist.[1] The question arises as to whether these or similar divisions survived the transplantation of logical empiricism to its new homes in exile, and if so, what the issues were around which they turned and how these issues were related to the changes that emigration and adaptation to the new cultural context brought about within logical empiricism itself.

The thesis I wish to present for consideration here focuses on just one aspect of the complex phenomenon in question. Relatively early in the history of logical empiricism's transplantation to North America, a self-conscious attempt was made by one of the emigrants to differentiate between the European proponents of logical empiricism. The point was to stress and highlight one particular orientation of and motivation for early logical empiricist philosophy, an orientation that was perceived to be in danger of becoming marginalized in the process of its Americanization. Needless to say, a certain amount of tact was called for if one wanted to avoid disruption to the personal and institutional aspects of the ongoing acculturation. In that respect, the intervention at issue was all too successful: Whereas the text in question has rightly become a classic account of the development of the movement, its interventionist subplot was lost.

That historical accounts of discipline formation and tradition are by no means wholly dispassionate but often carry a certain agenda has been shown for a variety of cases in the natural and social sciences.[2] Historical accounts in philosophy are, if anything, even more likely to be partisan, and the various accounts of the history of their own movement by members of the Vienna Circle and the Berlin Society for Empirical Philosophy can hardly be expected to be an exception. A fascinating study of them was undertaken by Alessandra D'Acconti in her still unpublished Cambridge Ph.D. dissertation (1995), whose second part, dealing with post–World War II developments, provides much material of relevance to the question of how the then new, now standard, image of logical empiricism was constructed. D'Acconti concentrated on the retrospective essay by Herbert Feigl (1969a), on Carnap's autobiography (1963), and on Hans Reichenbach's *Rise of Scientific Philosophy* (1951). One point of particular interest is that she noted these authors' virtual "erasure" (1995, 167) of the prehistory of the Vienna Circle of the Schlick years and the "anomaly" (169) in this respect of Frank's 1949 "Introduction: Historical Background" in *Modern Science and Its Philosophy*.[3] It is D'Acconti's speculative suggestion that Frank may have wished "to tell the other story of logical empiricism" (171) that I am concerned to explore here.

I will argue that Frank's "Introduction" was part of a long-standing but now forgotten campaign. Nowadays, Frank's "Introduction" is indeed best known for its valuable hints toward what Rudolf Haller (1985) has called the "first" Vienna Circle, the discussion group of Frank, Hahn, and Neurath that met from 1907 to 1912. My guiding question thus becomes, What was the function that Frank's mention of the early group was meant to serve and how was it effected? To answer this question, I will first undertake a brief rereading of Frank's text, then consider some relevant correspondence of Circle members, and finally note some pertinent aspects of Frank's work in his North American exile.

Frank's Histories of the Vienna Circle

Frank actually discussed the first Vienna Circle twice, both times in the introduction to collections of his essays (1941b, 1949b). Each of these collections begins with a translation of his "Kausalgesetz und Erfahrung" (1907) documenting his early appropriation of conventionalist themes at the time when the first Circle met. For our purposes, it is particularly interesting to note the subtle shifts of accent between Frank's two accounts.

Frank's First Account

Frank's preface of 1941, also called "Introduction: Historical Background" (1941a) sets out to provide a "contribution to the history of the development of 'logical empiricism,'" a development characterized, apparently with an eye to its then most recent stages, as the result of the "coöperation of Central-

Thomas E. Uebel

6
Philipp Frank's History of the Vienna Circle: A Programmatic Retrospective

It is no secret anymore that despite much effort to retain a common public front against "school philosophy," the Vienna Circle was riven by internal divisions. The best known is that between the Moritz Schlick–Friedrich Waismann wing of Wittgensteinian fellowship and the so-called left wing of the Circle, with Rudolf Carnap, Philipp Frank, Hans Hahn, and Otto Neurath, especially the rift between Schlick and Neurath; a more subtle one is the division within the left wing between Carnap as a rational reconstructionist and Neurath as a naturalist.[1] The question arises as to whether these or similar divisions survived the transplantation of logical empiricism to its new homes in exile, and if so, what the issues were around which they turned and how these issues were related to the changes that emigration and adaptation to the new cultural context brought about within logical empiricism itself.

The thesis I wish to present for consideration here focuses on just one aspect of the complex phenomenon in question. Relatively early in the history of logical empiricism's transplantation to North America, a self-conscious attempt was made by one of the emigrants to differentiate between the European proponents of logical empiricism. The point was to stress and highlight one particular orientation of and motivation for early logical empiricist philosophy, an orientation that was perceived to be in danger of becoming marginalized in the process of its Americanization. Needless to say, a certain amount of tact was called for if one wanted to avoid disruption to the personal and institutional aspects of the ongoing acculturation. In that respect, the intervention at issue was all too successful: Whereas the text in question has rightly become a classic account of the development of the movement, its interventionist subplot was lost.

That historical accounts of discipline formation and tradition are by no means wholly dispassionate but often carry a certain agenda has been shown for a variety of cases in the natural and social sciences.[2] Historical accounts in philosophy are, if anything, even more likely to be partisan, and the various accounts of the history of their own movement by members of the Vienna Circle and the Berlin Society for Empirical Philosophy can hardly be expected to be an exception. A fascinating study of them was undertaken by Alessandra D'Acconti in her still unpublished Cambridge Ph.D. dissertation (1995), whose second part, dealing with post–World War II developments, provides much material of relevance to the question of how the then new, now standard, image of logical empiricism was constructed. D'Acconti concentrated on the retrospective essay by Herbert Feigl (1969a), on Carnap's autobiography (1963), and on Hans Reichenbach's *Rise of Scientific Philosophy* (1951). One point of particular interest is that she noted these authors' virtual "erasure" (1995, 167) of the prehistory of the Vienna Circle of the Schlick years and the "anomaly" (169) in this respect of Frank's 1949 "Introduction: Historical Background" in *Modern Science and Its Philosophy*.[3] It is D'Acconti's speculative suggestion that Frank may have wished "to tell the other story of logical empiricism" (171) that I am concerned to explore here.

I will argue that Frank's "Introduction" was part of a long-standing but now forgotten campaign. Nowadays, Frank's "Introduction" is indeed best known for its valuable hints toward what Rudolf Haller (1985) has called the "first" Vienna Circle, the discussion group of Frank, Hahn, and Neurath that met from 1907 to 1912. My guiding question thus becomes, What was the function that Frank's mention of the early group was meant to serve and how was it effected? To answer this question, I will first undertake a brief rereading of Frank's text, then consider some relevant correspondence of Circle members, and finally note some pertinent aspects of Frank's work in his North American exile.

Frank's Histories of the Vienna Circle

Frank actually discussed the first Vienna Circle twice, both times in the introduction to collections of his essays (1941b, 1949b). Each of these collections begins with a translation of his "Kausalgesetz und Erfahrung" (1907) documenting his early appropriation of conventionalist themes at the time when the first Circle met. For our purposes, it is particularly interesting to note the subtle shifts of accent between Frank's two accounts.

Frank's First Account

Frank's preface of 1941, also called "Introduction: Historical Background" (1941a) sets out to provide a "contribution to the history of the development of 'logical empiricism,'" a development characterized, apparently with an eye to its then most recent stages, as the result of the "coöperation of Central-

European positivism with some groups representing American pragmatism" (1941a, 5). Frank's contribution to its history concerns the central-European positivist wing. The perspective outlined on central-European positivism is in essence that pioneered by Neurath (1936)—though that is not explicitly noted—but it is one that in the light of international political developments had gained an additional edge:

> The European movement had its origin in the ideas of the Austrian physicist Ernst Mach. At the beginning of the twentieth century it had a large following in the scientific circles of Austria, especially in Vienna and Prague. In spite of the common German language, this movement could find only a few adherents in the universities of the German Reich, because there the philosophy of Kant and his metaphysical successors reigned, being regarded as a world picture particularly suited to the German nation. (Frank 1941a, 6)

Next Frank explicitly referred to the group "from which the 'Vienna Circle' evolved twenty years later" (8) and developed its genealogy:

> About 1910 there began in Vienna a movement which regarded Mach's positivist philosophy of science as having great importance for general intellectual life, but was clearly aware of [its] shortcomings.... An attempt was made by a group of young men to retain the most essential points of Mach's positivism, especially his stand against the misuse of metaphysics in science. With regard to those points, however, where Mach stood in opposition to the present course of the development of science, they planned a reconstruction of his doctrines. To this group belonged the mathematician H. Hahn, the political economist Otto Neurath, and the author of this book, at the time an instructor in theoretical physics in Vienna. The attempts at this reconstruction were at first made rather gropingly; they were only preparations. We tried to supplement Mach's ideas by those of the French philosophy of science of Henri Poincaré and Pierre Duhem, and also to connect them with the investigations in logic of such authors as Couturat, Schröder, Hilbert, etc. The attitude toward the atomic theory was indicated to us by the ideas first of L. Boltzmann and then of A. Einstein. (6–7)

Frank's account is highly suggestive, but we cannot here follow him further into the intellectual world of the first Vienna Circle.[4] Frank also provided no further details on the first Circle, besides noting:

> At that time there was prevalent a strong aversion toward weaving into the philosophy of science any consideration of a moral, religious or political nature. Hence it was not realized that American pragmatism was a related movement, although at about this time a group of sociologists in Vienna came out in support of it. The ivory-tower attitude of the positivism of those days

is best seen from the fact that there was present in it even a certain appreciation of the vitalism of Hans Driesch. (1941a, 7–8)

This remark demands closer interpretation. Just what sort of mixing of science and *Weltanschauung* was rejected here? After all, Frank also lauded Mach's philosophy of science for its "importance for the intellectual life of our times" (1917, in the title).

In the preface of 1941, this cultural dimension remains somewhat underexposed. It seems plausible, however, to understand the "ivory-tower attitude" to denote the denial or at least the neglect of the existential and political valence of the questions, concepts, and distinctions pursued and developed in science and philosophy of science—just that valence that Neurath was to point to repeatedly from the early 1920s onward. The embarrassed retrospective reference to Driesch supports this interpretation (he was approvingly referred to in Frank's 1907 paper). For we may note, as Zilsel has already remarked, that Driesch's organistic and vitalistic conception came to serve as one of the foundations for the *völkisch*-German pseudo-philosophy of National Socialism (something that did not save Driesch from being himself barred from university teaching by the Nazis) (Zilsel 1933). The mixing of philosophy of science and "consideration[s] of a moral, religious or political nature" that the first Vienna Circle rejected may thus be taken to denote the attempt to "weave" *intrinsic* relations between them. It need by no means denote disinterest in moral, religious, or political questions as such.

Frank's Second Account

The agenda of Frank's introduction of 1949 is similar to that of the 1941 one, but whereas the latter comprised thirteen pages, the new one comprised fifty-two. This time, the first Vienna Circle appears right at the very beginning of the account, in section 1, "Discussions in a Vienna Coffee House." And this time, Frank stressed, on the one hand, that this was not a group of mere students and, on the other, that their discussions were not merely shoptalk:

> Although all three of us [Frank, Hahn, and Neurath] were at that time actively engaged in research in our special fields, we made great efforts to absorb as much information, methodology and background from other fields as we were able to get. Our field of interest included also a great variety of political, historical and religious problems which we discussed as scientifically as possible. Our group had at that time no particular common predilection for a certain political or religious creed. We had, however, an inclination towards empiricism on the one hand and long and clear-cut chains of logical conclusions on the other. (1949a, 1)

Note how Frank here fixed the reference for his later expressions "our group" (cf. 1949a, 3, 7, 8), "we/us" (1949a, 8, 9), and even more importantly "the old Vienna group" (33, 35), "the/our original Viennese group" (34, 36), "our Viennese group" (31), and similar ones that appear throughout his narrative. Without giving explicit expression to the thesis, Frank styled the group Hahn-Frank-Neurath as the *original core* of the Vienna Circle. Frank's claim is notable, I think, both for its subtlety and its boldness.

Note next that Frank ascribed to this group a practical ambition that had been left underdescribed in his own preface of 1941:

> The whole original Viennese group was convinced that the elimination of metaphysics not only was a question of a better logic but was of great relevance for the social and cultural life. They were also convinced that the elimination of metaphysics would deprive the groups that we call today totalitarian of their scientific and philosophic basis and would lay bare the fact that these groups are actually fighting for special interests of some kind. (1949a, 35)

Is this compatible with Frank's earlier remarks? Prima facie, he seems to be contradicting himself. Of central importance, however, is his stress on the rejection of metaphysics. Although the first Vienna Circle may have still neglected the existential and political dimension of scientific concepts and questions, it was already aware of the ideological dimension of metaphysical philosophical thinking. Therefore, the first Circle was not naively unpolitical, even though it does not yet seem to have possessed its later awareness and explicitly political orientation. Also to be noted is that they understood their own antimetaphysical efforts not only as politically but also as, as it were, existentially relevant:

> Our original Viennese group and particularly Neurath were not satisfied with ascribing to our new philosophical group mainly critical and analytical objectives. We knew well that man is longing for a philosophy of integration. If the new philosophy refuses to serve the cause of integration, a great many people, including even scientists, would rather return to traditional metaphysics than be restricted to a purely analytic attitude. (1949a, 36)

Here, then, we already find the origin of the "positive paradigm" (Hegselmann 1987) of logical empiricist thought, the unity of science of the later Vienna Circle. Criticism of metaphysics alone would not do if one wanted to "serve" the social and cultural "life" (Carnap, Hahn, and Neurath 1929, as it appears in 1973, 318) of one's place and time.

Frank's narrative develops the *Problemgeschichte,* the intellectual history of this original group, up to the heyday of the Vienna Circle around Schlick and its dispersion in exile. It begins with the discussion of the "crisis in science" (Frank 1949a, 4) that was widely felt after the turn of the century and explicitly topicalized as such by the French conventionalist Abel Rey as the "crisis of contemporary physics" (Frank 1949a, 3). The developments contributing to this crisis and its solution led, claims Frank, from the collapse of the mechanistic worldview and the apparent need for philosophical supplementations of scientific insights (via the critical reflection on the methodological legacy of Mach, encouraged by the study of Poincaré and Duhem) up to the first schematic attempts to integrate David Hilbert's and Albert Einstein's overturning of the foundations of geometry and physics with their categorical rejection of Kant's synthetic a priori. Frank called this the "new positivism" (11).

After World War I, "new men" were recruited to help realize the project of this Viennese group (Frank 1949a, 26). Frank mentions Schlick, Reichenbach, and Carnap. Two of Frank's remarks are particularly noteworthy: "The work done by Schlick and Reichenbach made a strong impression upon our Viennese group. In our search for a scientifically founded philosophy we were glad to find collaborators who attacked the task, I would say, from a more professional angle" (31). What made Schlick more "professional" was less his valued understanding of the physics of relativity—Frank himself had been publishing in this field since 1908 (and in 1908 had already coined the term "Galilean transformations")—than his recognized and distinguished position as a *philosopher*. Thus, Frank noted in his eulogy for Schlick, it was Schlick who connected with "the idea of Ernst Mach, Bertrand Russell, Henri Poincaré and others who had until then been accepted by German physicists only when they succeeded in watering down their teachings for purposes of interpretations of Kant and his successors" and who succeeded in freeing the understanding of the theory of relativity from these fetters (Frank 1937). (And although Reichenbach in the early 1920s was still striving for his own professional recognition, he nevertheless was also doing so within the same field as Schlick.)

Second, we note that Frank did not fail to notice that the Vienna Circle around Schlick constituted an expansion by means of new members whose thematic concerns were closely related to those of the original group but who were descended from a different philosophical tradition:

> The men who had expanded the new positivism into a general logical basis of human thought—Schlick and Carnap—came now into personal contact with the original Viennese group.... As a result of the developing coöperation the new philosophy became more and more different from the traditional German philosophy, to which both Schlick and Carnap were bound to have some sentimental ties originating from their training at German universities. (1949a, 34)

This time Frank explicitly accepted as correct Neurath's thesis of a distinct Austrian tradition in philosophy (47) and reproduced in its entirety the table of ancestors set out in the manifesto of 1929 with the laconic remark "Neurath was eager to trace the genealogic lineage of our movement" (39).

The Repositioning of the First Vienna Circle after the Dispersion of the Vienna Circle around Schlick

Frank's account raises the question of the identity of the Vienna Circle: Whose questions and issues are to be thought of as definitive? But it also raises the question of the identity of logical empiricism: Which questions and issues are to be thought of as definitive?

One point of particular interest in Frank's second report is his decisive accentuation of the role of the first Circle. Frank supported not only Neurath's thesis of the distinct Austrian philosophical tradition but also Neurath's broad remarks on that tradition's long-term dynamics as encompassing even the Circle around Schlick. As it happens, Frank even seems to *strengthen* Neurath's claim: "The" Vienna Circle virtually becomes an episode in the development of the "first" Vienna Circle. That is to say, Frank's accounts suggest that some kind of program for an autonomous philosophy of science—autonomous from metaphysical school philosophy—had already been developed by the first Vienna Circle. Before I am misunderstood, let me stress that Frank, like Neurath before him, was also very explicit about advances that were only made in the Circle around Schlick: It was the broad program, not its specific forms and solutions, that had been developed by the first Vienna Circle. Frank's claims were bold, but not hyperbolic.

To be sure, in his first account Frank had already noted, with an eye to Schlick, that "Hahn was looking for someone who could become the successor of Mach and work along the line of the new turn in the doctrine of positivism" (1941a, 8). That Frank stressed the role of the first Circle even further in his second account may well be explained by the changed situation in 1949. After the death of Neurath in 1945, Frank was the last living member of the original group. Thus, it fell to him to present its work and legacy in the proper historical light.

A second but related point of interest also emerges in Frank's second account. He emphasizes that the basic understanding of the issues and problems of philosophy of science was shared by the members of the first Circle, thereby providing circumstantial support for the view that their positions in the Circle around Schlick also agreed to a large extent. Even a certain tendency can be detected in this respect. Whereas in 1941 Frank still wrote of Schlick and Carnap that "these two Germans from the Reich . . .

were not able to find there any real echo for their ideas" (1941a, 9), in 1949 he could not help noticing, as we saw, their emotional attachment to the German tradition. It would seem that the question of the identity of the Vienna Circle had become an issue for Frank by 1949. As noted, the Circle around Schlick was not homogeneous, but neither, we begin to suspect, was the so-called left Vienna Circle, which, after all, included Carnap.

A third point of interest in Frank's accounts is provided by the expression of his hope that his collections of essays would help prevent logical empiricism from becoming a "narrow sect" and would encourage its integration in the "broader stream of intellectual and cultural development" (1941a, 16). Here Frank appealed to the wider cultural dimension to which, by his own account, the first Circle already saw its efforts directed. At just the time when logical empiricism was beginning to undergo an ever stricter professionalization in the course of its ascent in Anglo-American academic philosophy, Frank recalled and reaffirmed the existential and political ambitions of the original group.[5] However discreetly he went about it, Frank's opposition to the "watering down"—this time of the wider cultural ambitions of early logical empiricism—can no longer be overlooked.

A related point of interest here is that, on his own account, Frank returned to "old arguments in the Vienna coffeehouses around 1907 about Abel Rey, Ernst Mach, and Henri Poincaré" in order to overcome the "new crisis" of science that took its start from the quantum physics of the 1920s (1949a, 46), just as he returned to the "historical approach . . . familiar . . . since [his] student years from the meetings with [his] older friends" in beginning to teach philosophy of science at Harvard in J. B. Conant's new General Education Program (1949a, 50–51).

Finally, we may note that Frank gives the following assignment to historians of logical empiricism: "To understand this evolution precisely, it is not sufficient to follow the gradual alterations from the logical angle. We must also consider carefully the historical trend and background of the arguments" (1949a, 5). The connection with Mach's "historical-critical" method is clear, and again we are referred back, albeit implicitly, to the beginnings of the first Vienna Circle.

The Frank-Neurath Axis

Frank's histories of the Vienna Circle point not only backward but also forward.[6] To see still better the future direction of development that Frank would have preferred, it is necessary to deepen our appreciation of the agreement among the Hahn, Frank, and Neurath trio. Other documents allow us to extend the consequences we drew from Frank's account.

Frank's correspondence with Neurath after the latter's emigration to Holland, though hardly extensive, is revealing in several respects. In his first introduction, Frank had already remarked upon the divisions in the wider movement of logical empiricism, noting that the Berlin group around Reichenbach "never completely accepted the radical program of the Vienna positivism" (1941a, 10). This was not a new point of his (1929–30, as it appears in 1949b, 112), and it was, in any case, in agreement with Reichenbach's own concern to distinguish the Berlin Society for Empirical Philosophy from the Vienna Circle (Danneberg 1993, Hoffmann 1993). But soon Frank also had to note increasing differences among the members of the Circle itself. Thus, he wrote to Neurath from Cambridge, Massachusetts,

> Speaking about [our] movement, I am afraid to say that it has led into a certain impasse. This impasse comes from the lack of any real cooperation. Some people get more and more into purely logical formalism which means almost into a new scholasticism. Other ones who try to influence the real world profess ideas which have little connection with a general scientific approach. (Frank to Neurath, December 10, 1943, Vienna Circle Archive [hereafter WKA])

Whereas the last remark points to Charles W. Morris, the first seems aimed at Carnap. Needless to say, Frank's remark only confirmed Neurath's suspicions, which were fed by and voiced in his dispute over semantics with Carnap: "I have the feeling that the Viennese Circle people become formalists and less and less interested in empiricism as a living thing" (Neurath to Frank, June 16, 1945, WKA). Frank responded laconically, "I have not seen Carnap for five years and [have] not been in touch by writing [for] two years. I think that he drifts away from our group as most people do. It is probably very urgent to call a meeting of the 'inner circle' in order to straighten things out." But he added, "[The] prospect for Unified Science in the English speaking countries is not too bad. But it has to be stimulated by a broader approach" (Frank to Neurath, December 15, 1945, WKA). Whether Neurath received this letter before his death on December 23 is not known. Still, the divisions within the Vienna Circle in exile—even within its left wing—found clear expression here. (Consider also Frank's enigmatic later contribution to Paul Schilpp's Festschrift for Carnap [Frank 1963; cf. Danneberg 1996].)

But what precisely was "[logical] empiricism as a living thing"? It is easy enough to object to oversharp formalization as scholasticism, but what was the alternative? It is instructive here to consider Neurath's correspondence with Carl Gustav Hempel, originally a member of Reichenbach's

Berlin group. When Neurath received word that Hempel was about to enter into the Circle's protocol-sentence debate, he wrote,

> I am much looking forward to your paper. I hope I will not appear in it as someone in apostasy from Wittgenstein. In a certain sense I am glad that you now will have to read my older works in an, as it were, professional capacity. From these you will see that everything which I stand for nowadays dates back to before the common [Circle] period, the tremendous importance of which I certainly do not wish to discount. I just do not like it when you Germans *[Reichsdeutsche]* . . . think of Austria as some kind of appendage and fail to take account of the original French-English-Austrian development and only take cognisance of what happened since Wittgenstein. Thus people always underestimate Philipp Frank. I mean this in a historical sense, not to bluntly assert priorities [*prioritätlich* (sic)]. (February 2, 1935, WKA, my translation)

Neurath's admonition fell on unresponsive ears, and he had to repeat it (again without much appreciable effect), but here we need not reconsider the protocol-sentence debate or the relevance of this exchange with Hempel,[7] nor do we need to consider Neurath's related later reaction to A. J. Ayer's *Language, Truth, and Logic,* which, he wrote, gave him "goose pimples" (Neurath to Hempel, December 1, 1937, WKA, my translation). We may simply note that Neurath too insisted on the historical antecedence of what we now call the first Vienna Circle.

But it was not just the line of historical development that Neurath was concerned to stress. Consider, therefore, that in his exchanges with Hempel, Neurath also asked himself why it was he who saw "metaphysical developments much earlier than the most of us, even though I do not command better tools . . . and hardly am more clever 'as such'" (February 8, 1935, WKA, my translation). Notable in Neurath's own answer is not only that he expressed himself vividly or that he accentuated similar aspects as did Frank, but especially that he explicated the difference in philosophical procedure that separated not just him but all the members of the first Vienna Circle from Carnap and such followers as Hempel:

> I suspect it has to do with the fact that I proceed very pragmatically and less by way of logical analysis. This means the following. For instance, Carnap takes sentences which he wants to 'check' [*'prüfen'*]. Of course, these sentences are very coarse, mushy, like *Ballungen [ballig]*. So Carnap first has to make them precise and clean in order to properly kill them off or redeem them. I, by contrast—I am after all very sensitive about metaphysical or sloppy ways—I take one of these structures, roll it about in my palm, look at how one can get a grip on it, at what it does when one pinches it and

whether one can drive in nails with it. And soon this queer thing shows all sorts of strange properties. I look at these and then set about characterizing the thing itself—either in order to render it more precise myself or to delight in another's doing this. This is, I believe, the basis of the more historical approach *[historisierende Art]*—it is a pragmatic one. Philipp Frank has that very pronouncedly, probably even more so than I do, whereas Schlick, Carnap, and you too have a more logical approach *[logisierende Art]*. If you have a sentence before you, you know more than well with which others it stands in contradiction etc. . . . I am full of admiration for you in this. But when one gives you a sentence structure that is not yet ready for sharp analysis, your judgment is not too secure, and your prediction is often quite poor concerning just where a metaphysical tail will emerge that some devil will place on the altar of the chapel. . . . In this respect, you should not be too glad that Carnap is not very historical. For the pragmatic consideration of *Ballungen* is very important for all scientific work in the analytic sciences; this is quite a different thing from logical analysis. . . . For if one really wishes to engage with empirical science, with all of it, then one has to cultivate the pragmatic-historical approach. . . . The purely logical approach . . . can be misused. I can demonstrate this, as I remarked repeatedly, with examples from political economy. The result is BLUBO with Frege and Rickert with the calculus of relations. Be aware! That was the danger of scholasticism. And then one poured out the logic with the metaphysical bathwater.[8] (Neurath to Hempel, February 8, 1935, WKA, my translation)

This passage contains much valuable material. Neurath not only explained clearly why he insisted, in so many words, on the primacy of the physicalistic, everyday language as the basis for unified science, but he also contrasted the two tendencies within the Vienna Circle's thinking as "logical" and "historical" in a way that illuminates Frank's caustic remarks in his December 10, 1943, letter to Neurath (quoted above). "Empiricism as a living thing," to use Neurath's phrase, essentially depended on employing not only the logical but also the historical way. Empiricism was, as far as Frank and Neurath (and Hahn) were concerned, dead otherwise.

But in his correspondence, Neurath also illuminated the role played by Frank in the Vienna Circle. Thus we must note that it was by no means only in his North American exile that Frank supported Neurath's concern with "empiricism as a living thing" and his skepticism of the ever increasing formalization of their philosophical problematics. Thus, in 1935 Neurath could already write to Frank,

[H]ow correct is your overall attitude, which aims to state honestly and clearly that [one's own thought] still has uncertain margins and that despite this there is progress. Of course, I also said this, but inside I still had something that could not help trying to make things all too clear. The struggle against pseudo-rationalism also goes against something within me, perhaps against something in all of us. (January 18, 1935, WKA, my translation)

Clearly, Frank *confirmed* Neurath in his conviction that something like his encyclopedic conception of unified science had to be developed, that a philosophy of science had to be developed that, against the ideal of "the system," started from the given, fragmented practice of science. It even sounds as if Frank prompted Neurath's efforts in this direction; in any case, Neurath appreciated Frank's points. Soon Neurath wrote to Carnap, "Now I just want to know whether we too have our metaphysics but don't recognize it" (February 2, 1935, WKA, my translation). It becomes apparent here that the Vienna Circle's prima facie typically "modern" critique of metaphysics thus entered into the reflexive dimension that is normally reserved for postmodern discourse and that Frank was at one with Neurath in this respect.

These highlights from the correspondence of members of the first Vienna Circle from the time of the dispersion of Schlick's Vienna Circle in exile lead to two conclusions: First, Frank's retrospectives do not represent an idiosyncratic point of view; rather, they reflect the self-understanding of the earlier members of the first Vienna Circle at an advanced stage of their development. And second, it was not only Neurath's conception of the Vienna Circle as deeply indebted to and in fact embedded in the development of the distinct tradition of Austrian philosophy that found support from Frank, but also Neurath's pragmatic-historical approach to philosophy of science, and moreover, Frank had supported and even shaped Neurath's approach at least since the 1930s.

Into the 1950s

Although I cannot here consider much of Frank's varied work in America[9]— among which, in particular, his intellectual biography of Einstein deserves special mention for its (at the time) very unusual attention to the socio-historical and cultural context of Einstein's work (Frank 1947–49)—I may note that Frank's agreement with Neurath was not limited to their correspondence but extended far into Frank's own post–World War II work. What is especially noteworthy is that Frank developed the pragmatic-historical approach to philosophy of science along its sociological dimension particularly.

Consider Neurath's remark in the closing sections of his own contribution to the *International Encyclopedia of Unified Science*:

> [A]ltering our scientific language is cohesive *[sic]* with altering our social and private life. There is no extraterritoriality for sociologists, or for other scientists, and this is not always sufficiently acknowledged. Sociologists are not only outside their scientific field in arguing, deciding, and acting like other human beings; they also argue, decide, and act like other human beings within their scientific field. (1944, 46–47)

In his very last paper, Neurath stressed the point again:

> I do not think that one can distinguish between the problems of the scientists and the problems of the man in the street. In the end, they are more interlinked than people sometimes realize. Any synthesis of our intellectual life, any orchestration of various attempts to handle life and arguments should never forget these far reaching social implications. (1946, 79)

Neurath urged increased attention to historical, sociological, and educational studies of science—in support of the Enlightenment project.

Work in Neurath's spirit was carried on by Frank, who, in fact, gave much-needed substance to Neurath's suggestive but all-too-brief remarks. As I have noted elsewhere (1991a), Frank's American publications often address the very question that lies at the center of Neurath's thought: the importance of scientific knowledge not only in providing technological progress but also in transforming people's conception of themselves as epistemic agents (Frank 1946). And whereas orthodox logical empiricism sidestepped considerations of "external" influences on theory choice, Frank made it his business to promote research into so-called external determinants of theory choice (for example, Frank 1956).

In one of his contributions to the *Proceedings of the American Academy of Arts and Sciences* (whose volume 80 represents, as it were, the proceedings of the Cambridge, Massachusetts, Institute for the Unity of Science), Frank even presented what looks like the central argument of the radical, post-Mertonian sociology of knowledge. He asked, "[I]f a doctrine is confirmed 'by science,' meaning by observations and logical conclusions, how can its social usefulness have an influence on its acceptance?" and he answered,

> It is certainly possible that several scientific theories may account for the same observed facts. In this case, the "logical-empirical" criteria are compatible with two or more different theories. The authorities, state or church or public opinion, can select from among these doctrines the one which is the most useful for the training of "good citizens." . . . Many people have claimed that there cannot be any influence of state or church on scientific doctrines because no authority can "change observed facts." This is certainly

true. But an 'authority' can require the abandonment or acceptance of scientific *theories* without requiring a "change of the facts." (1951, 19)

Of course, ultraradical sociologists of knowledge would go further still and claim that due to the theory-ladenness of observation, even observational facts are accepted for at least partly external reasons. Even so, it seems noteworthy that the basic argument from evidential underdetermination to external influence on theory choice is unambiguously endorsed by Frank.

Moreover, he did not think this influence necessarily deleterious. Thus, he wrote in the conclusion of his 1957 textbook,

[T]he validity of a scientific theory cannot be judged unless we ascribe a certain purpose to that theory. The achievement of that purpose depends upon the degree to which the different criteria for the acceptance of a theory are satisfied, agreement with observed facts, simplicity and elegance, agreement with common sense, fitness to support desirable human conduct, etc. Hence, the validity of a theory cannot be judged by 'scientific' criteria in the narrower sense: agreement with observations and logical consistency. After application of all these criteria, there remains often a choice among several theories. However, if we mean by "science" not only physical science, but also the science of human behavior (psychology and sociology), then we can decide which among several physical theories achieves a certain human purpose in the best way. (1957, 359)

Not only was the appeal to external criteria of theory choice unavoidable, but theory choice so conceived could itself profit from a scientific approach. Far from rendering theory choice irrational, Frank's acceptance of external criteria as legitimate (presumably with qualifications) promised to increase the rationality of theory choice.

At this point, another significant agreement between Frank and Neurath comes into view.[10] Frank endorsed Neurath's ambitions for a unified science. These included the defense, or rather the fight for, the autonomy of a unified science. Frank approached the question "Can science be its own philosophy?" (1951, 29) in much the same spirit:

We can see that generally the powers that have been interested in controlling human conduct have envisaged an attitude of "humble positivism" among scientists. Scientists should simply collect material and leave the undecided questions of science to be resolved by the educators and rulers of men in such a way that the result will support an education for a "good life." But some scientists do not believe that "superscientific" methods of knowledge can answer the questions concerning which science has reached a stalemate. These scientists believe that where science has no answer, no one else has any answers either. One has to wait until a scientific solution is reached. The decisions encouraged by the social powers do not offer us any "truth." They are instruments of propaganda for a certain way of action. Whether one approves such an interpretation of science depends only upon whether one approves the way of life which is to be supported. This attitude is not "humble positivism" or "passive positivism" but we may call it "proud positivism" or "active positivism." (1951, 28)

Clearly, "proud positivism" affirms the autonomy of a unified science from philosophy—among whose enemies Frank named the "Communist Party in the Soviet Union" and "[o]rganizations like the Roman Church or the Third Reich in Germany," all of which encourage "humble positivism" (28).[11] It is important that "active positivism" does not defend this autonomy for the price of the naturalistic fallacy, by prescribing social goals on purely scientific grounds. Rather, it does so by pointing out the wider contexts into which questions of theory choice are embedded and how the relevant means-ends relations are to be assessed, by providing a scientific analysis of the choice situation.[12]

"Active positivism," accordingly, holds that "[t]he fact that no special science can . . . 'defend its own principles' does not lead to the conclusion that the system of *all* sciences cannot do so." Rather, it holds that "cooperation between the natural and social sciences would reach the objectives that were formerly reserved to philosophy" (1951, 30). By turning to the social sciences, theory choice may be explained *and* legitimated. Of course, given the pluralism of values and goals that is to be expected, theory choice will not be uniquely determined either, but that would surely be "hyperbolic positivism" (30). Frank aimed to follow a middle path. Thus, note also that the restriction of the relevant underdetermination to the theoretical level bars the path from the absence of what Neurath had called the "extraterritoriality of the scientist" (1944, 46) to the abandonment of all objectivity.

Last, one other point of continuity between Frank and Neurath may be mentioned. It was a point of particular satisfaction for Neurath to have won John Dewey as a contributor to the *International Encyclopedia of Unified Science*. Dewey's first contribution already touched upon the "human, cultural meaning of the unity of science" (Dewey 1938, 32), and in his own monograph for the series he returned to the matter: "The practical problem that has to be faced is the establishment of cultural conditions that will support the kinds of behavior in which emotions and ideas, desires and appraisals are integrated" (1939, 65). Frank was very concerned to continue this active cooperation with American pragmatism. Not only does Frank's first account of the history of the Vienna Circle's thought stress this cooperation as the way forward if logical empiricism was not to become "a narrow sect," but the very same point is echoed in his contribution to the American Academy I quoted above. "John Dewey," Frank concluded, "presents the case for 'active positivism' in a strong and impressive way," and he ended with a quotation from Dewey's *Reconstruction in Philosophy:*

> The history of philosophy will take on a new significance. What is lost from the standpoint of would-be science is regained from the standpoint of humanity. Instead of disputes of rivals about the nature of reality, we have the scene of human clash of social purpose and aspirations. (Dewey, quoted in Frank 1951, 30)

Or as Frank himself put it in his own terminology:

> The only way to include theory-building in the general science of human behavior is to refrain from ordering around scientists by telling them what they 'should' do and to find how each special purpose can be achieved by a theory. Only in this way can science as a human activity be "scientifically" understood and the gap between the scientific and the humanistic aspect be bridged. (1954, as it appears in 1956, 18)

Conclusion

My concern here has been to render explicit and plausible what I take to be the agenda enlivening Frank's histories: Behind the remarkable stress on the role of the first Vienna Circle—indeed, its elevation to the status of founding fathers—lay the desire to reaffirm the pragmatic-historical approach of this early group and its existential and social engagement at a time when the pressures of the academicization of logical empiricism pointed in the opposite direction. This interpretation gains further support from the personal communications of Frank and Neurath and from some of Neurath's contretemps with younger German members of the logical empiricist movement, such as Hempel, as well as from Frank's own work after World War II.

How successful was Frank's intervention in the struggle, as it were, for the soul of logical empiricism? Carnap hardly heeded his exhortation to consider the pragmatic aspects of science and stuck mainly to its logic. Similarly, Hempel concentrated on the development of a type of confirmation theory that betrayed little awareness of the legitimacy of external considerations, rediscovering the pragmatic approach only in his later years in the 1970s.[13] And in 1950, Frank had already had to cross swords with

Feigl over his attempt to reintroduce the realistic interpretation of scientific theory in place of the instrumentalist stance (Frank 1950). Thus, Frank's oldest colleagues in the movement already either neglected his "pragmatic" extension of logical empiricism or rejected his more traditionally "positivist" stances. In the long run, the wider philosophy of science community was no more encouraging.[14]

Neither Frank's historiographically disguised intervention concerning the future of the movement nor his own work, we can conclude, found the response among logical empiricists, old or new, that he hoped for. It was, of course, precisely the neglect of what Frank stressed as the pragmatic-historical approach of the first Vienna Circle and its existential and social engagement that prompted much of the antipositivistic criticism that has become all too familiar since the late 1960s.[15]

Notes

1. For an analysis of all three of these disputes in the context of the Circle's protocol-sentence debate, see Uebel 1992.

2. See Graham, Lepenies, and Weingart 1983; Woodward and Cohen 1991.

3. The exception here is, as D'Acconti noted, Feigl. One of his retrospective writings briefly mentions a "prehistoric Vienna Circle" and Hahn's role in Schlick's appointment in Vienna, but this is compensated by Feigl putting himself into the picture of the origin of logical empiricism as having suggested to Schlick, with Waismann, the regular meetings that became the Vienna Circle proper (1969a, in 1981, 58–60). In another, only the latter claim finds expression (1974, in 1981, 7), whereas the Viennese "prehistory" finds no mention there or in "The Origin and Spirit of Logical Positivism" (1969b) or in the "Historical Notes" of his "Logical Empiricism" (1943; dropped from the reprint).

4. For a detailed exploration of work of Frank, Hahn, and Neurath during their early years that substantiates Frank's account and Neurath's more coded remarks about the first Vienna Circle, see Uebel 2000; for the *Problemstellung* of the young Neurath, see also Uebel 1996.

5. On the pressures of the professionalization of logical empiricism, see Howard's chapter 2 in this volume.

6. But see note 4 above. Of course, for present purposes, it is not even required that Frank's reports of the first Vienna Circle approximate the truth; what is important is, rather, that Frank chose to tell the story of logical empiricism in the way that he did.

7. The issue concerns the appropriateness as a physicalist response, in Neurath's sense, to Schlick 1934 of Hempel 1935; see Uebel 1992, chap. 8.

8. "BLUBO" is short for *"Blut-und-Boden,"* the Nazi phrase "blood and soil."

9. For accounts rich with archive material, see Holton 1992, 1995; Galison 1998.

10. This agreement also seems to date back to the first Circle; see the long quotation from page 35 of Frank 1949a in the section "Frank's Second Account."

11. If the former reference sounds like a calculated political positioning of his view in the Cold War, note that Frank's equation of the role of the Catholic church and that of Nazi rulers was hardly likely to win him new friends in "God's own country," despite the Protestant bias in the United States.

12. This programmatic point of the Circle's "scientific world-conception" is explored in Uebel 1998.

13. On the trajectory of Hempel's development, see Friedman's chapter 3 in this volume.

14. On the need for the qualification, see again Howard's chapter 2 in this volume.

15. I wish to thank the participants of the 1998 Logical Empiricism in North America Conference for comments and criticisms, and the curator of the Vienna Circle Archive, now at the Rijksarchif Noord-Holland, for permission to quote from their holdings.

References

Archive of Scientific Philosophy (ASP). Correspondence of R. Carnap. Special Collections, Hillman Library, University of Pittsburgh.

Carnap, R. 1963. "Intellectual Autobiography." In *The Philosophy of Rudolf Carnap*, ed. P. A. Schilpp, 3–84. La Salle, Ill.: Open Court, 1963.

Carnap, R., H. Hahn, and O. Neurath. 1929. *Wissenschaftliche Weltauffassung: Der Wiener Kreis*. Vienna: Wolf. Translation: "The Scientific World-Conception: The Vienna Circle." In *Empiricism and Sociology*, ed. O. Neurath and R. S. Cohen, trans. P. Foulkes and M. Neurath, 299–318. Dordrecht and Boston: Reidel, 1973.

D'Acconti, A. 1995. "The Genealogies of Logical Positivism." Ph.D. diss., Cambridge University.

Danneberg, L. 1993. "Logischer Empirismus in Deutschland." In *Wien—Berlin—Prag: Der Aufstieg der wissenschaftlichen Philosophie*, ed. R. Haller and F. Stadler, 320–61. Vienna: Hölder-Pichler-Tempsky, 1993.

———. 1996. "Zu Brechts Rezeption des Logischen Empirismus." *Deutsche Zeitschrift für Philosophie* 44: 363–87.

Dewey, J. 1938. "Unity of Science as Social Problem." In *Encyclopedia of Unified Science*, by O. Neurath, R. Carnap, and C. W. Morris, vol. 1: 77–137. Chicago: University of Chicago Press, 1938.

———. 1939. *Theory of Valuation*. Chicago: University of Chicago Press.

Feigl, H. 1943. "Logical Empiricism." In *Twentieth Century Philosophy*, ed. D. D. Runes, 371–418. New York: Philosophical Library. Reprinted in *Readings in Philosophical Analysis*, ed. H. Feigl and W. Sellars, 3–26. New York: Appleton Crofts, 1949.

———. 1969a. "The *Wiener Kreis* in America." In *The Intellectual Migration*, ed. D. Fleming and B. Baylin, 630–73. Cambridge Mass.: Harvard University Press. Reprinted in *Inquiries and Provocations: Selected Writings, 1929–1974*, by H. Feigl, ed. R. S. Cohen, 57–94. Dordrecht, Boston, and London: Reidel, 1981.

———. 1969b. "The Origin and Spirit of Logical Positivism." In *The Legacy of Logical Positivism*, ed. P. Achinstein and S. F. Barker, 3–24. Baltimore: The John Hopkins University Press. Reprinted in *Inquiries and Provocations: Selected Writings, 1929–1974*, by H. Feigl, 21–37. Dordrecht, Boston, and London: Reidel, 1981.

———. 1974. "No Pot of Message." In *Mid-Twentieth Century Philosophy: Personal Statements*, ed. P. Bertocci, 120–59. New York: Humanities Press. Reprinted in H. Feigl, *Inquiries and Provocations: Selected Writings, 1929–1974*, ed. R. S. Cohen, 1–20. Dordrecht, Boston, and London: Reidel, 1981.

———. 1981. *Inquiries and Provocations*, ed. R. S. Cohen. Dordrecht, Boston, and London: Reidel.

Frank, Philipp. 1907. "Kausalgesetz und Erfahrung." *Annalen der Naturphilosophie* 6: 443–50. Translated as "The Law of Causality and Experience" in *Between Physics and*

Philosophy, by P. Frank, 17–27. Cambridge, Mass.: Harvard University Press, 1941. And "Experience and the Law of Causality" in *Modern Science and its Philosophy,* by P. Frank, 53–60. Cambridge, Mass.: Harvard University Press, 1949.

———. 1917. "Die Bedeutung der physikalischen Erkenntnistheorie Ernst Machs für das Geistesleben unserer Zeit." *Die Naturwissenschaften* 5: 65–72. Translated as "The Importance for Our Times of Ernst Mach's Philosophy of Science" in *Between Physics and Philosophy,* by P. Frank, 28–53. Cambridge, Mass.: Harvard University Press, 1941. And in *Modern Science and its Philosophy,* by P. Frank, 61–79. Cambridge, Mass.: Harvard University Press, 1949.

———. 1929–30. "Was bedeuten die gegenwärtigen physikalischen Theorien für die allgemeine Erkenntnislehre?" *Die Naturwissenschaften* 17 (1929): 971–77, 987–94; also *Erkenntnis* 1 (1930): 126–57. Translated as "Physical Theories of the Twentieth Century and School Philosophy." In *Between Physics and Philosophy,* by P. Frank, 54–103. Cambridge, Mass: Harvard University Press, 1941. And in *Modern Science and its Philosophy,* by P. Frank, 90–121. Cambridge, Mass: Harvard University Press, 1949.

———. 1932. *Das Kausalgesetz und seine Grenzen.* Vienna: Springer. Translated as *The Law of Causality and Its Limits.* Dordrecht: Kluwer, 1998.

———. 1937. "Nachruf auf Moritz Schlick." *Erkenntnis* 6: 291–91.

———. 1941a. "Introduction: Historical Background." In *Between Physics and Philosophy,* by P. Frank, 3–16. Cambridge, Mass: Harvard University Press, 1941.

———. 1941b. *Between Physics and Philosophy.* Cambridge, Mass.: Harvard University Press.

———. 1946. "Science Teaching and the Humanities." *ETC* 4: 3–28. Reprinted in *Modern Science and Its Philosophy,* by P. Frank, 260–85. Cambridge, Mass: Harvard University Press, 1949.

———. 1947-49. *Einstein: His Life and Times.* New York: Knopff. Enlarged ed. *Einstein: Sein Leben und seine Zeit.* Munich: List, 1949; reprinted Darmstadt: Wissenschaftliche Buchgesellschaft, 1979.

———. 1949a. "Introduction: Historical Background." In *Modern Science and Its Philosophy,* by P. Frank, 1–51. Cambridge, Mass.: Harvard University Press, 1949.

———. 1949b. *Modern Science and Its Philosophy.* Cambridge, Mass.: Harvard University Press.

———. 1950. "Comments on Realistic versus Phenomenalistic Interpretations." *Philosophy of Science* 17: 166–70.

———. 1951. "The Logical and Sociological Aspects of Science." *Proceedings of the American Academy of Arts and Sciences* 80: 16–30.

———. 1954. "The Variety of Reasons for the Acceptance of Scientific Theories." *Scientific Monthly* 79. Reprinted in *The Validation of Scientific Theories,* ed. P. Frank, 3–17. Boston: Beacon Press, 1956.

———, ed. 1956. *The Validation of Scientific Theories.* Boston: Beacon Press.

———. 1957. *Philosophy of Science: The Link between Science and Philosophy.* Englewood Cliffs, N.J.: Prentice Hall.

———. 1963. "The Pragmatic Components in Carnap's 'Elimination of Metaphysics.'" In *The Philosophy of Rudolf Carnap,* ed. P. A. Schilpp, 159–64. La Salle, Ill.: Open Court, 1963.

Friedman, M., 2003. "Hempel and the Vienna Circle." In *Logical Empiricism in North America,* ed. G. L. Hardcastle and A. W. Richardson, 94–114. Minneapolis: University of Minnesota Press, 2003.

Galison, P. 1998. "The Americanization of Unity." *Daedalus* 127: 45–73.
Graham, L., W. Lepenies, and P. Weingart, eds. 1983. *Function and Uses of Disciplinary Histories.* Dordrecht: Reidel.
Haller, R., 1985. "Der erste Wiener Kreis." *Erkenntnis* 22: 341–58. Translated as "The First Vienna Circle." In *Rediscovering the Forgotten Vienna Circle,* ed. T. E. Uebel, 95–108. Dordrecht: Kluwer, 1991.
Haller, R., and F. Stadler, eds. 1993. *Wien–Berlin–Prag: Der Aufstieg der wissenschaftlichen Philosophie.* Vienna: Hölder-Pichler-Tempsky.
Hegselmann, R. 1987. Introduction to *Unified Science,* ed. B. McGuiness, ix–xxi. Dordrecht: Reidel.
Hempel, C. G., 1935. "On the Logical Positivists' Theory of Truth." *Analysis* 2: 49–59.
Hoffmann, D. 1993. "Die Berliner Gesellschaft für empirische/wissenschaftliche Philosophie." In *Wien—Berlin—Prag: Der Aufstieg der wissenschaftlichen Philosophie,* ed. R. Haller and F. Stadler, 386–401. Vienna: Hölder-Pichler-Tempsky, 1993.
Holton, G. 1992. "Ernst Mach and the Fortunes of Positivism in America." *Isis* 83: 27–69. Reprinted in G. Holton, *Science and Anti-Science,* ed. Holton, 1–56. Cambridge, Mass.: Harvard, 1993.
———. 1995. "On the Vienna Circle in Exile: An Eyewitness Report." In *The Foundational Debate,* ed. W. DePauli-Schimanovich, E. Köhler, and F. Stadler, 269–92. Dordrecht: Kluwer, 1995.
Howard, D. 2003. "Two Left Turns Make a Right: On the Curious Political Career of North American Philosophy of Science at Mid-Century." In *Logical Empiricism in North America,* ed. G. L. Hardcastle and A. W. Richardson, 25–93. Minneapolis: University of Minnesota Press, 2003.
Neurath, O. 1936. *Le Developpement du Cercle de Vienne et l'avenir de l'Empiricisme logique.* Paris: Hermann et Cie. Translated by B. Treschmitzer and H. G. Zilian as "Die Entwicklung des Wiener Kreises und die Zukunft des Logischen Empirismus." In *Gesammelte philosophische und methodologische Schriften,* by O. Neurath, ed. R. Haller and H. Rutte, 673–703. Vienna: Hölder-Pichler-Tempsky, 1981.
———. 1944. *Foundations of the Social Sciences.* Chicago: University of Chicago Press.
———. 1946. "After Six Years." *Synthese* 5: 77–82.
———. 1973. *Empiricism and Sociology.* Ed. M. Neurath and R. S. Cohen, trans. P. Foulkes and M. Neurath. Dordrecht: Reidel.
———. 1981. *Gesammelte philosophische und methodologische Schriften.* Ed. R. Haller and H. Rutte. Vienna: Hölder-Pichler-Tempsky.
Neurath, O., R. Carnap, and C. W. Morris. 1938. *Encyclopedia of Unified Science.* Chicago: University of Chicago Press.
Reichenbach, H. 1951. *The Rise of Scientific Philosophy.* Berkeley and Los Angeles: University of California Press.
Schilpp, P. A., ed., 1963. *The Philosophy of Rudolf Carnap.* La Salle, Ill.: Open Court.
Schlick, M. 1934. "Über das Fundament der Erkenntnis." *Erkenntnis* 4: 79–99. Translated by P. Heath as "The Foundation of Knowledge." In *Philosophical Papers,* by M. Schlick, vol. 2, 1925–1936, ed. H. L. Mulder and B. van de Velde-Schlick, 370–87. Dordrecht: Reidel, 1979.
Uebel, T. E. 1991a. "Otto Neurath and the Neurath Reception: Puzzle and Promise." In *Rediscovering the Forgotten Vienna Circle,* ed. T. E. Uebel, 3–22. Dordrecht: Kluwer, 1991.
———, ed. 1991b. *Rediscovering the Forgotten Vienna Circle.* Dordrecht: Kluwer.
———. 1992. *Overcoming Logical Positivism from Within: The Emergence of Neurath's*

Naturalism in the Vienna Circle's Protocol Sentence Debate. Amsterdam and Atlanta, Ga.: Rodopi.

———. 1996. "On Neurath's Boat." In *Otto Neurath: Philosophy between Science and Politics,* ed. N. Cartwright, J. Cat, L. Fleck, and T. E. Uebel, 89–166. Cambridge: Cambridge University Press, 1996.

———. 1998. "Enlightenment and the Vienna Circle's Scientific World-Conception." In *Philosophers on Education: New Historical Perspectives,* ed. A. O. Rorty, 418–38. London and New York: Routledge, 1998.

———. 2000. *Vernunftkritik und Wissenschaft: Otto Neurath und der erste Wiener Kreis.* Vienna: Springer.

Vienna Circle Archive (WKA). Correspondence of O. Neurath. Rijksarchief Noord-Holland, Haarlem.

Woodward, W. R., and R. S. Cohen, eds. 1991. *World Views and Scientific Discipline Formation.* Dordrecht: Kluwer.

Zilsel, E. [pseud. R. Richter]. 1933. "Das Dritte Reich und die Wissenschaft." *Der Kampf* 26: 486–93.

Gary L. Hardcastle

7
Debabelizing Science:
The Harvard Science of Science
Discussion Group, 1940–41

Exile on Mass. Ave.: The Harvard Science of Science Discussion Group (SSDG)

In the fall of 1940, Harvard University witnessed a remarkable confluence of philosophically minded scientists and scientifically minded philosophers. A year before, on September 3–6, 1939, Harvard had hosted the Fifth International Congress for the Unity of Science, an event notable not only because it marked the gathering of an impressive array of scientists and philosophers[1] but because it took place against the backdrop of the German invasion of Poland, begun September 1, and the French and British declarations of war against Germany, which occurred on the conference's opening day. The flight of intellectuals to the United States that began in the 1930s thus reached a fever pitch by late 1940, and Harvard was a way station for the many, Jewish and non-Jewish alike, who sought in the United States the safety and stability by then completely unavailable to them in Europe.[2]

Consequently, many among the prominent philosophers and scientists in or around Cambridge in 1940 were émigrés. They included Rudolf Carnap, who had arrived in the United States from Prague in late 1935 and was visiting Harvard in 1940–41 from the University of Chicago in exchange for the Harvard logician Henry Sheffer (see Quine 1985, 149–59; and Carnap 1963, 34); Philipp Frank, Albert Einstein's successor at Prague, an émigré to the United States in April 1938, and a presence at Harvard from late 1938 until his death in 1966 (Holton 1995; 1993b, esp. 54–57, 64; and 1992, 41–44; and Jones 1984, 233); the psychiatrist Kurt Goldstein, newly appointed clinical professor of neurology at Tufts College Medical School following an appointment at Columbia University and a Rockefeller Fellowship given in the wake of his forced departure from the University of Berlin in 1933 (see Simmel 1968, 3–11); Alfred Tarski, the Polish logician whose residency in the United States began when he found himself unable to return to Warsaw from the Cambridge congress of 1939 (see Moore 1990, 893–96; Givant 1991, 25; and Jones 1984, 221); and Richard von Mises, the Austrian mathematician (and brother to economist Ludwig) who joined Harvard in applied mathematics in 1939 after six years at the University of Istanbul, where he had fled from the University of Berlin upon the rise of National Socialism in 1933 (see Mises 1963–64, vii–viii; and Birkhoff 1983, 283–84).[3]

What these émigrés found in Cambridge—and, in many cases, what brought them there in the first place—were scholars intellectually sympathetic to their scientific or philosophical outlook. In some cases, the sympathetic were German or Austrian emigrants who had come to the United States in the previous decade less for political than for professional reasons. Herbert Feigl, for example, came to the United States in 1930 to work with the Harvard physicist P. W. Bridgman, and he was present again at Harvard in 1940, visiting from the University of Iowa with support from the Rockefeller Foundation (see Feigl 1969, 643–51). The Italian historian of science Giorgio de Santillana had come to the United States from Paris in 1936 and soon after joined the faculty of MIT (de Santillana 1968). The Austrian economist J. A. Schumpeter, a former Austrian minister of finance who joined Harvard's Department of Economics in 1932, was also present at Harvard in 1940 (see Swedberg 1991, 108ff.). Among the "locally grown" sympathetic scholars were two luminaries already mentioned: Bridgman, whose 1927 book *The Logic of Modern Physics* fueled "operationism" movements across the sciences (see Walter 1990; and Moyer 1991a, 1991b), and Shapley, whose efforts brought Tarski and Frank to Harvard, the latter in a permanent half-time lectureship (Jones 1984). At least as significant as these two American natives was W. V. O. Quine, who in 1940 was entering his fifth year as an instructor in philosophy, following four years of postdoctoral support at Harvard (the first spent learning scientific philosophy in Vienna, Prague, and Warsaw in 1932 on a Sheldon Traveling Fellowship; see Quine 1985). Add to Bridgman, Shapley, and Quine the biochemist L. J. Henderson of Harvard's Medical School, whose interest in history and philosophy of science led him to teach Harvard's first history of science course in 1911 (see Parascandola 1970, 260–62); George Sarton, founder of the History of Science Society and editor of its journal, *Isis* (see Edsall 1994, Cohen 1994, and Kuhn 1994); and the young Harvard psychologist S. S. Stevens, whose sustained arguments for operationism in psychology and the social sciences formed a methodological backdrop for his research on hearing (see Hardcastle 1995, 404–24). In a crowning touch, at Harvard that fall was the man many took to be the intellectual father of scientific philosophy, Bertrand Russell, present to deliver the William

James Lectures, which were published that year under the title *An Inquiry into Meaning and Truth* (Russell 1940).

Such, then, was the magnitude and diversity of scientist-philosophers and philosopher-scientists gathered in Cambridge in the fall of 1940.[4] Indeed, this fall counts as an intellectual moment against which similar moments in the history of thought about science can be compared. It is, as we will see, an episode that can help us assess some recent claims about the development of science and its self-understanding in the twentieth century. For the philosophers and scientists in Cambridge that academic year tended not just to their own researches. Rather, a substantial portion (including many of those already mentioned) met regularly in organized discussions on the nature of science, discussions that they consciously identified as scientific. In mid-October 1940, an invitation penned by Stevens (and signed by Edwin Boring, Carnap, Feigl, Frank, Quine, Schumpeter, and Stevens) was distributed to many of the scholars around Cambridge. The invitation read,

> As an effort in the direction of debabelization the undersigned committee is organizing a supper-and-discussion-group to consider topics in the <u>Science of Science</u>. Scientists from various fields, together with some logicians and methodologists present this year at Harvard, will debate issues concerning the meaning and methods of science.
>
> The first topic:
> <u>Analysis of Fundamentals in Thermodynamics</u>
> will be introduced by Professor P. W. Bridgman.
> Place: Harvard Faculty Club
> Time: Thursday, October 31. Supper at 6:30. Discussion at 7:45.
> We hope to have you with us.
>
> (S. S. Stevens Papers [hereafter SSSP] 2.12, dated 1940, emphasis in original)

Among those sent the invitation were many of those mentioned above. Here is the full list, as it appears in the "Science of Science" folder in Stevens's papers:

H. Feigl	J. A. Schumpeter	Hans Kelsen
S. S. Stevens	P. W. Bridgman	G. Haberler
E. G. Boring	G. Birkhoff	C. I. Lewis
R. Carnap	G. D. Birkhoff	De Santillana
P. Frank	Bertrand Russell	K. Goldstein
W. V. O. Quine	R. von Mises	Karl Korsch

A. Wundheiler	J. H. Van Vleck	A. Rosenblueth
N. Goodman	G. Sarton	L. J. Henderson
K. S. Lashley	A. S. Coolidge	D. V. McGranahan
D. Struik	M. H. Stone	H. Shapley
J. Cooley	S. MacLane	Waelden
J. S. Brown	T. Parsons	E. V. Huntington
I. A. Richards	Alfred Tarski	W. H. Furry
V. Guillemin	G. K. Zipf	R. C. Jones
E. C. Kemble	C. Kluckhohn	Polya

The diversity of both discipline and rank reflected in this list is significant. Mathematics, biology, physics, psychology, physiology, economics, and law are all represented, and represented by graduate students and Harvard professors alike.

The SSDG's first topic, "Analysis of Fundamentals in Thermodynamics," was "introduced" by Percy Bridgman at the group's first meeting on October 31, 1940. Stevens moderated this event, and in his papers is the record of who attended, which reads (as it appears, again, from the documents among the SSSP) as follows:

Carnap
Goodman—logic <u>applied</u>
Boring—Probability
Coolidge—chem
Kemble—Quantum Mechanics
Brown
Richards—Meaning of Meaning
Rosenblueth—Physical-chem med
Frank—Math Physics—Vienna
Feigle—????—public relations
Kelsen—law Phil—
Huntington—First principles
Tarksi—Warsaw—Great Polish Sc.
Birkhoff—How me?? Of foreign Soc
Polya—Math Zurich

Mac Lane—ideals—
Zipf—German—Word Counts
Jones—Theor Phys.
Furry—" " Thermal diffusion
Shumpeter—Econ-Student of Sci.
Bridgman
Stevens
Oct 31—late
Wundheiler—Warsaw
Petliss—T
G. Birkhoff
Quine
Kluckholn
Van Vleck

(SSSP 2.12)

Of the approximately forty-five who were invited, at least twenty-nine attended. Russell, C. I. Lewis, and Sarton were apparently absent from this first meeting; Quine, Garrett Birkhoff, and a few others appear to have arrived late.

The SSDG held eight meetings at the Harvard Faculty Club, each on a Thursday following a 6:30 dinner—the same day and hour of the week, incidentally, favored by the Wiener Kreis. Within seven months of its inception, the SSDG had sponsored eight discussions of various topics, each "introduced" by an SSDG participant. From various notes and papers we can reconstruct the SSDG's year as follows:

October 31, 1940	P. W. Bridgman	"Analysis of Fundamentals in Thermodynamics"
January 9, 1941	G. D. Birkhoff	"Sufficient Reasons"
February 13, 1941	S. S. Stevens	"The Formal and Empirical Aspects of Scales"
March 6, 1941	L. J. Henderson	"The Study of Man: A Comparison of the Medical and the Social Sciences"
March 20, 1941	R. von Mises	"Recent Developments in Probability Theory"
April 10, 1941	H. Kelsen	"Causality and Retribution"
May 8, 1941	P. Frank	"The Fossilization of Scientific Theories"
unknown	O. Morgenstern	"Theory of Games"

This list should spark interest in anyone slightly curious about science and its philosophy in this century, for it depicts central figures in the natural and social sciences, mathematics, and philosophy discussing then-recent intellectual developments from an overarching point of view and in what we would now identify as an interdisciplinary context. But the historical instinct to locate these talks (preferably in typescript, but at least in their respective authors' later publications) and reconstruct what was said during a given SSDG gathering is destined to be frustrated, for the topics listed here were *topics,* not papers or talks. Consequently, there are neither typescripts nor ready identifications of SSDG topics in later publications. The SSDG was truly a *discussion* group.[5] And unlike the many significant groups in the history of the philosophy of science for which we have good documentation,[6] there are no minutes of SSDG meetings. What exactly occurred when Oskar Morgenstern led a discussion on the "Theory of

Games," for example, with Birkhoff, Carnap, Talcott Parsons, Russell, von Mises, and others in attendance, must be left to the historically informed imagination. Perhaps the only feature that *can* be gleaned from recollections of the SSDG is that it produced considerable debate. Carnap, for example, offered this account, some twenty years after the fact, of the discussion he led on mathematics and empirical science:

> During the academic year 1940–41, . . . I gave a talk on the relation of mathematics to empirical science in a large discussion group of faculty members interested in the foundations of science. My main thesis was that mathematics has no factual content and, therefore, is not in need of empirical confirmation. . . . I thought this was an old story and at any rate a purely academic question. But to my great surprise, the audience responded with vehement emotions. Even before I had finished my lecture, excited objections were raised. Afterwards we had a long and heated discussion in which several people talked at the same time. . . . Many others rejected my view. . . . [O]n the whole, the discussion was too vehement to permit a good mutual understanding. (Carnap 1963, 64–65)

It is in the Harvard SSDG that we find the crystallization of the heterogeneous aggregate of intellect present in Cambridge in the fall of 1940. Moreover, we find this crystallization in that very brief moment, before the dramatic changes in professional science wrought by World War II but just after scientific philosophy had begun to adapt to the American context. As such, the SSDG offers to our contemporary eye a glimpse of logical empiricism in transition.

In this chapter, I analyze the SSDG. I will center my analysis on the concept of "debabelization" mentioned in the SSDG's first invitation and on the notion of scientific unity that that term connoted and that the SSDG aimed to promote.[7] That analysis will lead us in turn through the views of Carnap and Stevens, the initiators of the SSDG, a competing construal of the unity of science from the late 1930s, and finally to the echo of the SSDG in two later groups, the 1944 Inter-Science Discussion Group and the 1947 Institute for the Unity of Science, both of which affected American thought about science through the 1960s.

I will examine the SSDG from the point of view of the notion of the unity of science it promoted, but I do not suggest that we find within the SSDG just one conception of the unity of science or that any of the senses of unity that are locatable within the SSDG are novel or striking. We will see, though, that one notion of the unity of science *is* central to the SSDG, is well captured by the term 'debabelization' and reflected in the SSDG's activities, and is significant in light of later groups devoted to the scientific study of science.

Before turning to these topics, a more general observation about the SSDG is in order. The nature—indeed, the very existence—of the SSDG speaks against accounts of logical empiricism and its reception in North America that foreground the intellectual exchanges between logicians and philosophers over technical matters (such as analyticity, semantics, and metalanguages) at the expense of exchanges embracing a far wider range of topics *across* scientific and humanistic disciplines. Here, for example, is how Quine, nominally an SSDG organizer, recalls the fall of 1940 at Harvard:

> Tarski had spent the year at Harvard on a meager make-shift appointment. Carnap, who had moved from Prague to Chicago, was at Harvard for the year as visiting professor. Russell came for the second half of 1940 as William James Lecturer. Halcyon days. Carnap, Tarski, and I met for discussion from time to time, together perhaps with I. A. Richards, Nelson Goodman, or John Cooley. By way of providing structure for our discussions, Carnap proposed reading the manuscript of his *Introduction to Semantics* for criticism. Midway in the first page, Tarski and I took issue with Carnap on analyticity. The controversy continued through subsequent sessions, without resolution and without progress in the reading of Carnap's manuscript. (Quine 1986, 19; for similar recountings see Quine 1985, 149–50; 1991, 267)

In this description of that fall (and in many others from Quine), there is no mention of the SSDG, even though Quine cosponsored the SSDG's initial invitation. A history of logical empiricism in North America gleaned from Quine's recollections would neglect the issues and people represented by the SSDG.

Mentions of the SSDG in retrospective writings by other SSDG participants amend Quine's picture slightly; recall Carnap's mention of the "large discussion group of faculty members interested in the foundations of science." Similarly, Feigl's recollections hint that the SSDG was made up of scientists as well as philosophers:

> There was a sort of revival of the Vienna Circle in 1940. This was at the time when Bertrand Russell gave his William James Lectures . . . at Harvard. Carnap and Tarski were at Harvard as visiting professors. I . . . was able to participate in fascinating regular discussions that took place in the Fall of 1940 and included, besides Russell, Philipp Frank, Richard von Mises, W. V. Quine, E. G. Boring, S. S. Stevens, P. W. Bridgman, and I. A. Richards, among the more active members. Issues in the foundations of logic and semantics, as well as of the theory of probability, were extensively discussed. (Feigl 1969, 660–61)

But Quine's authority and the relative accessibility and ubiquity of his writings suggest that the long-dominant image of logical empiricism (among philosophers, let alone other scholars) as a philosophical and logical endeavor entirely disengaged from actual science or scientists is in fact the product of an entirely Quinean lore that is at odds with what *actually* happened as logical empiricists adapted themselves to North America.[8]

Debabelizing Science: The Unity of Science in the SSDG

We can locate the notion of the unity of science central to the SSDG in the visions of its immediate founders, Carnap and Stevens. The SSDG emerged from a proposal Carnap brought to Stevens early in the fall semester of 1940. As Stevens later wrote, recollecting logical empiricism's arrival at Harvard,

> The philosophers inhabited the first floor of Emerson Hall, and it was convenient to take a seat at their lectures. A. N. Whitehead was the chief luminary in the early 1930s, but toward the end of the decade there was an influx from Europe, particularly from the Vienna Circle, which produced an enormous ferment. Rudolf Carnap climbed up to my laboratory one day and proposed that we should form a shop club to discuss the Science of Science. I sent out the invitations, and on Halloween 1940 we inaugurated monthly meetings by listening to P. W. Bridgman defend his solipsistic brand of operationism. Many notables followed. (Stevens 1974a, 408–9)[9]

The accuracy of this recollection is attested to by the assortment of invitations, consultations, and notes pertaining to the SSDG that we find among Stevens's papers. Stevens composed the SSDG's invitations, maintained its address lists, and arranged its meetings. His duties extended beyond the secretarial, for he introduced the participants in the first SSDG meeting to each other, moderated that meeting's discussion, and himself led an SSDG discussion.[10]

In his proposal to start what Stevens called a "shop club to discuss the Science of Science," Carnap probably had in mind the University of Chicago's Logic of Science Discussion Group, in which he had been involved. That group was organized by Charles W. Morris, then an assistant professor of philosophy at the University of Chicago, who had visited Carnap in Prague in 1934. Morris subsequently helped secure Carnap's appointment at the University of Chicago in 1936, where Carnap joined Morris in editing, with Otto Neurath, the initial monographs of the *International Encyclopedia of Unified Science,* eventually available as volumes 1 and 2 of *Foundations of the Unity of Science: Toward an International Encyclopedia of Unified Science* (see Reisch 1994 and 1995, chap. 3). Carnap's recollections of the Logic of Science Discussion Group are also corroborated by Morris's correspondence, for example, Morris's letters to Neurath of January 26, 1937, and June 20, 1937, in which Morris refers to a Logic of Science Discussion Group, and perhaps also a letter of September 17, 1936, to Neurath in which Morris invites Neurath to speak to a group of diverse faculty members at Chicago.[11] As Carnap would later recall, the Chicago Logic of Science Discussion Group "discussed questions of methodology with scientists from various fields and tried to achieve a better understanding among representatives of different disciplines and greater clarity on the essential characteristics of the scientific method." But its "productivity," Carnap added, "was somewhat limited by the fact that most of the participants, although interested in foundational problems, were not sufficiently acquainted with logical and methodological techniques" (Carnap 1963, 35). As we have seen, such was hardly the case at Harvard in the fall of 1940, a fact to which the invitation to the SSDG's initial meeting drew attention.

It is not surprising, then, that Carnap initiated a similar discussion group when he arrived at Harvard, nor that he approached Stevens for help. That fall, the thirty-four-year-old Stevens was entering his third year as an assistant professor in Harvard's Department of Psychology. He had taken a Ph.D. there under Boring in 1933 and had then spent the five intervening years in postdoctoral research fellowships in Harvard's Medical School and Department of Physics (with support from the National Research Council and the Rockefeller Foundation, respectively) and later as an instructor (and subsequently faculty instructor) in the Department of Psychology. It was in these years that Stevens conducted and published the research on audition for which he was well known by the late 1930s. This published work included his 1938 book with Hallowell Davis of Harvard's Medical School on hearing (Stevens and Davis 1938), featuring one of the first detailed mappings of the frequency response of the mammalian cochlea; a series of papers (Stevens 1934a, 1934b, 1936a, 1936b; Stevens and Davis 1936) reporting the experimental isolation of the four tonal psychological attributes of pitch, volume, density, and loudness; and toward the end of that decade, work on the measurement of sensation that led to the classification of nominal, ordinal, interval, and ratio scales of measurement (Stevens 1939b; see also Stevens 1974a and 1974b, and Woodward 1990). The topic of the SSDG discussion led by Stevens on February 13, 1941, was "The Formal and Empirical Aspects of Scales."

Stevens's scientific work was matched in impact at the end of the 1930s by four papers outlining his "operationism" (Stevens 1935a, 1935b, 1936c, and 1939a), a methodology of scientific concept formation that was, Stevens argued, exemplified in the most successful sciences and desperately wanted in the rest. In stark contrast to Bridgman's increasingly solipsistic "operational viewpoint,"[12] Stevens's operationism was founded on the view that public agreement between scientists was the defining characteristic of scientific knowledge. In "Psychology: The Propaedeutic Science" (1936c), written for the new journal *Philosophy of Science*, Stevens summarized this "social aspect of knowledge" and the subsequent importance attached to agreement. "[C]ertain results do get into the scientific textbooks," Stevens wrote,

> not because they satisfy or fall short of some absolute standard of merit, but simply because they meet with the approval or disapproval of other scientists.... The "true" value of a physical constant ... is true because physicists agree that it is true, and if someone convinces physicists that the value is not true, it will thereafter be false. (Stevens 1936c, 96–97; see also 1935a, 327; 1935b, 517; and 1939a, 227)

The greatest possible threat to science so understood was unresolvable *disagreement*. Such disagreements, Stevens suggested, typically revolved around the question of when a given concept applied in a particular situation. Thus, the proper scientific method would both prevent irresolvable disagreements over the application of a concept and settle the question (when it arose) of whether a particular concept applied in a particular case. That proper method was operationism. It required that every scientific concept be associated with a rule for its application, a rule that referred only to concrete, publicly available *operations,* like pointing. Deploring psychology's history of conceptual and theoretical revolution, Stevens heralded operationism as *the* means for eliminating controversy and promoting scientific progress. Operationism "ensures us against hazy, ambiguous and contradictory notions and provides the rigor of definition which silences useless controversy"; it was "the revolution that will put an end to the possibility of revolutions" (Stevens 1935a, 323, emphasis in original; see also 324–26). The arguments, or the rhetoric, were effective; by the late 1930s experimental psychologists (and social scientists in general) aligned themselves, if not their actual practice, with operationism.[13]

Stevens's operationism embraced a unity of science in one obvious sense, for operationism was in Stevens's hands a wholly general scientific prescription. It was applicable to all enterprises for which agreement was a defining characteristic, which is to say, for Stevens, all of science. Stevens's avowal of a unity of science goes deeper than its *method,* however. Scientific knowledge was, for Stevens, constituted in agreement. But agreement in turn is a matter of matching "concrete differential reactions [or discriminations] . . . to environmental states, either internal or external" (Stevens 1935b, 519), and from this Stevens drew two closely related conclusions: The first was that all scientific knowledge is of a single form; it is "obtained, conveyed, and verified by means of" discrimination (518). The second was that in this form inheres all the objective knowledge that could be had of knowledge itself; to reach beyond discrimination in the attempt to grasp *what* was discriminated was symptomatic of confusion over the very nature of knowledge. Discrimination was fundamental; it was the only starting point. The relatively applied consequence of this result was the expressibility of all science in its most fundamental terms, discrimination. In "Psychology and the Science of Science," a 1939 review of the intertwined strands of psychology, operationism, and logical empiricism—an interaction that Stevens characterized as the emerging "science of science"[14]—Stevens was only too happy to make this point via an identification of operationism with physicalism, in this context not the "metaphysical doctrine" that everything there is is physical but, rather, the assertion that any scientific claim could be put into "a language similar to contemporary physics." Thus, Stevens suggested, the protocol sentences sought by Neurath, Carnap, and Schlick, sufficient in principle for the reduction of all scientific sentences, were in fact propositions "relating to elementary discriminations." The sense in which science was unified on this basis was "obvious indeed," Stevens continued, for the reduction of "the esoteric jargons of all the separate sciences . . . to a single coherent language" demonstrates "a fundamental logical unity" (1939a, 240). With the form or structure of scientific claims taken to exhaust the objective knowledge we could have, a unity of form among scientific claims is fundamental indeed. In light of it, the distinctions we draw between disciplines are merely pragmatic; they are only one way among many of cutting up the different contexts in which our discriminatory capacities match up. Their artificiality becomes visible, however, only in the wake of an appreciation of the role of agreement and discrimination.

Return now to the SSDG and to debabelization. In the context of what the unity of science meant to Stevens, to 'debabelize' is not to invent a shared scientific language but to recover a common one. Debabelization is the elimination of the differences that have grown up between disciplines, differences that are superficial but have nonetheless invited the reification of distinctions between disciplines where no real distinctions exist. In his first operationism paper, Stevens implored psychologists to attend to their local conceptual environs in this fashion. "We must," he wrote,

> examine and sift the meanings of [psychological] concepts in accordance with operational procedure. We must determine as precisely as discrimina-

tion will permit what every term denotes.... [T]hen, with psychology stabilized on the operational basis, we must ... maintain constant vigil against the human tendency to read into a concept more than is contained in the operations by which it is determined. (1935a, 330)

Science so debabelized is not simply science tidied up but science optimized for agreement, and hence for progress; indeed, debabelization *was* progress. For making plain the operational basis of science forecloses the confusion and disagreement that had impeded every science at some point and in physics required Einstein to repair. Einstein's achievement (as Stevens read it, echoing Bridgman) was, after all, to recognize the operational basis of fundamental concepts in physics, in particular simultaneity, and then to reform other concepts accordingly. The ensuing resolution of conceptual anomalies in nineteenth-century physics and the articulation of special and general relativity was a brilliant achievement, to be sure, but one we would have been spared had we appreciated the operational basis of our concepts all along. Here was a point on which Stevens and Bridgman could agree: Debabelizing science would, in Bridgman's words, "render unnecessary the services of unborn Einsteins" (Bridgman 1927, 24).[15]

We can place Stevens's sense of unity in a somewhat wider frame. His emphasis on discrimination aside, his sense of the unity of science is of a piece with versions of that thesis familiar to this century's philosophers and scientists. To many, the unity of science thesis is the claim that the truths we find among the different sciences are all reducible to the truths of one sort—truths of physics, typically—and these in turn are reducible to the smallest set from which we can derive or otherwise reconstruct the rest. But for economy of expression and ease of pedagogy, we probably *would* formulate biology, chemistry, psychology, and all the other true sciences (and the true parts thereof) as these simplest truths, subject to no further reduction.

Perhaps such *archetypal reductionism* is now nowhere endorsed, if indeed it ever was (see Galison and Stump 1996), but more tenable visions of the unity of science share a significant feature with this one. What are these more tenable takes on the unity of science? First, notice that even a very strong sort of dependence of one theory or discipline upon another need not entail that the former is *ancillary* or eliminable in favor of the latter; reducing theories or disciplines may not eliminate the theories or disciplines they reduce. *Noneliminativist reductionism,* then, grants the dependency of all sciences upon one without entailing that the reduced sciences are to be eliminated, even in principle. Second, one may hold that no science is reducible to or dependent upon any other, but that, still, among the sciences there is but one method, aim, and even outcome. We saw one version of this *nonreductionist* sense of the unity of science thesis in Stevens's view that operationism was a universal method, and other versions are easy to imagine. What these three construals of the unity of science thesis share, among other things, is the idea that science's *multiplicity*—its division into disciplines or differences of method, aims, and outcomes—is illusory, or at most transient. The unity of science thesis in the broad sense that encapsulates these three positions holds that the sciences, and their methods, aims, and outcomes, are *one,* though from our perspective it may be difficult to see them as such. This is the whole, of which Stevens's unity and its attendant debabelization is a piece.

The sense in which Carnap endorsed the unity of science in the late 1930s was different from Stevens's, but not too different; it was another piece of the same whole. To be sure, Carnap's philosophical views in the years between the 1928 *Aufbau* and his turn to semantics and modal logic early in the 1940s shifted dramatically. In that period, Carnap abandoned epistemology as he had understood it, namely, as the demonstration of the possibility of objective knowledge (see Richardson 1997). But there are grounds to think that his vision of the unity of science in the 1920s was substantially unchanged in 1940 (see, for example, §16 of Carnap 1928, where Carnap identifies "a basic thesis of constitution theory" [cited in Richardson 1997, 8, Richardson's translation]). At any rate, Carnap's only sustained treatment of the unity of science, his *Logical Foundations of the Unity of Science* (1938), comes just two years before the SSDG.

"When we ask," Carnap wrote in 1938, "whether there is a unity to science, we mean this as a question of logic, concerning the logical relationships between the terms and the laws of the various branches of science." So understood, he added, "the question concerns scientists and logicians alike" (49). Carnap's answer to the question was affirmative—there was a unity to science—and consisted in arguments to the effect that any scientific claim could bear reduction to another claim that contained, beyond logical expressions like 'and' and 'not', terms drawn only from what Carnap called the "thing-language," that is, "terms which we use on a prescientific level in our everyday language, and for whose application no scientific procedure is necessary." Carnap offers, as exemplary members of the thing-language, terms like 'hot', 'cold', 'heavy', 'light', and 'red'. By means of reduction sentences effecting the elimination of one term for others, Carnap argued that all of science—or more accurately, any of the sentences of science—could be reduced to sentences of the thing-language. Proceeding from the physical language (with terms like 'temperature' and 'elasticity') through biology ('muscle'), psychology ('anger'), and social science, Carnap sketched how the reduction sentences could be prepared: "[T]he class of observable thing-predicates is a sufficient reduction basis for the whole of the language of science, including the cognitive part of the everyday language," he concluded (1938, 60).

Sixty years later, Carnap's construal and defense of the unity of science strikes modern ears as remarkably antique, a relic of a long-lost time when evidence was evidence, theory was theory, the good wore white, and life was generally simpler. But in the interest of distinguishing Carnap's actual view from its all-too-frequent caricatures, it is well to emphasize (as has Richard Creath [1996, 159] in just this context) that Carnap's is not a view we would now call foundationalist. The thing-language to which all scientific statements can in principle be expressed is only *a* language to which science can be reduced; indeed, it was one of the few themes of Carnap's entire philosophical career that any number of languages could serve our ends. It is not in the reduction *to the thing-language,* then, that the unity of science is demonstrated or achieved. It is in a reduction *period.*

At this point, it may seem that Carnap differs from Stevens on the matter of unity twice over, for although they agree that there is a unity to science, they disagree over the nature of the unity—for 'discrimination' is not a term of the thing-language but a term from psychology—and more deeply, perhaps, over the ways unity might be demonstrated. Stevens's operationism might well have struck Carnap as intolerant rather than progressive. Against these very real differences, however, a shared vision stands out. In Carnap's view, as in Stevens's, the differences between scientific disciplines are hardly as deep as we might have been led to think. The apparent differences between them—in method, content, and aims—are in fact consequences of differences between the languages they use, differences that Carnap, like Stevens, took to be demonstrably superficial. Similarly, 'debabelization' has in this context the sense of screening out these illusory differences so as to let a fundamental commonality stand forth.

With Stevens and Carnap united under the same Unity of Science banner despite their differences, one might well ask whether there was *any* sense of that thesis afoot in the 1930s to which these two could be opposed. Indeed there was, and we find it in Neurath, Carnap's peer in the Vienna Circle's so-called left wing.

It is, of course, richly ironic that this century's two premier expositors of the unity of science, Neurath and Carnap, would disagree, and disagree profoundly, over the sense in which science was or could be unified. The disagreement flourished partly due to Neurath's style of expression, which was at once cryptic, voluminous, dense, and polemical. It may well be that Carnap never grasped Neurath's notion of unity and so never understood the differences between them. With the "rediscovery" of Neurath's work in the 1990s, however, this and other aspects of Neurath's thought have been skillfully articulated (see, for example, Neurath 1973, Haller

1979, Stadler 1982, Uebel 1991, Cartwright et al. 1996, and Reisch 1997; for discussions of Neurath's philosophical vision, see Michael Friedman's chapter 3 in this volume and Thomas Uebel's chapter 6 in this volume). For our present, rather narrow interests, the key point is that on Neurath's understanding of the unity of science the distinctions we recognize between scientific disciplines are *not* necessarily illusory or transient. Unity is attained, rather, in the practical "real world" connections between scientific disciplines, and it is in turn mandated by our common practical need to predict and control our world. Thus, science is unified despite a multiplicity of methods, aims, and results; unified science is, Neurath wrote, "the stock of all connectable, and indeed logically compatible, laws" (Cat, Cartwright, and Chang 1996, 352). Such unity was, as Neurath and many others have pointed out (see, especially, Reisch 1994), the reason that the *International Encyclopedia of Unified Sciences* was an *encyclopedia,* rather than, say, an axiomatic system or a dictionary. In the *Encyclopedia*'s introductory monograph, "Unified Science as Encyclopedic Integration," Neurath wrote,

> Science itself is supplying its own integrating glue instead of aiming at a synthesis on the basis of a "super science" which is to legislate for the special scientific activities. The historical tendency of the unified science movement is toward a unified science departmentalized into special sciences, and not toward a speculative juxtaposition of an autonomous philosophy and a group of scientific disciplines. If one rejects the idea of such a super science as well as the idea of a pseudo-rationalistic anticipation of *the* system of science, what is the maximum of scientific coordination which remains possible? The answer given by the unity of science movement is: an encyclopedia of unified science . . . in contradistinction to an anticipated system or a system constructed a priori. (Neurath 1938, 20, emphasis in original)

Here, then, is the alternative to the sense of unity that was embraced by Stevens and Carnap and in which the SSDG was invested. Neurath, of course, did not survive the war; he died of natural causes in London in 1945. But, as we will see, his vision of the unity of science, embedded as it was in the *International Encyclopedia of Unified Science,* survived.

Unity after the War: The Inter-Science Discussion Group and the Institute for the Unity of Science

By the spring of 1941, the "remarkable confluence" that had been apparent eight months earlier around Harvard—and that had sparked the

SSDG—predictably dissolved. Visiting scholars, Carnap and Feigl among them, departed, and their émigré colleagues found employment in California, New York, Iowa, and other places. A few—notably, Frank—remained in Cambridge. The clamor across disciplines, intellectual and otherwise, faded, one comparatively small effect of totalitarianism's rise in the 1930s and Europe's plunge into war. Interdisciplinary discussions were squelched almost entirely with the official entrance of the United States into the war late in 1941. A moral compulsion to devote one's energies only to war-related matters, felt immediately among academics, was frequently complemented by Armed Service appointments and the accompanying security clearances and prohibitions on discussing war-related research with those lacking the requisite "need-to-know." By late 1941, for example, Stevens's administration of Harvard's Psycho-Acoustic Laboratory, in which research on acoustics, hearing, and communications was being turned toward the improvement of flight crew communications in combat aircraft, precluded him from discussing the lab's research beyond the basement walls of Memorial Hall, which housed it (see Stevens 1974a, Galison 1998, 62–63). Quine, to take another example, began learning cryptanalysis via a Navy correspondence course in June 1941 so as to make himself "useful" in the fight against Nazism; his three subsequent war years were spent as a Navy lieutenant analyzing decoded German submarine communications (Quine 1985, 151, 159, 181–90). The SSDG's time had clearly passed.

While the transformation of scientific practice in the United States during World War II is enormously profound and truly complex (see Kevles 1995, Galison 1997, and Lowen 1997), my interests in this final section are not in change but in a particular thread of continuity. A little over three years after its last session, with mobilization achieved and U.S. anxieties, at least among academics, eased, the SSDG was reborn. Sometime in 1944, under the direction of Frank, an Inter-Science Discussion Group (ISDG) was formed at Harvard. The ISDG's inception, workings, and ultimate effects have been described by its postwar secretary, Gerald Holton, then a graduate student in Harvard's Department of Physics. Holton's description makes clear the similarities between the ISDG and the SSDG, well beyond their temporal and geographic proximity. Like the SSDG, the ISDG met in the evening at the Harvard Faculty Club following a dinner at 6:30. Invitations to participate in an ISDG meeting resembled earlier SSDG invitations; one invitation to the ISDG read in part as follows:

> Our group consists of persons in different fields who feel that the extreme specialization within science demands as its corrective an interest in the

entire scientific edifice. We plan to hold meetings from time to time in which discussions of different topics will be led by competent scholars. . . .
Sincerely Yours,
The Committee:
Percy W. Bridgman
Walter Cannon
Philipp Frank
Philippe LeCorbeiller
Wassily W. Eeontief
Harlow Shapley
George Uhlenbeck
(SSSP, "Inter-Scientific Discussion Group" invitation, December 30, 1944)

There was also a considerable overlap of participants: Aside from Frank, Bridgman, Edwin Kemble, Parsons, Richards, Santillana, Schumpeter, Shapley, and von Mises regularly attended meetings of both the SSDG and the ISDG. The wide range of expertise among the participants led to an equally wide range of topics, for, as the invitation of December 30 noted, ISDG members believed that "extreme specialization within science demands as its corrective an interest in the entire scientific edifice." Thus, the ISDG's meetings featured topics that ranged from "Psychoanalysis and the Theory of Social Systems," led by Parsons, to "The Brain and the Computing Machine," led by Norbert Weiner, to "General Education," led jointly by Kemble, Ivor Richards, Samuel Beer, Philippe LeCorbeiller, and E. S. Castle (Holton 1995, 271–72).

There is another connection between the two groups, less firm perhaps, but far more suggestive. Two years after the inception of the ISDG, in the wake of the war's end and amid efforts to reclaim prewar projects, Frank sought funds from the Rockefeller Foundation to create within Boston's American Academy of Arts and Sciences an Institute for the Unity of Science (IUS), ostensibly on the basis of the ISDG.[16] In his appeal to the Rockefeller Foundation, Frank enthusiastically described the emergence of "several domains of special science." But "unfortunately," he wrote,

> the terminologies in different "fields" are not coordinated with each other. . . . [S]everal fields of study use the same words in totally different meanings and, what is still worse, in slightly different meanings. Therefore, if we attempt to put the pieces together we get ourselves into many apparently "insoluble problems." The situation reminds [us of] the biblical story of the tower of Babel. Because of the confusion in human language the tower of science cannot grow toward the heavens. (Frank 1947, 161–62)

In light of the emphasis I have placed on debabelization in my discussion of the SSDG, this passage raises some obvious questions. Did the ISDG and its successor, the IUS, embrace the vision of scientific unity we located in the SSDG? Did 'debabelization', a notion I have argued was characteristic of Americanized scientific philosophy in the late 1930s, survive the war to influence American thought and policy about science through the 1950s?

No. Although there can be little doubt of the historical connection between these two uses of 'debabelization', the sense Frank attached to scientific unity in his Rockefeller Foundation appeal (and in the context, presumably, of the ISDG) was very different from that around which the SSDG was centered. As Peter Galison (1998) has recently shown, the notion of unity that informed the IUS and the ISDG was not founded on locating a single method, language, or set of concepts in which all sciences could be expressed or to which they all could be reduced. Instead, it was a unity made up "piecewise" of "cross-connections"; it was a "unification through localized sets of common concepts, not through a global metaphysical reductionism" (Galison 1998, 67), and one "predicated on assembling diverse methods, professions, and patterns of work into the production of pragmatic solutions to immediate problems" (64). At its heart, then, was the "formation of entirely new combinations of disciplines" like physical chemistry, biophysics, information theory, and behavioral psychology, disciplines that had in many cases just emerged from wartime labs like Stevens's Psycho-Acoustic Laboratory. The notion of unity for which Frank proposed an institute was, Galison concludes, "a philosophical outlook squarely located in the scientific concerns of an age of computers and nuclear power" (1998, 46).

A glance back at Frank's report confirms a portion of Galison's conclusion and provides us with a sense of debabelization very different from what we find in the SSDG. Frank did not always fully grasp the distinctive character of his own brand of unity;[17] more often, he saw exciting and unexplored possibilities in the pursuit of a scientific unity created by connections between disciplines. "Hybrid fields like 'mathematical biophysics' or 'mathematical economics,'" wrote Frank, no doubt recalling ISDG meetings on those very topics,

> are no longer isolated cells where some queer professors may enjoy their strange fancies, but ... the roots of new developments leading towards the integration of human knowledge and human behavior. These cross-connections become the vanguards of the science of the future. (Frank 1947, 166)

'Debabelization' in the context of this sense of unity is not, as it was for the SSDG, a ruthless excision of linguistic accretions that have concealed the form or content common to all scientific sentences and therefore blinded us to science's true unity. It is, instead, the maintenance, and in some cases the creation, of languages that mediate between distinct disciplines. Among its tasks, as we saw above, is the prevention of instances in which "several fields of study use the same words in totally different meanings and, what is still worse, in slightly different meanings." Indeed, in Frank's appeal is an implicit criticism of scientific unity and debabelization as the SSDG pursued these aims. For scientists ignorant of the true nature of scientific unity and its maintenance, Frank claimed

> there seems to be a dilemma: either to be satisfied with an exact exploration and presentation of special fields isolated from each other with no bridges between them, or with an integration by a logically consistent philosophical system which is a superstructure imposed upon science and formulated in a terminology which is only vaguely connected with the terminology of the scientists. (Frank 1947, 162)

In Frank's IUS, by contrast, scientists would realize the unity they desired *by the formation of bridges among the sciences*; the dilemma "between Specialization and Superficiality" (162) would dissolve. The vision of science as a tower growing toward heaven is a powerful one, and it is one that Frank, like Stevens, employed to considerable effect. But Frank's notion of unity and Stevens's were far apart.

But was Frank's sense of the unity of science novel? Galison, we have seen, has claimed it was—a product, namely, of a tumult of disciplinary interconnections forged in the often frantic problem solving of American laboratories in the early years of World War II. The discussion at the end of the previous section, however, suggests a slightly different conclusion. Although evidence of the historical connection between Neurath and Frank on this score is lacking in Frank's proposal, his notion of unity resembles Neurath's. One of Neurath's favored illustrations of the unity of science—indeed, one employed by Galison in his characterization of prewar notions of the unity of science—makes clear the coincidence of Frank's and Neurath's ideas. "Certainly different kinds of laws can be distinguished," wrote Neurath, but

> it can *not be said of a prediction of a concrete, individual process that it depend[s] on one definite kind of law only*. For example, whether a forest will burn down at a certain location on earth depends as much on the weather as on whether human intervention takes place or not. This intervention, however, can only be predicted if one knows the laws of human behavior.

That is, under certain circumstances, it must be possible to connect all kinds of laws with each other. Therefore all laws, whether chemical, climatological, or sociological, must be conceived as *parts of a system*, namely of *unified science*. (Neurath, cited in Cat, Cartwright, and Chang 1996, 352)

We can well imagine Frank assenting to this account of unified science, but not to those endorsed by Carnap or Stevens. I will not attempt a more detailed tracing of the apparent influence from Neurath to Frank on this matter but will rest content with the conclusion that, *contra* Galison, the sense of unity embraced by the ISDG was not solely a product of World War II.

Conclusion

I have in this paper traced what I earlier characterized as a unique and untainted sample of logical empiricism American style, existing in a brief historical moment before World War II yet after the Americanization of scientific philosophy. We see in the Harvard SSDG the embodiment and implementation of a particular notion of the unity of science, an implementation that called for the active participation of powerful and talented scientists from apparently disparate backgrounds to join together in ridding their common enterprise of babel, so that its true form—a tower reaching to heaven—would be manifest. Likewise, the SSDG's postwar instantiations, the Harvard ISDG and the ensuing IUS, consisted of powerful, talented scientists of disparate backgrounds, ostensibly engaged in debabelization as well. But it was unity and debabelization of a different, albeit not a novel, sort. It was in the Neurathian development of interdisciplines and the forging of connections among extant disciplines that Frank and his fellow scientists understood themselves to be *building* a tower of science to the heavens. We are left, in the end, not just with an array of senses in which the phrases 'unity of science' and 'debabelization' could be and were put, but with a demonstration of the malleability, adaptability, and power of the ideas behind these phrases.

Notes

All passages from the S. S. Stevens Papers (SSSP) are reproduced by permission of Harvard University Archives.

1. The participants in the 1939 Congress included P. W. Bridgman, Rudolf Carnap, Alonzo Church, Herbert Feigl, Philipp Frank, Carl Hempel, Kurt Goldstein, L. J. Henderson, Werner Jaeger, Felix Kaufmann, S. C. Kleene, Susanne Langer, Kurt Lewin, R. B. Lindsay, Henry Margenau, Richard von Mises, Charles W. Morris, Ernest Nagel, Otto Neurath, F. C. S. Northrop, Talcott Parsons, W. V. O. Quine, Hans Reichenbach, Giorgio

de Santillana, George Sarton, S. S. Stevens, Alfred Tarski, and Friedrich Waismann, among others. A listing of the papers delivered there appears in volume 8 of *Erkenntnis*. The original listing of expected speakers (from the papers of the Congress's organizer Bridgman) is reproduced in Holton 1995; see also Joergensen 1951, 47–48. Due in part no doubt to the outbreak of war, some papers (e.g., those of J. Lindenbaum-Hosiasson, Leon Chwistek, Karl Dürr, Goldstein, and K. Reach) were read by persons other that their authors, who did not attend. As the *Erkenntnis* report makes clear (371), the Congress proceedings were to appear in volume 9 of *Erkenntnis* under its new title, *Journal of Unified Science*. The German invasion of Holland in May 1940, however, apparently rendered publication of that number impossible (see Quine 1985, 141).

It is this 1939 Congress that Quine later recognized as "the Vienna Circle, with accretions, in international exile" (Quine 1986, 19; see also Quine 1985, 140–41). For a discussion of the significance of the 1939 Congress in the context of the transmission of logical empiricism to North America, see Holton 1992, 1993b, 1995. The 1939 Congress is not the only event to be later christened a "Vienna Circle in exile." Feigl, for example, described this paper's topic, the 1940 Science of Science Discussion Group, as "a sort of revival of the Vienna Circle" (1969, 660), and Quine described meetings of the American Academy of Arts and Sciences organized by Frank in 1949 as "somewhat in the way of a Vienna Circle in exile" (1985, 219).

2. For recent discussions of intellectual migration from Europe to the United States in the 1930s, see Heilbut 1997 and Jackman and Borden 1983. Other useful sources include Coser 1984; Fermi 1971, 364–72; and Fleming and Bailyn 1969. Jones 1984 addresses the efforts of Harvard astronomer Harlow Shapley, an SSDG participant, to aid displaced European intellectuals.

3. In his flight to the revitalized University of Istanbul, von Mises was joined by several other German scientists and philosophers, notably, Hans Reichenbach, who learned English and wrote his *Experience and Prediction* (1938) in Istanbul. On the significance of the University of Istanbul for European refugees, see Jones 1984, 208, n. 5.

4. That this collection was a relative novelty even there is evidenced by a comment Bridgman made to Frank in a letter of March 30, 1938, two and a half years before the SSDG. In response to an overture from Frank of the month before, Bridgman replied, "I am afraid you will not find Cambridge the center of activity with regard to the questions of interest to you. . . . My work is done practically alone. I have no students [in philosophy of science], and have practically no contact with members of the department of philosophy, and, in fact, most of them are not at all sympathetic with our point of view. The only young philosopher here whom I have particularly interested is Dr. Quine" (quoted in Holton 1993, 65). Bridgman's isolation may reflect his own iconoclasm. But even on a weak reading of this description, by the fall of 1940 the situation in Cambridge, and at Harvard in particular, had changed drastically.

5. In fact, at least one scholar declined to participate in the SSDG on just these grounds. In a letter to Stevens dated October 19, 1940, Sarton wrote, "I am very sorry not to be able to attend these interesting discussions. I am not the right man for them. I prefer to *read* and *write* than to *listen* and *talk,* and can do the former much better than the latter. It is true I teach but only because I must; I would prefer not to teach orally but only by example—in coöperative work, even as a master craftsman teaches apprentices. I am sorry that I cannot be of use in your very laudable undertaking—Vale" (SSSP 2.12, emphasis in original).

6. See, for example, Moritz Schlick's instantiation of the Vienna Circle; the similar records of the Inter-Science Discussion Group and its successor, the Institute for the Unity

of Science, in Cambridge, Massachusetts, from 1944 through 1956 (see Holton 1995); and the extensive records (including audio tapes) of the discussions and conferences held at the Minnesota Center for Philosophy of Science in the 1950s.

7. My focus upon this one word from the SSDG's initial invitation and its meaning may seem unnecessarily restrictive, or at least too slender a basis on which to construct an analysis of the SSDG. My focus is warranted in part because 'debabelization' begs comment in this context. The "babelization" described in Gen. 11:1–9 is, after all, God's introduction of confusion (in the form of multiple languages) in an attempt to frustrate the hubristic attempt to build a tower to the heavens. *De*babelization might therefore be understood at a first glance as the *embrace* of hubris. The term's rich meaning would not have escaped Stevens (who probably coined it), for Stevens taught a Mormon Sunday school for adults in Cambridge in the early 1930s and was given to the playful and irreverent adoption of religious and biblical terms for academic purposes (see Stevens 1974, 395–420, 401; Hardcastle 2000). 'Debabelization' was therefore probably a term coined precisely to convey to the wider Cambridge community the nature and aims of the SSDG.

My focus is motivated also by the fact that Frank, seven years after the SSDG, described the "background and purpose" of his own Institute for the Unity of Science with reference to the Tower of Babel (Frank 1947, 161–62); see the discussion later in the text. Far from being a salient but offhand neologism of the SSDG, 'debabelization' is thus a natural point from which to approach the nature and aims of the SSDG.

8. The disparity between what was once common lore about scientific philosophy or logical empiricism and what the historical facts are is getting to be old news, as historical scholarship on logical positivism and scientific philosophy has come into its own; see, for example, Giere and Richardson 1996 and Friedman 1999. For analyses of logical empiricism in North America especially, see Holton 1992 and 1995.

9. See also Stevens 1966, where he writes: "[P]erhaps a little too shy to do the task himself, [Carnap] suggested one day that I should try to gather a monthly discussion group composed of those whose interests could be said to overflow the fences surrounding their narrow professional specialties" (1966, 215).

10. I have found no evidence in the papers of Stevens, Boring, or Bridgman or in the published work of Boring, Stevens, Frank, Bridgman, Quine, Carnap, Schumpeter, or Feigl that the inception or maintenance of the SSDG was ever in the hands of anyone other than Carnap or Stevens. The archival evidence suggests that the SSDG was primarily Stevens's project. Accordingly, Boring, Frank, Schumpeter, and the rest were apparently enlisted as cosigners on the invitation (as suggested in Stevens 1966, 215), and though they participated in discussions, they were apparently not active in initiating or planning the SSDG.

11. George Reisch, personal communication.

12. Bridgman 1936 makes evident the divergence of Bridgman's views from the views of those who had taken up his 1927 book as heralding operation*ism,* a neologism Bridgman deplored. Where operationists saw a means to render science public and objective, Bridgman described science as essentially and intensely private: "[I]n the last analysis," he wrote, "science is only my private science" (Bridgman 1936, 13).

For analyses of Bridgman's philosophical position, see Koch 1992, Moyer 1991a and 1991b, and Walter 1990, chap. 7; see also Bridgman 1959, which is Bridgman's own assessment of his 1927 book. For discussions of operationism, see Hardcastle 1995, Rogers 1989, and Smith 1986, esp. chap. 7.

13. Stevens was not the only early advocate of an operational outlook. B. F. Skinner and E. C. Tolman also promoted arguably operational viewpoints in the 1930s, Skinner in his Harvard Ph.D. dissertation and Tolman in his "An Operational Analysis of 'Demands'"

(1936). (The influence accorded by T. B. Rogers in "Operationism in Psychology" [1989] to the educational testing movement and to Boring's views is exaggerated; cf. Hardcastle 1995). This lends some credence to Boring's oft-cited comment that operationism "was there all along" in psychology (see Boring 1950, 656). However, Stevens's views were articulated in far greater detail and were in general far more sophisticated than the positions outlined by either Tolman or Skinner. Stevens's work became the reference point for favorable and unfavorable discussions of operationism (see, for example, Bergmann and Spence 1941, Israel and Goldstein 1944, and MacCorquodale and Meehl 1948).

14. Others in the 1930s used the phrase 'science of science' to characterize their discussions about science, most significantly for our purposes, Charles Morris. In the "introductory prospectus" widely distributed by the University of Chicago Press to publicize the forthcoming *International Encyclopedia of Unified Science,* Morris wrote, "The concern throughout the world for the logic of science, the history of science, and the sociology of science reveals a comprehensive international movement interested in considering science as a whole in terms of the scientific temper itself. A science of science is appearing" (Morris, cited in Reisch 1994, 153). It is very likely that this is where Stevens got the phrase, for in his "Psychology and the Science of Science" (1939a), he cited Morris as providing the "fullest account" of the science of science and proceeded to outline Morris's "semiotic."

15. Debabelization also made the science that addressed discrimination especially significant—in Stevens's words, "propaedeutic." "Does it not appear," he asked, "that the Science of Science must go directly to psychology for an answer to many of its problems? . . . The psychologist works out the laws under which different stimuli evoke equivalent reactions. . . . The entire activity of the scientist as a sign-using organism constitutes, therefore, a type of behavior for which behavioristics seeks the laws. If there is a sense in which psychology is the propaedeutic science, . . . it is undoubtedly in its ability to study the behavior, *qua* behavior, of the science-makers" (1939a, 250). Stevens recognized that psychology so described was in the vertiginous position of having itself as an object of study, and he addressed this in a passage that echoed Neurath in Stevens's own time and prefigured the Quinean view to come; see Hardcastle 1995.

16. Frank's appeal for Rockefeller funds was successful. In late 1947, a year after Frank's request, the foundation awarded the IUS $9,000 over a three-year period (Resolution RF 47131, RG 1.1, 100 Unity of Science, Box 35, Folder 281, Rockefeller Foundation Archives; cited in Galison 1998, 69). The IUS was granted a provisional charter by the State of New York on July 31, 1947, although owing to unclarities concerning its tax status, the Rockefeller money was not available until July 1, 1949 (Galison 1998, 69). The IUS remained active through 1958. Then, under the direction of its secretary, Robert Cohen, it was dissolved and its funds were officially transferred to the Philosophy of Science Association. It was reborn in 1959 under the direction of Cohen and Marx Wartofsky as the Boston Colloquium for Philosophy of Science. As Holton notes, "Frank and other members of the IUS contributed regularly, and thereby smoothed the transition" (Holton 1995, 279). The Boston Colloquium continues to this day, thus continuing a tradition of interchange among scientists, philosophers, historians, sociologists, and other intellectuals that can be traced back more than sixty years to Morris's Logic of Science Discussion Group at the University of Chicago.

17. For example, Frank aligns the aims of the IUS with those of the prewar Unity of Science movement led by Carnap and Neurath, which held that synthesis of the sciences would take place through "logical-empirical or semantical analysis" and "the propositions of all the sciences become . . . life" (Frank 1947, 165). On the view presented here, Frank is simply not aware of the new notion emerging.

References

Bergmann, G., and K. Spence. 1941. "Operationism and Theory in Psychology." *Psychological Review* 48: 1–14.
Birkhoff, G. 1983. "Richard von Mises's Years at Harvard." *Zeitschrift Für Angewandte Mathematik und Mechanik* 63: 283–84.
Bridgman, P. W. 1927. *The Logic of Modern Physics*. New York: Macmillan.
———. 1936. *The Nature of Physical Theory*. New York: Dover.
———. 1959. "P. W. Bridgman's 'The Logic of Modern Physics' after Thirty Years." *Daedalus* 88: 518–26.
Carnap, R. 1928. *Der logische Aufbau der Welt*. Berlin: Weltkrcis. Translated by R. George as *The Logical Structure of the World*. Berkeley and Los Angeles: University of California Press, 1967.
———. 1938. "Logical Foundations of the Unity of Science." In *Foundations of the Unity of Science: Toward an International Encyclopedia of Unified Science,* ed. O. Neurath, R. Carnap, and C. Morris, 42–62. Chicago: University of Chicago Press, 1969.
———. 1963. "Intellectual Autobiography." In *The Philosophy of Rudolf Carnap,* ed. P. A. Schilpp, 3–84. La Salle, Ill.: Open Court Press, 1963.
Cartwright, N., J. Cat, L. Fleck, and T. Uebel, eds. 1996. *Otto Neurath: Philosophy between Science and Politics*. Cambridge: Cambridge University Press.
Cat, J., N. Cartwright, and H. Chang, eds. 1996. "Otto Neurath: Politics and the Unity of Science." In *The Disunity of Science: Boundaries, Contexts, and Power,* ed. P. Galison and D. Stump, 347–69. Stanford, Calif.: Stanford University Press.
Cohen, I. B. 1994. "A Harvard Education." *Isis* 75: 13–21.
Coser, L. A. 1984. *Refugee Scholars in America: Their Impact and Their Experiences*. New Haven, Conn.: Yale University Press.
Creath, R. 1996. "The Unity of Science: Carnap, Neurath, and Beyond." In *The Disunity of Science: Boundaries, Contexts, and Power,* ed. P. Galison and D. Stump, 158–69. Stanford, Calif.: Stanford University Press.
de Santillana, G. 1968. *Reflections on Men and Ideas*. Cambridge: MIT Press.
Edsall, J. T. 1994. "Lawrence J. Henderson and George Sarton." *Isis* 75: 11–13.
Feigl, H. 1969. "The *Wiener Kreis* in America." In *The Intellectual Migration: Europe and America, 1930–1960,* ed. D. H. Fleming and B. Bailyn, 630–73. Cambridge, Mass.: Harvard University Press.
Fermi, Laura. 1971. *Illustrious Immigrants; The Intellectual Migration from Europe, 1930-41,* 2nd ed. Chicago: University of Chicago Press.
Fleming, D. H., and B. Bailyn, eds. 1969. *The Intellectual Migration: Europe and America, 1930–1960.* Cambridge, Mass.: Harvard University Press.
Frank, P. 1947. "The Institute for the Unity of Science: Its Background and Its Purpose." *Synthese* 6: 161–62.
Friedman, M. 1999. *Re-evaluating Logical Positivism*. Cambridge: Cambridge University Press.
———. 2003. "Hempel and the Vienna Circle." In *Logical Empiricism in North America,* ed. G. L. Hardcastle and A. Richardson, 94–114. Minneapolis: University of Minnesota Press.
Galison, P. 1997. *Image and Logic: A Material Culture of Microphysics*. Chicago: University of Chicago Press.
———. 1998. "The Americanization of Unity." *Daedalus* 127: 45–71.
Galison, P., and D. Stump, eds. 1996. *The Disunity of Science: Boundaries, Contexts, and Power.* Stanford, Calif.: Stanford University Press.

Giere, R., and A. Richardson, eds. 1996. *Origins of Logical Empiricism*. Minneapolis: University of Minnesota Press.

Givant, S. 1991. "A Portrait of Alfred Tarski." *Mathematical Intelligencer* 13: 16–32.

Haller, R. 1979. *Studien zur Österreichische Philosophie*. Amsterdam: Rodopi.

Hardcastle, G. L. 1995. "S. S. Stevens and the Origins of Operationism." *Philosophy of Science* 62: 404–24.

———. 2000. "The Cult of Experiment: The Psychological Round Table, 1936–1941." *History of Psychology* 3: 344–70.

Heilbut, A. 1997. *Exiled in Paradise: German Refugee Artists and Intellectuals in America, from the 1930s to the Present*. Berkeley and Los Angeles: University of California Press.

Holton, G. 1992. "Ernst Mach and the Fortunes of Positivism in America." *Isis* 83: 27–60. Reprinted in *Science and Anti-Science,* by G. Holton 1–55. Cambridge, Mass.: Harvard University Press, 1993.

———. 1993a. *Science and Anti-Science*. Cambridge, Mass.: Harvard University Press.

———. 1993b. "From the Vienna Circle to Harvard Square: The Americanization of a European World Conception." In *Scientific Philosophy: Origins and Developments,* ed. F. Stadler, 47–73. Boston: Kluwer.

———. 1995. "On the Vienna Circle in Exile: An Eyewitness Report." In *The Foundational Debate: Complexity and Constructivity in Mathematics and Physics,* ed. W. DePauli-Schimanovich, E. Köhler, and F. Stadler, 269–92. Dordrecht: Kluwer.

Israel, H., and B. Goldstein. 1944. "Operationism in Psychology." *Psychological Review* 51: 177–88.

Jackman, J. C. and C. M. Borden, eds. 1983. *The Muses Flee Hitler: Cultural Transfer and Adaptation, 1930–1945*. Washington, D.C.: Smithsonian Institute Press.

Joergensen, J. 1951. "The Development of Logical Empiricism." In *Foundations in the Unity of Science: Toward an International Encyclopedia of Unified Science,* ed. O. Neurath, R. Carnap, and C. Morris, 847–936. Chicago: University of Chicago Press.

Jones, B. Z. 1984. "To the Rescue of the Learned: The Asylum Fellowship Plan at Harvard, 1938–1940." *Harvard Library Bulletin* 32: 205–38.

Kevles, D. J. 1995. *The Physicists: The History of a Scientific Community in Modern America*. Cambridge, Mass.: Harvard University Press.

Koch, S. 1992. "Psychology's Bridgman versus Bridgman's Bridgman." *Theory and Psychology* 2: 261–90.

Kuhn, T. K. 1994. "Professionalization Recollected in Tranquility." *Isis* 75: 29–32.

Lowen, R. 1997. *Creating the Cold War University: The Transformation of Stanford*. Berkeley and Los Angeles: University of California Press.

MacCorquodale, K., and P. E. Meehl. 1948. "Hypothetical Constructs and Intervening Variables." *Psychological Review* 55: 95–107.

Mises, R. 1963–64. *Selected Papers*. Vol. 1, *Geometry, Mechanics, Analysis*. Ed. P. Frank and G. Birkhoff. Providence, R.I.: American Mathematical Society.

Moore, G. 1990. "Alfred Tarski." In *Dictionary of Scientific Biography,* ed. C. C. Gillespie, 893–96. New York: Scribner.

Moyer, A. E. 1991a. "P. W. Bridgman's Operational Perspective on Physics. Part I: Origins and Development." *Studies in the History and Philosophy of Science* 22: 237–58.

———. 1991b. "P. W. Bridgman's Operational Perspective on Physics. Part II: Refinements, Publications, and Reception." *Studies in the History and Philosophy of Science* 22: 373–97.

Neurath, O. 1938. "Unified Science as Encyclopedic Integration." In *Foundations of the*

Unity of Science: Toward an International Encyclopedia of Unified Science, ed. O. Neurath, R. Carnap, and C. Morris, 1–27. Chicago: University of Chicago Press.
———. 1973. *Empiricism and Sociology.* Ed. R. S. Cohen and M. Neurath. Dordrecht: Reidel.
Neurath, O., R. Carnap, and C. Morris, eds. 1938. *Foundations of the Unity of Science: Toward an International Encyclopedia of Unified Science.* Chicago: University of Chicago Press.
Parascandola, J. 1970. "Henderson, Lawrence Joseph." In *Dictionary of Scientific Biography,* ed. C. C. Gilliespie, 260–62. New York: Scribner.
Quine, Willard V. O. 1985. *The Time of My Life: An Autobiography.* Cambridge: MIT Press.
———. 1986. "Autobiography of W. V. Quine." In *The Philosophy of W. V. Quine,* ed. L. E. Hahn and P. A. Schilpp, 1–46. La Salle, Ill.: Open Court.
———. 1991. "Two Dogmas in Retrospect." *Canadian Journal of Philosophy* 21: 265–74.
Quine, W. V. O., and R. Carnap. 1990. *Dear Carnap, Dear Van: The Quine-Carnap Correspondence and Related Work.* Edited and with an introduction by R. Creath. Berkeley and Los Angeles: University of California Press.
Reichenbach, H. 1938. *Experience and Prediction: An Analysis of the Foundations and the Structure of Knowledge.* Chicago: University of Chicago Press.
Reisch, G. 1994. "Planning Science: Otto Neurath and the International Encyclopedia of Unified Science." *British Journal for the History of Science* 27: 153–75.
———. 1995. "A History of the International Encyclopedia of Unified Science." Ph.D. diss., University of Chicago.
———. 1997. "How Postmodern Was Neurath's Idea of Unity of Science?" *Studies in History and Philosophy of Science* 28: 439–51.
Richardson, A. 1994. "Introduction: Origins of Logical Empiricism." In *Origins of Logical Empiricism,* ed. R. Giere and A. Richardson, 1–13. Minneapolis: University of Minnesota Press.
———. 1997. *Carnap's Construction of the World: The Aufbau and the Emergence of Logical Empiricism.* Cambridge: Cambridge University Press.
Rogers, T. B. 1989. "Operationism in Psychology: A Discussion of Contextual Antecedents and an Historical Interpretation of Its Longevity." *Journal of the History of the Behavioral Sciences* 23: 139–53.
Russell, B. 1940. *Inquiry into Meaning and Truth.* London: Allen and Unwin.
Simmel, M. L., ed. 1968. *The Reach of Mind: Essays in Memory of Kurt Goldstein.* New York: Springer.
Smith, L. 1986. *Behaviorism and Logical Positivism: A Reassessment of the Alliance.* Stanford, Calif.: Stanford University Press.
S. S. Stevens Papers (SSSP). Harvard University Archives. Box HUG FP 2.12.
Stadler, F. ed. 1982. *Arbeiterbildung in der Zwischenkriegszeit: Ausstellungskatalog mit Forschungsteil.* Vienna: Österreichisches Gessellschafts- und Wirtschaftsmuseum.
Stevens, S. S. 1934a. "The Attributes of Tones." *Proceedings of the National Academy of Science* 20: 457–59.
———. 1934b. "Tonal Density." *Journal of Experimental Psychology* 17: 585–92.
———. 1935a. "The Operational Basis of Psychology." *American Journal of Psychology* 47: 323–30.
———. 1935b. "The Operational Definition of Psychological Concepts." *Psychological Review* 42: 517–27.

———. 1936a. "The Psychophysiology of Hearing." *Proceedings of the American Society of the Hard of Hearing* 17: 30–35.
———. 1936b. "A Scale for the Measurement of a Psychological Magnitude: Loudness." *Psychological Review* 43: 405–16.
———. 1936c. "Psychology: The Propaedeutic Science." *Philosophy of Science* 3: 90–103.
———. 1939a. "Psychology and the Science of Science." *Psychological Bulletin* 36: 221–63.
———. 1939b. "On the Problem of Scales for the Measurement of Psychological Magnitudes." *Journal of Unified Science* 9: 94–99.
———. 1966. "Quantifying the Sensory Experience." In *Mind, Matter, and Method: Essays in Philosophy and Science in Honor of Herbert Feigl*, ed. P. K. Feyerabend and G. Maxwell, 215–33. Minneapolis: University of Minnesota Press.
———. 1974a. "S. S. Stevens." In *A History of Psychology in Autobiography*, ed. G. Lindzey, 395–420. New York: Appleton, Century, Crofts.
———. 1974b. *Psychophysics: Introduction to Its Perceptual, Neural, and Social Prospects*. Ed. G. Stevens. New York: Wiley and Sons.
Stevens, S. S., and H. Davis. 1936. "Psychophysiological Acoustics: Pitch and Loudness." *Journal of the Acoustical Society of America* 8: 1–13.
———. 1938. *Hearing: Its Psychology and Physiology*. New York: Wiley and Sons.
Swedberg, R. 1991. *Schumpeter: A Biography*. Princeton, N.J.: Princeton University Press.
Tolman, E. C. 1936. "An Operational Analysis of 'Demands.'" *Erkenntnis* 6: 383–90.
Uebel, T., ed. 1991. *Rediscovering the Forgotten Vienna Circle*. Dordrecht: Reidel.
———. 2003. "Philipp Frank's History of the Vienna Circle: A Programmatic Retrospective." In *Logical Empiricism in North America*, ed. G. L. Hardcastle and A. Richardson, 149–69. Minneapolis: University of Minnesota Press.
Walter, M. L. 1990. *Science and Cultural Crisis: An Intellectual Biography of Percy Williams Bridgman (1882–1961)*. Stanford, Calif.: Stanford University Press.
Woodward, W. R. 1990. "Stanley Smith Stevens." In *Dictionary of Scientific Biography*, vol. 18, *Supplement 2*, ed. F. L. Holmes, 869–75. New York: Scribner.

George Reisch

8
Disunity in the International Encyclopedia of Unified Science

The "disunity" in my title is the disunity among Otto Neurath, Rudolf Carnap, and Charles W. Morris, the editors of the *International Encyclopedia of Unified Science*. There are also stories to tell about disunity among the *Encyclopedia*'s monographs and among their authors and about disagreements between the editors and their publisher, the University of Chicago Press. But these stories are subordinate to the many intellectual and personal differences among the editors, for the latter explain why the *Encyclopedia* failed after only a few years of success. I will detail this explanation here and also defend my view that the failure of the *Encyclopedia* ultimately caused the demise of logical empiricism some thirty or so years after it arrived in America.

My explanation for the failure of the *Encyclopedia* and logical empiricism is different from others, and there is a historiographic component to my paper that I would like to make explicit. First, my explanation cuts across the tempting distinction between the internal, "philosophical" development of logical empiricism and the external, "political" or "social" roles it consciously or unconsciously played. It focuses on dynamics internal to this philosophical project, dynamics that are simultaneously and in different measures intellectual, interpersonal, and political. Second, my account avoids the related mistake of construing the events that drove the history of logical empiricism as a microcosmic recapitulation of larger political and cultural conditions. Carnap and Neurath were indeed Europeans and Morris was indeed an American. But these facts do not entail that the *Encyclopedia* was a site where European and American intellectual cultures clashed. In particular, I will analyze the striking differences between the role played by Morris in my account and in Peter Galison's (1996) in order to sketch some methodological guidelines and warnings for writing the history of logical empiricism.

What Was "Logical Empiricism"?

In order to substantiate my claim that the history of the *Encyclopedia* has much to do with the history of logical empiricism, I must emphasize that the *Encyclopedia* was regarded by the early logical empiricists as the central focus of their activities in America. We tend to think of logical empiricism as it flourished in the 1950s as an intellectual program—a philosophical paradigm, perhaps—consisting of certain beliefs (or dogmas), unsolved problems, and styles of argumentation and inquiry. But in the mid-1930s, Neurath, Carnap, and Morris banded together to support a constructive scientific project, namely, the unification of the sciences. There was much debate about what terms were appropriate for the philosophical banner under which they would work—"logical empiricism," "empirical rationalism," "scientific empiricism," for example (Dahms 1997). As Neurath put it, "[E]ach baby needs a name."[1] If one were to ask Neurath, at least, this baby was to be raised more as a scientist than as an analytic philosopher of science.

The *Encyclopedia*'s primary goals were scientific. It would organize scientists and philosophers to work collectively toward unifying the sciences so that they could be better used for understanding and shaping the modern world. It was to be structured as a series of seventy-page monographs, in volumes of ten monographs each, written by experts in science, philosophy, or logic, and it would serve—Neurath hoped—as a living reflection of these ongoing efforts. It would not aim to present any final picture of the sciences or a blueprint for how the sciences should be unified in the future. The future of science was open, and the *Encyclopedia* would reflect that openness by avoiding any overarching organizational scheme or agenda.

The *Encyclopedia* was also a new project. The European publications in scientific philosophy at the time—namely, *Erkenntnis, Einheitswissenschaft,* and *Schriften zur Wissenschaftliche Weltauffassung*—were to be called to America and reconstituted as subordinate counterparts to the new encyclopedia. Their titles also signal a shift away from philosophy, epistemology, knowledge, and *Weltauffassungen* and toward unified science. *Erkenntnis* was for a short while refashioned as the *Journal of Unified Science,* and its editors also planned to establish a monographic series to be called the Library of Unified Science. As these titles suggest, when Morris brought both the *Encyclopedia* and Carnap to Chicago in 1936, he did not make that city a center of scientific philosophy or of logical empiricism. It became a center of unified science and of the larger Unity of Science movement, which consisted of the International Congresses for the Unity of Science (held annually from 1935 to 1940), Neurath's Institute for the Unity of Science, a host of organizing and editorial committees, the above-mentioned publications, and a short-lived "Unity of Science Forum" that appeared in *Synthese*.

Why Did the *Encyclopedia* Fail?

Why did the *Encyclopedia* fail? This is not a simple matter, I believe, and new possibilities continue to appear, such as those suggested by Don Howard's (1996) analyses of how the journal *Philosophy of Science* evolved and by John McCumber's (1996) examination of the American Philosophical Association's responses (or lack thereof) to McCarthyism in the 1950s. Clearly, the *Encyclopedia* and logical empiricism were buffeted by political winds. But it is possible to explain, or at least, to begin to explain, their demise by placing most explanatory weight on the dynamics among Neurath, Morris, and Carnap.[2]

The *Encyclopedia* was successful as a publishing venture only in its first few years.[3] It was first advertised as a subscription series in 1937, and by the beginning of 1940, eight of the first twenty monographs had appeared. The editors had hoped that the monographs would appear at a rate of one per month, but this shortfall was relatively minor. During and after the war, save for a burst of activity in 1951 and 1952 during which four monographs were published, the monographs appeared very infrequently. The last of the original twenty appeared in 1970.

By the "failure" of the *Encyclopedia,* I refer specifically to the facts that the project never recaptured the success it enjoyed before the war and that only these twenty monographs appeared. Neurath had once envisioned hundreds. My explanation for this consists in the intersection of three circumstances. First, there were pressures external to the project. The war frustrated all the main actors—the editors, the authors, the University of Chicago Press, and the *Encyclopedia*'s subscribers—and created pressure for the editors to produce (and sell) monographs quickly. Sometimes this forced them to compromise on their choices of authors and topics. In addition, the editors never found external funding for the *Encyclopedia*—funding that would have relieved some of this pressure.

The second circumstance was Neurath's death in late 1945. Although Neurath was more that just the Vienna Circle's "indefatigable organizer," as Joergen Joergensen put it (1951), it is true that he was an organizational and motivational wizard. He made the encyclopedia project go. Once he died, it began to wither.

The third circumstance was the disunity that developed within the project. Neurath, Carnap, and Morris had very different conceptions of what unified science was, of what the goals and methods of their *Encyclopedia* should be, and of the role of philosophy in the unification of the sciences. These differences fed several disputes, at least one of which turned personal and ugly.

These internal and external circumstances interacted to cause the downfall of the encyclopedia project: Once Neurath was gone, someone had to assume his leading, organizational role. Philipp Frank tried, but without much success. In the late 1940s and early 1950s, he operated the Institute for the Unity of Science in Boston while he lectured at Harvard. He obtained six years of funding from the Rockefeller Foundation, and the meetings of this new institute brought together the usual cast of characters, including Morris and Carnap. Frank's mission explicitly included supporting and continuing the *Encyclopedia,* but not much came of that support. Four monographs appeared in the early 1950s, but so far as I can tell, Frank never worked hands-on to reinvigorate the project. After the period of Frank's leadership, the Unity of Science movement effectively became moribund, and in 1973, the Institute for the Unity of Science was finally absorbed into the Philosophy of Science Association.

The crucial question is why neither Morris nor Carnap, nor both together, did not assume Neurath's role and tend to the growth of the *Encyclopedia*. They were in situ, and they were most familiar with the project. The answer lies in the first and third circumstances: The *Encyclopedia* was more work than they anticipated, and it went very slowly (especially because of the war). And the disunities within the editorial circle had taught them how difficult it would be to live up to the ideal of constructive, intellectual cooperation that the project celebrated. In short, by the time of Neurath's death, their original hope and optimism for the *Encyclopedia* had given way to frustration and disappointment.

Disunity in the Editorial Circle: Neurath and Carnap

There were two main axes of conflict among the editors: the relationships between Neurath and Carnap and those between Neurath and Morris. The disputes between Neurath and Carnap began shortly after the publication of Carnap's *Aufbau* (Uebel 1992) and continued, sporadically, and in one form or another, until Neurath's death. Roughly, they had different philosophical methods that led them to different conclusions about issues such as protocol sentences, metaphysics, and semantics. Carnap, the logician, sought to build languages in which sciences and the relations among them could be reconstructed and analyzed.[4] Neurath endorsed, instead, a kind of naturalism that would not privilege any one part of our knowledge (such as logic) over any other (such as empirical science). From his point of view, Carnap's methods appeared tinged with a kind of Platonism or absolutism that presumed—falsely, from Neurath's point of view—that laws of contradiction or entailment, for instance, are more real, robust, or enduring than ordinary sentences of everyday life.

For my purposes here, the relevant chapter in their ongoing disputes began in 1942, just after Carnap's *Introduction to Semantics* (1942) appeared. Neurath read it and renewed his charge that semantic theory entailed unacceptable "metaphysical," "ontological," or "absolutistic" beliefs about the world. In short, Neurath reasoned that theories about language and their relations to the world require independent, nonlinguistic access to the world. But this is not available. Thus, Neurath diagnosed semantics as resting upon a dangerous "duplication" or "reduplication," through which our always-instrumental knowledge of the world inflates into realistic claims about the way the world really and truly is—claims that no true empiricist would ever defend.[5]

With Neurath in England and Carnap in the United States, their ensuing dispute lasted for about three years and continued—at increasing pitch and intensity—until Neurath's sudden death in late 1945. Charges and countercharges included that Neurath was prone to violent, intimidating outbursts, like a "volcano"; that Carnap was treating Neurath as a less than top-notch thinker; that Neurath was not tolerant of views he did not like; that Carnap was not tolerant of views he did not like; that Carnap had stolen ideas from Neurath; that Neurath could not express his own ideas clearly in publications; and that Carnap made Neurath and the *Encyclopedia* look bad when he placed a disclaimer on Neurath's 1944 monograph *Foundations of the Social Sciences,* saying that Carnap did not share in the editorial responsibility for the monograph (which he did not).[6]

Are these merely routine difficulties among scholars and coeditors, on which little weight should be placed when explaining momentous changes in history of philosophy? No. In the spring and summer of 1945, Carnap was losing sleep. He wrote to Neurath, "Do you realize how distressing and disturbing your attitude towards me in [your last] two letters has been to me? If you knew how many sleepless hours at night they have caused me, and how much inability to work in the daytime!" (Carnap to Neurath, August 23, 1945 [ASP 102-55-09]). Uncharacteristically, Carnap ended that sentence with an exclamation point. One day later, Ina Carnap wrote to Neurath and noted that her husband was not exaggerating: "You should see how he sits and broods whenever one of your accusing letters arrives—for days he is unable to settle down to his work!" (Ina Carnap to Neurath, August 24, 1945 [ASP 102-55-10]). Neurath was shaken by the dispute, as well. He wrote the letters that so upset Carnap when he was in—very uncharacteristically—"a somewhat hopeless mood." Three months later, he noted, "this hopeless mood is continuing. . . . I feel really helpless and somewhat hopeless."[7]

From Carnap's point of view, things were bleak. The future of his collaboration on the *Encyclopedia* was in doubt, for he feared he would not be able to work with Neurath any longer if their dispute went unresolved and if Neurath (the "volcano") ever again erupted as he had on a few occasions in the past (Carnap to Neurath, August 23, 1945 [ASP 102-55-09]). Second, if any dispute could be settled by the kind of peaceful resolution by philosophical analysis that Carnap always promoted, then surely so should one between two committed logical empiricists. But this dispute worsened, and kept Carnap awake at night. Third, when Neurath suddenly died, Carnap knew that a great and once philosophically productive (if turbulent) friendship had ended in ruin.

Disunity in the Editorial Circle: Neurath and Morris

Morris's relationship with Neurath was less dramatic but no less difficult. First, it is necessary to see that Morris worked tirelessly for the movement. Besides reading and editing monographs, he wrote the *Encyclopedia*'s promotional materials (and revised them sometimes repeatedly at Neurath's request); he was the project's contact with prospective sources of funding and with the University of Chicago Press (to whom he routinely had to deliver bad news about the pace of the project); he organized the Fifth International Congress for the Unity of Science held at Harvard in the fall of 1939; and he worked hard on behalf of the *Journal of Unified Science* and Library of Unified Science.

Consequently, Morris must have been dismayed as he increasingly came to learn—sometimes years into their collaboration—that Neurath had very little respect for his own substantive views about science, philosophy, and the Unity of Science movement. Morris is remembered by a student as having a demeanor of "zen-placidity" (Sebeok 1991, 79), and he was a lifelong devotee of Buddhism. So one rarely finds Morris's ghost in the archives explicitly complaining about this situation. But the facts point toward disappointment: Morris devoted the best years of his life to a project that was controlled by a powerful leader who appreciated his labor but not his ideas.

As a young professor of philosophy, Morris's ambitions were much like Neurath's. In 1932, for example, before he had connected with Carnap and

Neurath, he proposed to University of Chicago president Robert Maynard Hutchins that the university establish an "institute of philosophy" whose mission would be to apply pragmatism's tools to the world's problems. "Such an Institute," he explained, "would make it possible, perhaps for the first time, for philosophy to perform its proper function in civilization." It would host visiting scholars from around the world, attempt to interpret and synthesize new scientific findings, sponsor public forums and radio shows, and produce an "Encyclopedia of the Philosophical Sciences."[8] Like Neurath's, Morris's vision included internationalism, unification of knowledge, and encyclopedias.

Morris's ambitions became even more similar to Neurath's as their collaboration began. He adopted a mild antimetaphysical attitude by referring, for example, to "the hidden metaphysical and conventional elements" in modern thought and to "the verbiage of most philosophical speculation" (Morris 1937, 3). And in his essay of 1934, *Pragmatism and the Crisis of Democracy,* he argued that the tradition of Charles Sanders Peirce, William James, John Dewey, and George Herbert Mead, Morris's teacher, was fully aligned with the Vienna Circle's crusade to build a scientific philosophy. In fact, he wrote that pragmatism simply *was* "a form of positivism or empiricism."[9] Here, Morris also endorsed a version of the technological modernism espoused by the Circle's left wing and analyzed recently by Peter Galison (1990, 1996). Whereas Neurath and Carnap looked to science, technology, and architecture as the main pillars of a new, modern form of life, Morris looked to the broader, Deweyan notion of "intelligence" as a "formative principle of reconstruction." This "intelligence" embraced science and technology, as well as philosophy, art, and design. It even served ethical and religious thought, all of which could be understood and practiced, he emphasized, in an empirically controlled, nonmetaphysical way (1934, 16). Finally, Morris shared Neurath's and Carnap's sympathies with socialism. Pragmatism showed how it would be possible, he explained, for America to build a democracy "contain[ing] as its essence the moral ideal of a classless functional society" (1934, 7). Privately, Morris favored this outcome, at least enough to have contributed to the Expansion Fund of the Socialist Party (Norman Thomas to Morris, March 16, 1933 [PEP]).

These shared enthusiasms explain why Morris was so energized by Neurath's project and why he nearly equaled Neurath as a cheerleader for unified science. In a review of the proceedings of the Prague Vorkonferenz, the conference at which Neurath began to assemble his new Unity of Science movement, Morris spoke of "the inauguration of a new era in the history of science and philosophy" (1936d, 130). At the First International Congress for the Unity of Science, held in Paris in 1935, he read a short paper, "Remarks on the Proposed Encyclopedia," depicting the project as expressing "a new spirit which is sweeping over the ranks," as presenting "an opportunity that may assume great historical importance," and as "something valuable enough to enlist our most earnest efforts and our strongest devotion" (1936a, 71, 74).

Morris most admired the collaborative and cooperative character of the movement: "the opportunity it present[ed] for carefully planned cooperative work" (1936a, 70). His self-appointed role was to articulate a broad philosophical framework to facilitate this interdisciplinary collaboration. In a series of papers from 1935 and 1936 (later collected in Morris 1937), he explained how philosophers and scientists of all stripes could work together under the umbrella of "scientific empiricism" and nourish the growth and unification of the sciences.

Ten years later, the *Encyclopedia* was up and running. Yet it did not command Morris's "strongest devotion." After Neurath's death in late 1945, he tended the *Encyclopedia* as occasion demanded until the last monograph officially appeared in 1970, and he participated in the Unity of Science meetings organized by Frank into the early 1950s. But he did not attempt to revive the project by recruiting new authors, pursuing the plans the editors had made for a third volume of monographs (to be called *Methods of Science*), or planning more international congresses. This was because even before Neurath died, Morris had begun to turn away from his collaboration with Neurath and Carnap. He turned away from natural science and more aggressively pursued his interests in social science, psychology, and religion.

In 1942, Neurath's eyebrows rose when Morris published a book titled *Paths of Life: Preface to a World Religion* (1942). Morris's theory held that the dominant religions of the world correspond to different points in what is (effectively) a three-dimensional personality space. Here, Morris relied heavily on the work of William Sheldon and his theory of "somatotypes," which holds that personality types are closely correlated with empirically specifiable body types. (This marks Morris's behaviorist turn, one that still dismays those of his followers who admire Morris the semiotician.) The ideal religion, Morris theorized, would correspond to the middle of this personality space: If we were to adopt a religion, or "path of life"—that is, a constellation of values and attitudes—that avoided the intellectual and temperamental excesses of the dominant religions (of which Morris identified six, one at the extremes of each dimension), the world would be more peaceful and satisfying.

Morris was confident about his analysis and its public importance. He promoted it by writing a popular article in *Fortune* magazine (Morris 1943), he contacted an agent who represented "distinguished personalities" (Charles S. Pearson to Morris, September 27, 1944 [PEP]), and he explained to a radio-show producer that "what I have to say is so relevant to the present crucial moment in American and world history that it should be said to a large audience" (Morris to Bryson, September 28, 1943 [PEP]). He also approached the Rockefeller Foundation and the Institute of Current World Affairs for funding to visit India, China, and Russia, where he would promote his theory and facilitate "the international exchange of ideas and ideals" (Morris to Walter S. Rogers, March 21, 1943 [PEP]). He did eventually tour China and Southeast Asia for several months in 1948–49 (Morris to Rossi-Landi, January 21, 1951, cited in Petrilli 1992, 47).

Why did Morris not pursue these goals primarily within the context of the Unity of Science movement? Why did he not go to these countries explicitly to promote the *Encyclopedia* and to recruit more collaborators? Based as it was on Sheldon's empirical classification of body types, his theory of religion was scientific. Morris had not turned his back on science or its unification. Instead, he believed that the Unity of Science movement and the *Encyclopedia* were not paying enough attention to "sociohumanistic" studies, to value theory, or to projects like his analysis of religion. Since there was little interest within the movement for these areas, Morris decided to pursue and promote it himself, outside the movement.

Yet Morris did not go outside the movement quietly. Four years before his book appeared, in 1938, he addressed his colleagues about this problem in *Synthese*'s "Unity of Science Forum." The movement was neglecting nonnatural sciences, he explained, and were it to realize its potential for transforming and improving culture, it must follow American pragmatism and pursue "the integration of socio-humanistic studies with the wider corpus of scientific knowledge and procedure." Were it to remain narrowly wedded to the natural sciences and to continue to shun research on values, then "creative cultural forces" could never gain strength and prestige from an alliance with natural science. Instead, these forces would remain suppressed in a scientific age that mistakenly regarded studies of "mind, value, art, and moral behavior" as necessarily unscientific (Morris 1938b, 28).

But the situation was even worse. From Morris's point of view, the movement was neglecting more than socio-humanistic areas of inquiry. It was also neglecting the best tool for both appreciating and achieving this integration of natural and socio-humanistic science. That tool was Morris's general theory of signs, or "semiotic." From the beginning of his association with Neurath and Carnap, Morris promoted semiotic as an obvious and viable generalization of Carnap's syntactic logic of science (Morris 1936c, 1938a). As a tripartite theory embracing syntactics, semantics, and pragmatics, it would allow us to see our knowledge and our language in their totality. With such a theoretical handle, we might better unify the sciences, incorporate value theory, and wisely harness these "creative cultural forces."

Carnap accepted the outlines of Morris's vision, and he adopted the term "semiotic" to designate a complete theory of language.[10] Neurath hated it. He did not interfere with Morris's two main writings for the *Encyclopedia*, both of which outlined and defended semiotic. But this was probably because he could not risk angering so helpful a collaborator. When Neurath and Morris discussed matters privately, however, Neurath was blunt. He criticized Morris's terminology (cautiously, since he was not a native speaker of English), and he could not abide Morris's aggressively vague and passive style of writing. Subject matters often flow from Morris's pen into an impressionistic collage of 'aspects' and 'dimensions' and 'emphases', all of which can easily be 'reconciled' and 'integrated'.[11] Neurath feared that Morris's vagaries about signs and symbols were too sympathetic to traditional metaphysics and not sufficiently empirical and scientific.[12]

The core idea of semiotic fared no better. Neurath's main objection was that Morris was attempting to unify the sciences by leaning heavily on this tripartite theory of language. Semiotic was the skeleton of Morris's program of scientific empiricism. For Neurath, however, any top-down, theory-driven approach was unacceptable for the same reasons that Neurath rejected talk of unified science as a "system"-building project: It was a prioristic and architectonic (Reisch 1997a). For Neurath, the movement should not risk being a slave to present-day expectations about what the future of science, or the future of language, might be like. He complained, for instance, that the tripartite structure of semiotic revealed a "Kantian inclination": "Whether an architecture with three-parts comes out or not cannot concern us. . . . One can indeed say all the beautiful things without ARCHITECTURE" (Neurath to Carnap and Morris, January 4, 1938 [USM, Box 2, Folder 10], capitalization in original). Nor was Neurath impressed by Morris's assurances that semiotic was fully empirical and scientific. Instead of articulating his new theory about signs and symbols, Morris should have been writing specifically about actual sciences: "[I try] to remain as 'open' as possible and to refer to the sciences. I would be very happy if you would share this attitude" (Neurath to Morris, January 28, 1938 [USM, Box 2, Folder 10]). "Would it not be possible to establish more contact with science?" (Neurath to Carnap and Morris, January 4, 1938 [USM, Box 2, Folder 10]).

Morris responded to Neurath's complaints, but he often did not understand them. In particular, he failed to appreciate Neurath's concern for terminology and the metaphysical baggage that words and phrases carry.

(This component of Neurath's program undergirds his proposal for establishing an *index verborum prohibitorum* [Reisch 1997b].) Morris proposed, for example, that the third volume of the *Encyclopedia* be titled "General Science" or perhaps "Science of Science"—phrases against which Neurath had strong doubts *("die weitest gehenden Bedenken")*. Neurath explained that such terms added to the impression that the *Encyclopedia* was prematurely reaching for an overarching architecture or system embracing all the sciences (Neurath to Carnap and Morris, January 10, 1938 [USM, Box 2, Folder 10]). Morris's marginalia in this letter show his defensiveness and his puzzlement: "I want nothing metaphysical—merely a name for semiotic, logic, induction, etc. etc."; "What [is] his attitude to method?"; "Difficulty here?"; "Does he oppose logical syntax?"

These were two minds *not* engaged in a productive dialogue. Morris believed that developing semiotic would help a movement grow and fulfil its cultural promise; Neurath believed that Morris's vague and architectonic formulations would hurt the movement. Over the course of their collaboration, they usually put their differences aside and dealt with the business of running the *Encyclopedia* and the movement.

Against this backdrop, consider how the encyclopedia project looked to Morris upon Neurath's death. He was growing increasingly tired of administrative work ("I am beginning to rebel at so much," he had told Neurath years before [Morris to Neurath, September 29, 1941 (USM, Box 2, Folder 1)]); he surely regretted how nasty the dispute between Neurath and Carnap had become; his best ideas (as he saw them) went unappreciated by the movement's leader; and his vision of the Unity of Science movement embracing semiotic, value theory, and "socio-humanistic" studies was not shared widely. Instead of attempting to assume Neurath's leading role in the movement, Morris not surprisingly pursued his own projects until the end of his career. In 1958, he took a job as research professor at the University of Florida in Gainesville, where he pursued his behavioristic and often questionnaire-based studies of value and esthetics. "My time," he wrote to Ferruccio Rossi-Landi, perhaps Morris's greatest fan, "will be completely at my own disposal." "This will give me a chance to finish the 3 or 4 books which still seem to be inside me" (Morris to Rossi-Landi, July 12, 1958, cited in Petrilli 1992, 95).

Morris's comment suggests the main failing of the *Encyclopedia*'s editorial engine. The point of the project, as Morris once emphasized, was to collaborate with others, *not* to publish writings that express an individual's private, subjective theories and proposals. Consensus was to emerge from regular, organized debate and discussion. Even *Encyclopedia* monographs subsequent to the original twenty were to be circulated and revised prior to publication. "Carefully planned collaborative work," as Morris noted in his remarks at the Paris congress in 1935, would replace egotism, ideological disputes, and the eternal conflicts between rigid schools of thought. But the editors who produced the *Encyclopedia* and who articulated and defended the project the most vigorously could not live up to this ideal of intellectual cooperation. Their collaboration proved difficult, frustrating, and tiring. After Neurath died, as Morris and Carnap contemplated the possible future of the *Encyclopedia,* the idea of working individually—on value theory, in Morris's case, or on semantics and inductive logic in Carnap's case—had more appeal than trying to revive the *Encyclopedia* and to rekindle worldwide discussion about the unification of the sciences.

Qualifications, Comparisons, and Historiographic Lessons

The clashes of intellect and personality I have described occurred alongside other disputes that I have not recounted here. These also contributed to the frustrations each of the editors felt over the course of their collaboration. In the wake of Neurath's 1944 monograph *Foundations of the Social Sciences,* for example, Neurath and Carnap argued over what precisely the *Encyclopedia*'s editors were responsible for (Neurath to Morris and Carnap, 18 November 1944 [PEP]). Earlier, in 1941, Morris and Carnap had become very upset with Neurath because he effectively ruined an opportunity to receive $20,000 from the Rockefeller Foundation to support the *Encyclopedia* and the movement. This was because Neurath, at the last minute, had scuttled long-negotiated plans to bring the *Journal of Unified Science* to the University of Chicago Press.[13]

There are also signs that these difficulties were noticed by scholars outside of the editorial circle. As Ernest Nagel put it, the publication of Neurath's 1944 monograph was a "black eye" for the *Encyclopedia*. The poor quality of the monograph suggested that the editor-in-chief was not subject to the same editorial control as other writers. Some of Nagel's colleagues, he reported, were "simply dismayed" by it (Nagel to Morris, November 16, 1944 [PEP]). If Nagel was right, the project was losing respect and prestige as early as 1944. This circumstance may help explain why other leading supporters of the *Encyclopedia,* such as Herbert Feigl, Carl Hempel, or Nagel himself, did not rush to fill Neurath's shoes after he died.

Incomplete as it is, however, my explanation avoids what the economic historian Donald McCloskey has called "the dogma of large-large" explanation. This is the historiographic assumption that (in some sense) "large" results or effects (in this case, the eventual demise of logical empiricism in North America) must have been brought about by (in some sense) "large" causes (McCloskey 1991, 32). My explanation counts as a "small-large" explanation because it traces this "large" change in North America's philosophical landscape to relatively "small" causes. Not coincidentally, my explanation is also Neurathian: It is physicalistic, insofar as it focuses on specific people, their specific experiences, and the specific decisions they made, and it is sympathetic with Neurath's claims for the unpredictability of historical events that are caused by small, unobserved, or seemingly insignificant factors (Reisch 2001).

In contrast to my account, there are at least two species of "large-large" explanations of the demise of logical empiricism. The first holds that the movement was defeated on conceptual and philosophical terrain. Candidates for its Achilles' heel are the observational-theoretical distinction (attacked by Thomas Kuhn and Norwood Russell Hanson) and the analytic-synthetic distinction (exposed as unworkable "dogma" by W. V. O. Quine). These were significant moments in the history of logical empiricism. But for at least two reasons, they fail to count as sufficient reasons for its demise. First, it is now understood that logical empiricism was not a monolithic program. The "received view" of logical empiricism, popularized in Frederick Suppe's *The Structure of Scientific Theories* (1977), is often regarded as a final snapshot of a coherently developing philosophical movement. Yet fundamental positions—such as the viability of rational reconstructions of theories, of epistemic foundationalism, or of overarching philosophical theories of knowledge—were in dispute since the early days of the Vienna Circle. This complexity makes it difficult to specify what logical empiricism was, and more difficult to specify when and how it was refuted.

Second, there is a surprising degree of agreement among leading logical empiricists and those who allegedly refuted the program. Kuhn's famous model of scientific change, for example, in which sciences lurch discontinuously from paradigm to paradigm, can be almost without residue translated into a sequence of Carnapian language frameworks (and this explains why Carnap liked Kuhn's *Structure of Scientific Revolutions* very much [Reisch 1991]). And Neurath promoted a version of epistemological naturalism long before Quine (Uebel 1991).

My account escapes these difficulties by focusing on this earlier, constructive, and scientific phase of logical empiricism. It was conveyed into and throughout North America largely on the back of the encyclopedia project—a vehicle that broke down as it became increasingly frustrating and difficult for the editors to manage. As the *Encyclopedia* lost its vitality in the 1940s and 1950s, logical empiricism, then decoupled from the scientific task of unification, evolved into the abstract, ahistorical, and politically neutral set of theories about science and techniques of analysis that make up the "received view." (If McCumber [1996] is right, this academicization of philosophy of science was partly driven by the conservative climate of McCarthyism.) If we look back and recover logical empiricism's early encyclopedic and scientific goals, however, we can see that it first succumbed to clashes of intellect and personality in the 1930s and early 1940s, however much its later evolution may have been guided by these political currents and the attacks led by Kuhn and Quine.

Another kind of "large-large" explanation holds that logical empiricism foundered mainly for cultural reasons. Peter Galison has promoted this line of inquiry by tracing the long roots of logical empiricism from Carnap's classic *Logische Aufbau der Welt* (1928) to modernist architecture to Austro-Marxist dreams of utopian technocracy. The common nutrient flowing through these diverse enterprises was the goal of completely reforming and modernizing public and private life (Galison 1990, 1996). When the European scholars emigrated, however, these long roots atrophied, because American cultural soil could not nourish such a symphony of meanings and socialistic ambitions. In North America, the Unity of Science movement would have to be strictly scientific and "politically homogenized and neutral" (1996, 38). But with its cultural roots so hemmed in, the movement was bound to atrophy, as it eventually did. In general, Galison writes, "people move across oceans with relative ease, complexes of ideas do not" (1996, 35).

One constraint on any historical explanation is that the actors in question must in some way perceive the forces and junctures central to the explanation. Otherwise, the historian risks producing a spurious reconstruction of events that would appear foreign and clumsy to those that actually experienced them. If Galison's account is right, someone must have recognized that the Unity of Science movement's social, political, and cultural agenda would have to be reigned in upon coming to America. Enter Charles Morris.

Morris was always deferential and somewhat awestruck by the talents of his new European colleagues. At the First International Congress for the Unity of Science in Paris, 1935, for example, he read a formal statement as "a representative of the United States." He defended the philosophical reputation of North American philosophy (as he would again do later in his "Unity of Science Movement and the United States" [1938b]) mainly by praising the breadth and quality of Peirce's work. Still, Morris clearly portrayed Americans as playing second or third fiddle to the Europeans: "We can thus appear before this congress not as beggars—though we do have much to learn and overemphases to correct—but as representatives of those who have helped to sow the seeds from which this Congress is itself one growth" (Morris 1936b, 22). This tone rings throughout Morris's writings, including his correspondence with his coeditors of the *Encyclopedia*. He routinely presented Neurath with important considerations and recommendations, but he knew and accepted that Neurath would make all the decisions about titles, topics, authors, textual formatting, publishing contracts, and so on.

Galison recognizes Morris's "tentative approaches" (1996, 37) to his new colleagues. But he sees a genuine shift in Morris's role once Carnap and the encyclopedia project had landed in Chicago: "Morris's position in the movement changed drastically." He "began to instruct his collaborators on how they must act in the United States" (1996, 37), and he actively attempted to reign in Neurath's wide-ranging technocratic, socialist ambitions. Citing comments from letters written in 1936 and 1937, Galison depicts several exchanges between the two in which Morris "resolutely resisted," "deprecated," and "would [not] countenance" (37) various plans or decisions that Neurath had made. Galison writes that Morris issued "quite evidently a warning to Neurath that his political views . . . would have no place" in the publications of the Unity of Science movement. If the *Encyclopedia* were to be successful in North America, it needed "*more science* and *more scientists*" in the ranks, "not more politics and philosophy" (37, emphasis in original).

This picture of Morris's role in the enterprise is very different from the one I have offered here, largely because it overlooks both Morris's reticent demeanor and his affinities with Neurath's internationalist ambitions. As a result, Morris's role in the history of the *Encyclopedia* is presented almost backward. As I showed above, Morris urged that the movement needed *more* emphasis on areas of study relevant to culture and religion, while Neurath, on the other hand, urged Morris to be *more scientific* in his thinking and writing.[14]

The disunity that plagued the *Encyclopedia* was not a clash of cultures. It was merely a clash of specific intellects and personalities. Still, it shaped the course of philosophy of science in North America.

Notes

Archival sources cited in this chapter are abbreviated as follows: PEP: Peirce Edition Project, Morris Papers, Indiana University–Purdue University, Indianapolis; USM: Unity of Science Movement Papers, Regenstein Library, University of Chicago; PP: Presidents' Papers, Regenstein Library, University of Chicago; ASP: Archives of Scientific Philosophy, Hillman Library, University of Pittsburgh; ONN: Otto Neurath Nachlass, Wiener Kreis Archiv, Rijksarchief in Noord-Holland, Haarlem, Netherlands. These letters are quoted by permission. All rights reserved.

1. Neurath to Morris and Carnap, November 18, 1944, PEP.
2. For a view that places much more weight on McCarthyism in understanding the development of logical empiricism during the Cold War, see Reisch, forthcoming.
3. The specific information presented in this paper about the history of the encyclopedia project is extracted from Reisch 1995.

4. See Uebel 1991 and the chapters by Uebel (chapter 6), Richardson (chapter 1), and Friedman (chapter 3) in this volume for more on the differences between Neurath and Carnap.

5. Neurath discusses "reduplication" more generally in "Unified Science and Psychology" (1932, 22) and "Foundations of the Social Sciences" (1944, 10, 16). He indirectly accuses Carnap and others of succumbing to reduplication in letters to Carnap dated January 15, 1943 (ASP 102-55-02) and June 16, 1945 (ASP 102-55-11).

6. For a detailed examination of this debate, see Reisch 1995, chap. 5; 2001.

7. Neurath to Carnap, September 22, 1945 (ONN, Folder 223). Neurath evidently did not send this letter to Carnap, for he marked it "not used."

8. Morris, "Institute of Philosophy," March 1, 1934 (PP, Box 106, Folder 14). The cover letter accompanying this proposal refers to an earlier proposal of 1932.

9. He described positivism as "the doctrine that propositions about existence are to be regarded as significant in proportion as they arise out of and are controllable by direct confrontation of the object or kind of object meant" (Morris 1934, 10).

10. See Morris's quotation from a revised version of Carnap's "Testability and Meaning" in Morris 1963, 88. In "Foundations of Logic and Mathematics," Carnap outlines his philosophy of science on the basis of a pragmatics-semantics-syntax schema (1939, 4).

11. Neurath once politely approached this tendency of Morris when reviewing one of his contributions to the *Encyclopedia:* "I prefer the concrete expressions over the abstract, if permissible" (Neurath to Carnap and Morris, January 4, 1938 [USM, Box 2, Folder 10]).

12. About Morris's first contribution to the *Encyclopedia,* Neurath wrote, "One shows that a certain metaphysical question-complex appears acceptable in a new framework. I find this part still somewhat problematic for an introductory monograph" (Neurath to Morris and Carnap, January 4, 1938 [USM, Box 2, Folder 10]).

13. In September 1941, Morris wrote to Neurath with good news. The Rockefeller Foundation's Warren Weaver had finally come around. A grant of $20,000, Morris surmised, would be offered to the encyclopedia project if Neurath submitted to Weaver "a concrete plan under reliable and careful leadership" according to which the *Journal* would come to the University of Chicago Press (Morris to Neurath, September 29, 1941 [USM, Box 2, Folder 1]). Neurath, then living in England, agreed to this arrangement. But at the time, he was also seeking to find an English publisher for the new journal and its new sister series, the Library of Unified Science. Evidently, Neurath, who chose not to emigrate to America, did not want all the movement's publications based there. Soon Neurath announced that he had secured an arrangement with Blackwell in England (Neurath to Morris, December 1, 1941 [USM, Box 2, Folder 14]), and he warned his coeditors, who were very much against his plan, that the matter was not negotiable. As Morris reported to Carnap, Frank, and Hans Reichenbach, they had no choice but to agree to Neurath's maneuver: "Unless we are willing to break off amicable relations with Neurath no other action seems possible. He has taken the matter very personally" (Morris to Carnap, Frank, and Reichenbach, December 26, 1941 [USM, Box 2, Folder 1]). Three months later, Morris wrote to Weaver to explain these developments and to say that the editors would postpone their official request for funds until things settled down and their plans were "more definite" (Morris to Weaver, March 25, 1942 [USM, Box 2, Folder 2]).

14. There are several mistakes in Galison's account that help explain the difference between my account of Morris and his. First, Galison describes the *International Encyclopedia of Unified Science* as "the successor journal" to *Erkenntnis* (1996, 36; 1990, 747). But the *Encyclopedia* was, instead, a new project that was to be supplemented by *Erkenntnis,* which the editors and the University of Chicago Press tried (unsuccessfully) to purchase from its publisher, Felix Meiner, and rename the *Journal of Unified Science* (see note 13

above). Second, Galison introduces Morris's suddenly increased organizational authority just after claiming that Morris's efforts to secure $20,000 from the Rockefeller Foundation were successful (1996, 36). But this money was never granted (see note 13 above). Third, the exchanges cited between Morris and Neurath do not evidence any genuine change in Morris's posture vis-à-vis Neurath. Throughout their collaboration, Morris provided his colleagues with a constant flow of suggestions and ideas for how to run their shared projects. But rarely do these amount to confrontations or stern warnings. For example, Galison writes that "Morris deprecated Neurath's title for the proposed encyclopedia," because it "began with the word 'international'" (1996, 37). Morris himself did not object to the word. He first told Neurath that "some people" had these objections, and later noted that they "were not my own" (Morris to Neurath, June 20, 22, 1937 [USM, Box 1, Folder 16]). Morris, in fact, had earlier coached Neurath to change the title from "International Unity of Science Encyclopedia" to "International Encyclopedia of Unified Science" (Morris to Neurath, February 21, 1937 [USM, Box 1, Folder 16]). English usage was one area where Neurath did routinely accept Morris's advice.

References

Archives of Scientific Philosophy (ASP). Hillman Library. University of Pittsburgh.

Carnap, R. 1939. "Foundations of Logic and Mathematics." In *International Encyclopedia of Unified Science,* ed. O. Neurath, R. Carnap, and C. W. Morris, vol. 1, no. 3. Chicago: University of Chicago Press.

———. 1942. *Introduction to Semantics.* Cambridge, Mass.: Harvard University Press.

Dahms, H. J. 1997. "Positivismus, Pragmatismus, Enzykopeodieproject, Zeichentheorie." *SB* 1: 25–73.

Friedman, M. 2003. "Hempel and the Vienna Circle." In *Logical Empiricism in North America,* ed. G. L. Hardcastle and A. W. Richardson, 94–114. Minneapolis: University of Minnesota Press, 2003.

Galison, P. 1990. "Aufbau/Bauhaus: Logical Positivism and Architectural Modernism." *Critical Inquiry* 16: 709–52.

———. 1996. "Constructing Modernism: The Cultural Location of Aufbau." In *Origins of Logical Empiricism,* ed. by R. Giere and A. Richardson, 17–44. Minnesota: University of Minnesota Press, 1996.

Howard, D. 1996. "Philosophy of Science and Social Responsibility: Some Historical Reflections." Philosophy of Science Association Annual Meeting.

Joergensen, J. 1951. "The Development of Logical Empiricism." *International Encyclopedia of Unified Science,* vol. 2, no. 9. Chicago: University of Chicago Press.

McCloskey, D. 1991. "History, Differential Equations, and the Problem of Narration." *History and Theory* 30, no. 1: 21–36.

McCumber, J. 1996. "Time in the Ditch: American Philosophy and the McCarthy Era." *Diacritics,* Spring, 33–49.

Morris, C. 1934. *Pragmatism and the Crisis of Democracy.* Public Policy Pamphlet, no. 12, ed. H. G. Gideonse. Chicago: University of Chicago Press.

———. 1936a. "Remarks on the Proposed Encyclopedia." In *Actes du Congrès International de Philosophie Scientifique Actualitès Scientifiques et Industrielles* 388 [fasc. 2] *Philosophie Scientifique et Empirisme Logique,* 71–74. Paris: Hermann et Cie.

———. 1936b. "Opening Speech (for the American Delegates)." In *Actes du Congrès International de Philosophie Scientifique Actualitès Scientifiques et Industrielles* 388 [fasc. 1] *Philosophie Scientifique et Empirisme Logique,* 22. Paris: Hermann et Cie.

———. 1936c. "Semiotic and Scientific Empiricism." In *Actes du Congrès International de Philosophie Scientifique Actualitès Scientifiques et Industrielles* 388 [fasc. 1] *Philosophie Scientifique et Empirisme Logique,* 42–56. Paris: Hermann et Cie; in *Logical Positivism, Pragmatism, and Scientific Empiricism,* 56–71. Paris: Hermann et Cie, 1937.

———. 1936d. Review of *Einheit der Wissenschaft: Prager Vorkonferenz der Internationlen Kongresse für Einheit der Wissenschaft, 1934.* In *Philosophy of Science* 3: 127–30.

———. 1937. *Logical Positivism, Pragmatism, and Scientific Empiricism.* Actualitès Scientifiques et Industrielles, vol. 449. Paris: Hermann et Cie.

———. 1938a. "Foundations of the Theory of Signs." In *International Encyclopedia of Unified Science,* vol. 1, no. 2. Chicago: University of Chicago Press.

———. 1938b. "The Unity of Science Movement and the United States." *Synthese,* November, 25–29.

———. 1942. *Paths of Life: Preface to a World Religion.* Chicago: University of Chicago Press.

———. 1943. "Freedom or Frustration: Can Americans Find an Answer in Their Own Age-Old Traits?" *Fortune,* September, 148.

———. 1963. "Pragmatism and Logical Empiricism." In *The Philosophy of Rudolf Carnap,* ed. P. A. Schilpp, 87–98. La Salle, Ill.: Open Court.

Neurath, O. 1932. "Unified Science and Psychology." In *Unified Science,* ed. B. McGuinness, 1–32. Dordrecht: Reidel, 1987.

———. 1944. "Foundations of the Social Sciences." In *International Encyclopedia of Unified Science,* vol. 2, no. 1. Chicago: University of Chicago Press.

Otto Neurath Nachlass (ONN). Wiener Kreis Archive. Rijksarchief in Noord-Holland, Haarlem, Netherlands.

Peirce Edition Project (PEP). Morris Papers. Indiana University–Purdue University, Indianapolis.

Petrilli, S., ed. 1992. "The Correspondence between Charles Morris and Ferruccio Rossi-Landi." *Semiotica* 88, no. 1/2: 32–122.

Presidents' Papers (PP). Regenstein Library. University of Chicago.

Reisch, G. 1991. "Did Kuhn Kill Logical Empiricism?" *Philosophy of Science* 58: 264–77.

———. 1995. "A History of the *International Encyclopedia of Unified Science.*" Ph.D. diss., University of Chicago.

———. 1997a. "How Postmodern Was Neurath's Idea of Unified Science?" *Studies in History and Philosophy of Science* 28: 439–51.

———. 1997b. "Epistemologist, Economist . . . and Censor? On Otto Neurath's Infamous Index Verborum Prohibitorum." *Perspectives on Science* 5: 452–80.

———. 2001. "Against a Third Dogma of Logical Empiricism: Otto Neurath and 'Unpredictability in Principle.'" *International Studies in the Philosophy of Science* 15: 199–209.

———. Forthcoming. "From 'The Life of the Present' to the 'Icy Slopes of Logic': Logical Empiricism, the Unity of Science Movement, and the Cold War." In *Cambridge Companion to Logical Empiricism,* ed. A. Richardson and T. E. Uebel.

Richardson, A. 2003. "Logical Empiricism, American Pragmatism, and the Fate of Scientific Philosophy in North America." In *Logical Empiricism in North America,* ed. G. L. Hardcastle and A. W. Richardson, 1–24. Minneapolis: University of Minnesota Press, 2003.

Sebeok, T. 1991. *Semiotics in the United States.* Bloomington: Indiana University Press.

Suppe, F. 1977. *The Structure of Scientific Theories.* Urbana: University of Illinois Press.

Uebel, T. 1991. "Neurath's Programme for Naturalistic Epistemology." *Studies in History and Philosophy of Science* 22: 623–46.
———. 1992. *Overcoming Logical Positivism from Within: The Emergence of Neurath's Naturalism in the Vienna Circle's Protocol Sentence Debate*. Atlanta, Ga.: Rodopi.
———. 2003. "Philipp Frank's History of the Vienna Circle: A Programmatic Retrospective." In *Logical Empiricism in North America,* ed. G. L. Hardcastle and A. W. Richardson, 149–69. Minneapolis: University of Minnesota Press, 2003.
Unity of Science Movement Papers (USM). Regenstein Library. University of Chicago.

Friedrich K. Stadler

9
Transfer and Transformation of Logical Empiricism: Quantitative and Qualitative Aspects

There is a general consensus in twentieth-century historiography that modern philosophy of science *(Wissenschaftstheorie)* has, since World War II, been strongly influenced by the direct and indirect contributions of logical empiricism, including the Vienna Circle and the Berlin Group (Danneberg, Kamlah, and Schäfer 1994; Haller and Stadler 1993; Stadler 1997b). Less well explored, however, are the ways this transfer of knowledge from Europe to North America took place as a consequence of forced emigration. It is also not clear what impact this so-called sea change (Hughes 1975) has had. The way the reception of the internationally dominant philosophy of science, established in the Anglo-American countries, influenced the scientific culture of the German-speaking countries in the decades after 1945 has also not yet been adequately explored (Fischer and Wimmer 1993). In this chapter, I describe and assess the transfer, transformation, and (reciprocal) reception of the scientific philosophy that originated in central Europe, especially in Austria, and became associated with the term *Wissenschaftslogik* ["logic of science"] in the interwar years. This intellectual process will be reconstructed with a special focus on the Vienna Circle's conception of philosophy of science.

Wissenschaftslogik—Philosophy of Science—*Wissenschaftstheorie:* A Survey

It is only in the 1990s that the history of logical empiricism (alternatively called logical positivism or neopositivism) has been appreciated in studies of emigration from the German-speaking countries (Stadler 1997b, 1998a, 1998b). This transfer of scientific knowledge has been ignored for many years due, on the one hand, to the deficits of postwar German historiography and, on the other, to the ideologies of the so-called positivism debate of the early 1960s (Dahms 1994). The ahistorical account of the Vienna

Circle in the historiography of German philosophy since 1945 has resulted in a rather one-dimensional, clichéd image (Marcuse 1964). As a result, the development of logical empiricism considered as a typical emigration phenomenon or a phenomenon of cultural exodus (Stadler 1987–88) has been obscured. This deficient historical reconstruction has led to a widespread ignorance of the institutional and theoretical development of logical empiricism as a pluralist movement, leaving the impression that it is intellectually both diffuse and exhausted—for example, as reflected in discussions of the "received view" (Suppe 1977).

In the 1990s, new research has spurred a historical and critical rediscovery of the Vienna Circle in its social context, first in Austria (Uebel 1991) and, more recently, in North America (Giere and Richardson 1996). These studies reflect two different lines of approach: (1) exile and emigration research, and (2) history and philosophy of science. Both lines will be further explored in the following assessment of the history of reception. First, however, a brief survey of the history of emigration and acculturation of the logical empiricists is in order.

The Emigration of the Vienna Circle: Background and Development

Even at first glance, the forced emigration of the Vienna Circle, with its many Jewish members, can be recognized as a real exodus. The "liberalism," "Austro-Marxism," and "scientism" denounced by Austrofascism and national socialism, as well as their racism and anti-Semitism, had already begun to trigger a slow, continuous intellectual exodus at the beginning of the 1930s (Stadler and Weibel 1995). There were cognitive reasons in addition to these external reasons for the exodus. The "scientific conception of the world" (Carnap, Hahn, and Neurath 1929), which was developed against the background of a late Enlightenment in "Red Vienna" and motivated by empiricism, as well as the employment of linguistic analysis to "eliminate metaphysics through logical analysis of language" (Carnap 1932), represented stimulating alternatives to the fundamentalist holisms and traditional systemic philosophies found in Catholic or German national *Weltanschauungen*. Moreover, the personal engagement of both the so-called left wing of the Vienna Circle (Hans Hahn, Philipp Frank, Rudolf Carnap, Otto Neurath, and Edgar Zilsel) and the less radical individualist liberals of the movement (Moritz Schlick, Friedrich Waismann, Karl Menger, and Felix Kaufmann) was reason enough for the academic and political elite in Vienna to brand the logical empiricism of the Vienna Circle as a whole a degenerate syndrome of democratic and cosmopolitan culture.

A historical account of the emigration has been given elsewhere (Dahms 1987, Hegselmann 1988, Stadler 1998a, 1998b). In contrast to these investigations, the common image, of an exile beginning only in 1938 and of successful acculturation in the Anglo-American world, fails to reflect the various temporal, personal, and sociocultural factors in this process of diffusion, for example, by focusing exclusively on successful life stories (such as those of Gustav Bergmann, Carnap, Herbert Feigl, Frank, and Hans Reichenbach). In fact, there was an intensifying Austro-American exchange of ideas between *Wissenschaftslogik* and philosophy of science starting in the early 1920s that is yet to be fully explored. This more detailed history of reception has led to assessments that assign varying degrees of success to personal acculturation and reception in North America.[1]

Given the social conditions in central Europe, the brain drain set in relatively early, peaking after the *Anschluss* in 1938. The prewar internationalization of logical empiricism, along with its break with German philosophical culture, led to the emigration of all but a very few members of the movement. These factors also influenced the emigrants' decisions not to return to their native countries (in contrast to the members of the Frankfurt school, whose ties to the culture of the Germany remained intact in exile and who, therefore, returned after 1945).

A survey of the members and sympathizers of the Vienna Circle suggests the extent of the losses. Starting in 1931, thirteen members of the twenty core participants of the Vienna Circle emigrated for political, economic, or cultural reasons, most notably because of increasing racism. The list of those who emigrated from Vienna, Berlin, and Prague (Haller and Stadler 1993) reads like a who's who of twentieth-century philosophy of science: We have Gustav Bergmann (1939, to the United States), Rudolf Carnap (1936, United States), Herbert Feigl (1931, United States), Philipp Frank (1938, United States), Kurt Gödel (1939, United States), Olga Hahn-Neurath (1934, Netherlands), Felix Kaufmann (1938, United States), Karl Menger (1937, United States), Richard von Mises (1933, Turkey; 1939, United States), Otto Neurath (1934, Netherlands; 1940, Great Britain), Rose Rand (1939, Great Britain), Josef Schächter (1938, Palestine), Olga Taussky (1937, Great Britain; 1947, United States), Friedrich Waismann (1937, Great Britain), and Edgar Zilsel (1939, United States). Viktor Kraft and Heinrich Neider were forced into so-called inner emigration, with all its repressive circumstances, while Belá Juhos managed to live through the Nazi period in economic independence without any professional obligations.

The murder of Schlick on the *Philosophenstiege* of the University of Vienna in 1936 was undoubtedly symbolic of the cultural climate and the subsequent "cultural exodus from Austria" (Stadler and Weibel 1995). In the Austrian press, this tragedy was, for the most part, legitimized as a consequence of Schlick's "pernicious philosophy" of positivism. This event and Hahn's early death in 1934 contributed to the dissolution of the Circle.

At the same time, a considerable internationalization of logical empiricism can be noted from 1930 on. The convergence of the movement—most notably in the Anglo-American realm—with pragmatism, behaviorism, and the British analytic philosophy of language (itself influenced by Ludwig Wittgenstein) became manifest in six International Congresses for the Unity of Science held in France, Denmark, England, and the United States. Thus, an intellectual and institutional platform for the cultural transfer to the New World had been created. Such forms of communication had already been practiced in exemplary individual instances in philosophy generally, especially in scientific philosophy, since the turn of the century (for example, the communication of Ernst Mach with the group around *The Monist* and with William James or the contacts between Wittgenstein, Bertrand Russell, and Frank Ramsey), making a history that emphasizes such processes of transfer and transformation—accelerated and reinforced by external factors—appear even more plausible (Holton 1993).

If we take even a cursory glance at the situation in the most important host country, the United States, we can note a strong intellectual and institutional manifestation of what Herbert Feigl (1969) called "The Vienna Circle in America." In the wake of the International Congresses for the Unity of Science (from 1935 through 1941), the ambitious correlated publication project, the *International Encyclopedia of Unified Science,* was launched in 1938 under the aegis of Neurath, Carnap, and Charles Morris (Reisch 1995, Dahms 1999). The project remained incomplete because of World War II and because of theoretical differences of opinion among the editors (see Reisch's chapter 8 in this volume). The project ended in 1962, when nineteen monographs were published together under the title *Foundations of the Unity of Science* (Neurath, Carnap, and Morris 1970–71).

At the same time, the immigrant philosophers of science at Harvard participated in an interdisciplinary discussion of the "science of science" from 1940 on (see Hardcastle's chapter 7 in this volume). Such a discussion took place later at the Institute for the Unity of Science (1947–58), founded and headed by Frank. Jointly with other North American organizations, the institute sponsored conferences and proceedings on the theme of "Science and Culture," which were published as issues of the *Proceedings of the American Academy of Arts and Science, Isis,* and *Daedalus.*

After Frank retired from Harvard, a younger generation of philosophers of science who had been socialized there, such as Gerald Holton and Robert S. Cohen, sought to further develop what had become the discipline of history and philosophy of science: Cohen "inherited" the Institute for the Unity of Science, and in 1960, he established the still existing Boston Colloquium for the Philosophy of Science. In conjunction with this, he, together with Marx W. Wartofsky, has edited the series Boston Studies in the Philosophy of Science, numbering two hundred volumes to date.

A few years earlier, in 1953, Feigel, by then a naturalized U.S. citizen, had founded the Minnesota Center for the Philosophy of Science in Minneapolis (after an interim stop in Iowa). This center has become one of the leading locations for scholars interested in an expanded understanding of history, psychology, sociology, and philosophy of science. The conditions of excellence upheld by an open "Republic of Scholars" (to use a favorite phrase of Neurath's) in the spirit of the Vienna Circle are reflected in the series Minnesota Studies for the Philosophy of Science, as well as in the journal *Philosophical Studies,* founded by Feigl and Wilfrid Sellars. Natural science had been the focus of the Harvard Circle, while the Boston center featured a blend of natural science, social science, and humanities. In Minnesota, a psychological and social scientific focus, and an interest in foundations and applications were cultivated.

The Center for Philosophy of Science at the University of Pittsburgh, modeled after Feigl's Minnesota center, was established in 1960 under Adolf Grünbaum, Andrew Mellon Professor of Philosophy of Science and research professor of psychiatry. There, annual lectures—published in the University of Pittsburgh Series in the Philosophy of Science—were given by Paul Feyerabend, Hempel, Ernest Nagel, Michael Sciven, Sellars, and Nicholas Rescher. In 1978, the Archives for Scientific Philosophy were established at the University of Pittsburgh, which currently contain the papers of, among others, Carnap, Reichenbach, Hempel, Ramsey, and Feigl (on microfilm by courtesy of the University of Minnesota). Meanwhile, the University of Pittsburgh has embarked on an archival and publishing cooperation with the University of Konstanz in Germany. Their joint forum organizes the biennial Pittsburgh-Konstanz Colloquium in the Philosophy of Science, providing the Pittsburgh-Konstanz Series in the Philosophy and History of Science.

Philosophy of social science had, meanwhile, established itself above all at the emigrants' university, the New School for Social Research in New York (Rutkoff and Scott 1986, Krohn 1987). There, the "phenomenologist of the Vienna Circle," Felix Kaufmann, was, until his death in 1949, active with his friend from his Vienna days, Alfred Schuetz, in promoting a convergence of phenomenology, logical empiricism, and pragmatism (in the tradition of John Dewey) in the context of the Unity of Science movement. As coeditor of *Philosophy and Phenomenological Research,* he created a platform for such inter- and transdisciplinary debates. Kaufmann's intellectual mediation between the topoi of understanding and explanation in his *Methodenlehre der Sozialwissenschaften* (1936), modified in exile and published as the *Methodology of the Social Sciences* (1944), has only recently been addressed in philosophy of science (Stadler 1996, Zilian 1990, 1997). Another, independent development in the field of phenomenological research is associated with the "Vienna Circle dissident," Gustav Bergmann, who formed a sort of intellectual school at the Department of Psychology at the University of Iowa, where a specific "Iowa tradition" was established (Kendler 1989, Hochberg 1991). A similar convergence of philosophical movements could be described for central-European and British philosophy of science, although it goes beyond the scope of this paper, which is focused on the United States.

Even at first glance, we see a very complex institutional and theoretical development that cannot be described only in terms of "exile studies" or any "input-output" model applied in history of science. Accordingly, the emigration and exile of the Vienna Circle was manifold and conflicting, involving both success and failure, acculturation and disintegration, diffusion and isolation. Qualitative research on the reception of the Circle in exile must take into account its members' sociocultural contexts in addition to traditional concerns with the development of their philosophical viewpoints. Such work requires the resources of emigration research and discipline-oriented historiography of science, as well as general migration research and specific sociology of science (since mobility is a characteristic of the scientific community). Thus, the transition from *Wissenschaftslogik* to philosophy of science (see below), together with the already mentioned intense intellectual interaction between Europe and America between the wars, featuring the coevolution of scientific philosophy, is an independent field of investigation complementary to emigration studies (Jackman and Borden 1983). Thus, analyzing translations, relevant journals, and book series—for example the *Journal of Philosophy, Philosophy of Science, Erkenntnis/Journal of Unified Science, Social Research, Philosophy, and Phenomenological Research,* and *Synthese (Unity of Science Forum),* to name a few—prevents one from attempting overly simplistic (symmetric) loss-gain calculations.

The network of relationships in such cases of knowledge transfer is hard to discern, but it can be made transparent in some instances. One example is the transfer of knowledge in formal logic, game theory, and decision theory, through Gödel, John von Neumann, and Oskar Morgenstern at the Institute for Advanced Study and at Princeton University. Their contributions already figured in an international context in the 1930s and 1940s, and not just as a typical emigrant science (cf. Regis 1987, Dawson 1997). Nonetheless, they could *also* be described, without ethnocentric or nationalist intentions, as central-European, or "Austro-Hungarian," contributions to an open-minded cosmopolitan culture of science (Stadler 1997, Weibel 1998). There, we have integrated the internationalized modern logic and mathematics and the specific features of local scientific subcultures, as reflected in Menger's Mathematical Colloquium (1928–36) (Dierker and Sigmund 1998).

Such a differentiated study of impact is also required when we look at the outsiders who failed in exile because of external conditions. Whereas the mathematician Mises was able to "transfer" his lifework successfully in the realm of (applied) mathematics and theory of science (see Stadler's introduction to Mises 1990), the philosopher and sociologist of science Zilsel stands as a rather tragic case of failed transfer of knowledge. In spite of the Merton school, which emerged independently, Zilsel's work in the history and sociology of modern science has only become evident in the 1990s in the wake of the Karl Popper–Thomas Kuhn–Paul Feyerabend controversy and through reprints of his works (Haller and Stadler 1993, Zilsel 2000).

Similarly, from today's perspective, the Austrian/European contribution to the foundations of psychology and psychoanalysis in California—specifically the influence of Egon Brunswik and Else Frenkel-Brunswik (an intermediary of the Buehler school and the Vienna Circle)—seems almost marginal (Frenkel-Brunswik 1996), despite the increased convergence of Austro-American neobehaviorism and logical empiricism after 1930 (Smith 1986) and the increased notice of so-called ecological psychology in the German-speaking countries (Fischer and Stadler 1997).

On the Reception and Influence of Logical Empiricism

The United States was one of the most important host countries for what Mitchell Ash and Alfons Soellner (1996) have called a "forced migration" and Friedrich Stadler and Peter Weibel (1995) have termed a "cultural exodus" from Austria. The groundwork that has been done so far in emigration studies enables us to assess carefully the reception of the emigrants' contributions, taking into account the phenomena of integration, disintegration, acculturation, and the corresponding theoretical dynamic (*Oesterreicher im Exil* 1995). The central issue of the (direct and indirect) influence of external factors on the theoretical core of disciplines and movements can be studied by focusing on three basic aspects: transfer, transformation, and influence (Kroener 1988, Strauss 1991). A decisive factor in the emigration was that, given the restrictive quotas and the necessary affidavits of support, only those scientists or scholars who had been offered a position at a U.S. university could emigrate. They could become more or less integrated into American academic culture through the assistance of private organizations (such as the Rockefeller Foundation). Moreover, good general living and working conditions were crucial for successful active acculturation. A number of factors figured significantly in the emigrant experience, including the moment of emigration, age, sex, Jewish background, political orientation, and the status of the academic discipline in question in the host country. Against the background of these complex sociocultural factors, the specific details of intellectual emigration can be determined from émigrés' publications as well as from their research and teaching activities. Did a "deprovincialization" of intellectual life in the United States actually take place as a consequence of the immigration of scholars and scientists there, as was asserted many times in the early studies after 1945? And what features can we detect in the transfer, transformation, and influence of the philosophy of science originating in Austria and central Europe?

If one adopts Lewis Coser's perspective (1984, 1988), differences between German and Austrian emigration and subsequent typical academic careers can shed light on the transfer of logical empiricism. This is really an argument for differentiation within the population of German-speaking intellectual emigrants and immigrants.

In his "Acculturation of Thought," Barry Katz (1991) indicates that after a phase of affirmative post-1945 historiography and autobiographical success stories (see, for example, Carnap 1963, Feigl 1969, and Frank 1949), a more critical look back at the process of reciprocal acculturation (as opposed to the one-sided melting-pot theory) has been dominant in recent years. Accordingly, as new qualitative features of an Austrian "social theory in a new context," H. Stuart Hughes (1983) mentions such factors as the emigrants' increased metatheoretical reflection and methodological work in absence of a grand theory mediating 'positivism' and 'empiricism'. This may be applied to the story of logical empiricism. However, these surveys underestimate the importance of international and interdisciplinary communication both before and within the period of emigration among those in philosophy of science (for example, the visits in the 1930s of Morris, Nagel, E. C. Tolman, and W. V. O. Quine to Vienna or Prague). And these early *prewar* interactions point to a revision of traditional accounts of the Austrian emigration focusing on intellectual migration (for example, Spaulding 1968, Fermi 1971, Johnston 1972, Timms and Robertson 1995).

Only the most recent historiographical studies have focused specifically on the issue of the cultural transfer from Europe (in particular, from Vienna, Berlin, and Prague) to the United States, critically examining the linear image of what Reichenbach (1951) later called the "rise of scientific philosophy." (On the transfer from Europe to the United States, see

Fleming and Bailyn 1969, Pabisch 1988, and Hoelbling and Wagnleitner 1992.)

Based on the above-cited publications on intellectual migration, one can better, and critically, examine the transfer of knowledge and the influence of the Vienna Circle on history and philosophy of science in North America.

Explaining change in science due to the forced emigration from 1933 not in terms of costs and benefits but in terms of contingent processes of change presents a challenge. This account seems so difficult because such research is dependent on a methodically and factually well-founded historiography of science. Here, the plausible "denationalization of science" via emigration is a necessary but not sufficient explanation if, given the transnational character of science, one wants to avoid a circular explanation.

I will focus on a single case study. The influence of the Vienna Circle in U.S. social sciences, especially regarding empirical social research, is certainly not to be described as a success story. It has been correctly stated that there was a superficial reception based on legitimizing citations and not on substantial considerations (Platt and Hoch 1996). Yet the range of study must be extended beyond citation studies. Both the exclusion of outsiders such as Kaufmann and the false claim that no one in the Circle except for Neurath was interested in the social sciences limit the value of such a rather external history of science. Also, the view that there was no philosophy of science in the United States before 1938 is contradicted by well-known facts about the American scientific landscape (for example, the founding of the journal *Philosophy of Science* in 1934).

In the innovative anthology *Origins of Logical Empiricism,* Ronald Giere (1996) provides an alternative account of the development of philosophy of science from *Wissenschaftliche Philosophie*. According to this study, the prewar years were decisive for the emergence of philosophy of science in the United States. Here, the roots and subsequent developments following the *Anschluss* are analyzed within a historico-genetic context. Giere suggests, plausibly, that the members of the Vienna Circle decided what writings were deemed "fit for America." This, I think, could be described as part of the transition or modification from a "hard" to a "weak" philosophy of science, in accord with Hans-Joachim Dahms's (1987) account of the relevant changes brought on by emigration, including depolitization, liberalization, and academization of logical empiricism. Dahms explains these phenomena as resulting from the loss of a cultural context for a militant and popularizing "scientific world conception." However, Dahms's investigations are undermined by an analysis of philosophical texts that reflect internationalized communication between the wars (Stadler 2001). Giere closes with a counterfactual *Gedankenexperiment* to which no definitive answer is given, namely, imagining that the National Socialists had not come to power in Germany and Austria. He has inadvertently highlighted the relevant research program, namely, one that examines the influence and impact of the transfer and transformation in history and philosophy of science brought about by logical empiricism's emigration and exile.

On the Emergence of Philosophy of Science: A Reevaluation of *Wissenschaftslogik* before 1938

The discipline known today as *Wissenschaftstheorie,* or philosophy of science, resulted from the scientification of philosophy after the turn of the twentieth century. The programmatic "manifesto" of the Vienna Circle (Carnap, Hahn, and Neurath 1929) sought to replace the autonomous discipline of philosophy with an antimetaphysical and physicalist unified science. This idea was systematically elaborated in the 1930s, most notably in Carnap's writings (Richardson 1998). In the manifesto, reference had been made to Carnap's *Logische Aufbau der Welt* (1928). A few years later, the position he took in his *Logische Syntax der Sprache* (1934a) found acceptance. The logic of science (in *Die Aufgabe der Wissenschaftslogik* [1934b]) Carnap presented as the study of science as a whole or its disciplines:

> The concepts, propositions, proofs, theories appearing in the various realms of science are analyzed—less from the perspective of the historical development of science or of the sociological and psychological conditions of its functioning, but more from a logical perspective. This field of work for which no generic term has been able to gain acceptance, could be called theory of science or to be more precise logic of science. Science is understood as referring to the totality of accepted propositions. This does not just include the statements made by scholars but also those of everyday life. There is no clear boundary line drawn between these two areas. (Carnap 1934b, 5)

Here, the emancipation from traditional philosophy becomes salient. This new discipline is interested not in providing propositions about objects, the realm of the empirical disciplines, but, rather, in understanding "science itself as an orderly structure of propositions" (6). The "sense" of the propositions and the "meaning" of concepts are understood in the logical sense, and the domain of meaningful propositions is limited to the analytic propositions of logic/mathematics and the empirical propositions of real science *("Realwissenschaften").* Thus, in Carnap's view, "the propositions of the logic of science are propositions of the logical syntax of language. And these propositions lie within the boundaries drawn by Hume, for logical syntax is . . . nothing other than mathematics of language" (6).

The logic of science as a language-critical discipline had already been promulgated in the nineteenth century, for instance in Mach's work (Mach 1883 [1982 printing, 457–58]; see also Mises 1951, chap. 2). His writings were commonly used as an alternative to classical philosophy (Losee 1980). Nevertheless, Carnap, in his "Überwindung der Metaphysik durch logische Analyse der Sprache" (1932), seems to be the first to envision an overcoming of metaphysics through the logical analysis of language. He elaborated this program of unified science in his *Logische Syntax der Sprache* (1934a) and, as part of the internationalization of the Vienna Circle under way since 1929, in two small books that appeared almost at the same time in England, *The Unity of Science* (1934c) and *Philosophy and Logical Syntax* (1935), in the series Psychè Miniatures published by Kegan Paul. The former was an edition of the 1932 essay on scientific language, reworked by the author and translated by Max Black. The latter united three lectures that Carnap had given at the University of London in October 1934: "The Rejection of Metaphysics," "Logical Syntax of Language," and "Syntax as the Method of Philosophy." This attempt to popularize his ideas in the Anglo-Saxon world was completed by the translation of *Logische Syntax der Sprache,* which appeared in 1937 in a reworked and expanded edition, from the same English publisher, as *The Logical Syntax of Language* (1937). *Logical Syntax* was influenced by Polish and American logicians and philosophers of science (notably Alfred Tarski, Quine, and Morris) to expand the field of "logic of science" (Koehler and Wolenski 1998). In addition to the syntactic dimension, Carnap cited the semantic and pragmatic dimensions as future fields of work. Accordingly, he described the logic of science in his preface to the English edition as the "Analysis and Theory of the Language of Science":

> According to the present view, this theory comprises, in addition to logical syntax, mainly two further fields, i.e., semantics and pragmatics. Whereas syntax is purely formal, i.e., only studies the structure of linguistic expression, semantics studies the semantic relationship between expressions and objects or concepts.... Pragmatics also studies the psychological and sociological relations between persons using the language and the expressions. (Carnap 1937 [1968 printing, vii])

With this conceptualization of *Wissenschaftslogik,* which was already in place before the transfer to the United States, we have also defined the field of philosophy of science as well as the terminological structure for the Unity of Science movement. Of course, logical positivism had no universally agreed understanding of the relations of logic of science to philosophy. Here, however, I have focused on the elements that were to prove most relevant later in the Anglo-American world. In this context, therefore, I cannot dwell on other issues, such as the protocol-sentence debate within the Vienna Circle in the 1930s, in which various positions on the basic issue of foundations of empirical knowledge were elaborated, although at the time, this involved a heated debate of crucial questions in philosophy of science and in epistemology (Uebel 1992).

The Differentiation of History and Philosophy of Science and the Emigration of the Logical Empiricists

From the early 1930s, which mark the beginning of the transformation of logical empiricism in North America, it is possible to reconstruct the intellectual conditions of the convergent development of central European and North American philosophy of science.

In the contemporary *Dictionary of Philosophy* (Runes 1944), we find relevant discussions of the time summarized in various short entries. The contributions concerning philosophy of science were written by Carnap, Hempel, and Heinrich Gomperz. Carnap presents "philosophy of science" here as

> that philosophic discipline which is the systematic study of the nature of science, especially of its methods, its concepts and presuppositions, and its place in the general scheme of intellectual disciplines. No very precise definition of the term is possible since the discipline shades imperceptibly into science, on the one hand, and into philosophy in general, on the other. A working division of its subject-matter into three fields is helpful in specifying its problems, though the three fields should not be too sharply differentiated or separated. (Carnap, in Runes 1944, 284)

The three fields addressed are

1. A critical study of the method or methods of the sciences, of the nature of scientific symbols, and of the logical structure of scientific symbolic terms. . . .
2. The attempted clarification of the basic concepts, presuppositions and postulates of the sciences, and the revelation of the empirical, rational, or pragmatic grounds upon which they are presumed to rest. . . .
3. A highly composite and diverse study which attempts to ascertain the limits of the special sciences, to disclose their interrelations one with another, and to examine their implications so far as these contribute to a theory either of the universe as a whole or of some aspect of it. (Carnap, in Runes 1944, 284–85)

In a preceding section, Carnap had already subsumed "philosophy of science" under "science of science," which he described as "the analysis and description of science from various points of view, including logic, methodology, sociology, and history of science" (284). In this connection, he refers to his entry on scientific empiricism and the Unity of Science movement, which was, he said, "a wider movement, comprising besides logical positivism other groups and individuals with related views in various countries" (286). Carnap thus identified "The Unity of Science" with internationalization, and "Scientific Empiricism" is introduced as a transformation of logical positivism. With this understanding, a development that had been preceded by two decades of intellectual exchange between Europe and the United States was institutionalized and developed.

To be concrete: The wider concept of logical positivism as characterized via Carnap's comprehensive concept of philosophy of science (with the inclusion of syntax, semantics, *and* pragmatics) was developed after the peak of emigration in 1938. And this theoretical dynamic has to be coordinated with the classical emigration and exile studies in an international and institutional context. Therefore, innovative functional descriptions and interdisciplinary sketches may be developed in place of linear causal explanations. According to this approach, the transfer and transformation of logical empiricism will appear complex, both historically and philosophically.

The transfer, transformation, and impact of central-European, especially Austrian, philosophy of science in the period 1930–60 did not take place abruptly. Rather, it involved a continuous brain drain that was reinforced by the mass exodus that set in around 1938. The early ties to Anglo-American philosophy of science prepared the ground for a pronounced convergence between *Wissenschaftslogik* and history and philosophy of science in the United States. In the 1920s, there was already a trend toward internationalization. With the dominance of (neo)pragmatism in U.S. philosophy, "The *Wiener Kreis* in America" (Feigl 1969) initially appeared relatively successful.[2]

These findings suggest a need to bring together exile and emigration studies with science studies (including psychology and sociology of science) in order to illuminate the recent history of philosophy of science as well (Stadler 1996). History and philosophy of science must be regarded as a large, complex, self-organizing project within a sociopolitical context. The histories of the careers of the above-mentioned persons and institutions, along with a comparative account of the international development of the disciplines, provide the basic elements for a historical study of philosophy of science. This research, however, cannot dispense with an understanding of the theoretical work produced by the logical empiricists.

The complementarity of text and context is thus postulated for a representative study of philosophy of science.

In this sense, historiography of logical positivism is seen as a constitutive part of interdisciplinary research. It is a history that must address the disparate human, social, and natural sciences as cultural phenomena. If one views this cultural exodus in very general terms, it becomes clear how importantly factors such as migration, mobility, and internationality influenced the development of knowledge and science. Thus, the development from *Wissenschaftslogik,* via the philosophy of science, to today's *Wissenschaftstheorie* must be contextualized. This is necessary for fully understanding the state of the art in history and philosophy of science. We are just at the beginning of this comprehensive enterprise, but we have some promising prospects.

Notes

This paper is part of a larger study, "*Wissenschaftslogik*—Philosophy of Science—*Wissenschaftstheorie:* Transfer, Transformation, and Influence of Central European Theory of Science," in preparation. Thanks to the Austrian Fulbright Commission for supporting my research scholarship in the United States in 1998–99.

1. The Berlin counterpart to the Vienna Circle, the Gesellschaft für empirische Philosophie in Berlin (Berlin Society for Empirical Philosophy), with Reichenbach, Carl G. Hempel, Olaf Helmer, and Paul Oppenheim, together with the Berlin Gestalt psychologists Wolfgang Koehler, Kurt Koffka, and Kurt Lewin, experienced a similar fate. Only recently has it been the subject of study (Danneberg, Kamlah, and Schäfer 1994; Rescher 1996).

2. It is striking that the criticism of the "received view" in the 1960s had been anticipated in the development of history and philosophy of science since the internationalization—long before Kuhn's theory of paradigm shifts. In Kuhnian philosophy of science, one finds a number of themes and methodological principles more or less anchored in the program of the *International Encyclopedia of Unified Science,* from 1930 to 1960, especially the pragmatization, historization, and naturalization of philosophy of science. There was even a movement toward pluralism and relativism, for example, in Frank's *Relativity: A Richer Truth* (1950). For comparison with the Vienna Circle in Great Britain, see Stadler 2002.

References

Ash, M., and A. Soellner, eds. 1996. *Forced Migration and Scientific Change: Émigré German-Speaking Scientists and Scholars after 1933.* Cambridge: Cambridge University Press.

Carnap, R. 1928. *Der logische Aufbau der Welt.* Berlin-Schlachtensee: Weltkreis-Verlag.

———. 1932. "Überwindung der Metaphysik durch logische Analyse der Sprache." *Erkenntnis* 2: 219–41. English translation in *Logical Positivism,* ed. A. J. Ayer, 60–81. Glencoe, Ill.: Free Press, 1959.

———. 1934a. *Logische Syntax der Sprache.* Vienna: Springer.

———. 1934b. *Die Aufgabe der Wissenschaftslogik.* Vienna: Verlag Gerold und Co.

———. 1934c. *The Unity of Science*. Trans. Max Black. London: Kegan Paul.
———. 1935. *Philosophy and Logical Syntax*. London: Kegan Paul.
———. 1937. *Logical Syntax of Language*. London: Kegan Paul, 1968.
———. 1963. "Intellectual Autobiography." In *The Philosophy of Rudolf Carnap*, ed. P. A. Schilpp, 3–84. La Salle, Ill.: Open Court.
Carnap, R., H. Hahn, and O. Neurath. 1929. *Wissenschaftliche Weltauffassung: Der Wiener Kreis*. Vienna: Artur Wolf. English translation in *Empiricism and Sociology*, ed. M. Neurath and R. S. Cohen, 299–318. Dordrecht: Reidel, 1973.
Coser, L. 1984. *Refugee Scholars in America: Their Impact and Their Experiences*. New Haven, Conn.: Yale University Press.
———. 1988. "Die oesterreichische Emigration als Kulturtransfer Europa-Amerika." In *Vertriebene Vernunft: Emigration und Exil österreichischer Wissenschaft*. Vol. 2, ed. F. Stadler, 93–101. Vienna and Munich: Jugend und Volk.
Dahms, H.-J. 1987. "Die Emigration des Wiener Kreises." In *Vertriebene Vernunft: Emigration und Exil österreichischer Wissenschaft*. Vol 1, ed. F. Stadler, 66–122. Vienna and Munich: Jugend und Volk. English translation in *The Cultural Exodus from Austria*, ed. F. Stadler and P. Weibel. Vienna: Springer, 1995.
———. 1994. *Positivismusstreit: Die Auseinandersetzungen der Frankfurter Schule mit dem logischen Positivismus, dem amerikanischen Pragmatismus und dem kritischen Rationalismus*. Frankfurt: Suhrkamp.
———. 1995. "The Emigration of the Vienna Circle." In *The Cultural Exodus from Austria*, ed. F. Stadler and P. Weibel, 57–79. Vienna: Springer.
———. 1999. "Otto Neurath's *International Encyclopedia of Unified Science* als Torso." In *Otto Neurath: Rationalität, Planung, Vielfalt*, ed. E. Nemeth and R. Heinrich, 184–227. Vienna and Berlin: Oldenbourg-Akademie Verlag.
Danneberg, L., A. Kamlah, and L. Schäfer, eds. 1994. *Hans Reichenbach und die Berliner Gruppe*. Braunschweig-Wiesbaden: Vieweg.
Dawson, J. W. 1997. *Logical Dilemmas: The Life and Work of Kurt Goedel*. Wellesley, Mass.: A. K. Peters.
Dierker, E., and K. Sigmund, eds. 1998. *Karl Menger: Ergebnisse eines Mathematischen Kolloquiums*. Vienna and New York: Springer.
Feigl, H. 1969. "The *Wiener Kreis* in America." In *The Intellectual Migration: Europe and America, 1933–1960*, ed. D. Fleming and B. Bailyn, 630–74. Cambridge, Mass.: Harvard University Press.
Fermi, L. 1971. *Illustrious Immigrants: The Intellectual Migration from Europe, 1930–41*. Chicago: University of Chicago Press.
Fischer, K. R., and F. Stadler, eds. 1997. *Wahrnehmung und Gegenstandswelt: Zum Lebenswerk von Egon Brunswik (1903–1955)*. Vienna: Springer.
Fischer, K. R., and F. M. Wimmer, eds. 1993. *Der geistige Anschluß: Philosophie und Politik an der Universität Wien, 1930–1950*. Vienna: WUV Universitätsverlag.
Fleming, D., and B. Bailyn, eds. 1969. *The Intellectual Migration: Europe and America, 1933–1960*. Cambridge, Mass.: Harvard University Press.
Frank, P. 1949. *Modern Science and Its Philosophy*. Cambridge, Mass.: Harvard University Press.
———. 1950. *Relativity: A Richer Truth*. Foreword by A. Einstein. Boston: Beacon Press.
Frenkel-Brunswik, E. 1996. *Studien zur autoritären Persönlichkeit: Ausgewählte Schriften*, ed. D. Paier. Graz and Vienna: Nausner und Nausner.
Giere, R. 1996. "From *Wissenschaftliche Philosophie* to Philosophy of Science." In *Origins of Logical Empiricism*, ed. R. N. Giere and A. W. Richardson, 335–54. Minneapolis: University of Minnesota Press.

Giere, R. N., and A. W. Richardson, eds. 1996. *Origins of Logical Empiricism.* Minneapolis: University of Minnesota Press.
Haller, R., and F. Stadler, eds. 1993. *Wien—Berlin—Prag: Der Aufstieg der wissenschaftlichen Philosophie: Zentenarien Rudolf Carnap, Hans Reichenbach and Edgar Zilsel.* Vienna: Hölder-Pichler-Tempsky.
Hardcastle, G. L. 2003. "Debabelizing Science: The Harvard Science of Science Discussion Group, 1940–41." In *Logical Empiricism in North America,* ed. G. L. Hardcastle and A. W. Richardson, 170–96. Minneapolis: University of Minnesota Press, 2003.
Hegselmann, R. 1988. "Alles nur Mißverständnisse? Zur Vertreibung des Logischen Empirismus aus Österreich und Deutschland." In *Vertriebene Vernunft II: Emigration und Exil österreichischer Wissenschaft,* ed. F. Stadler, 188–203. Vienna and Munich: Jugend und Volk.
Hochberg, H. 1991. "Gustav Bergmann." In *Handbook of Metaphysics and Ontology,* ed. H. Burkhardt and B. Smith, 82–86. Munich: Philosophia.
Hoelbling, W., and R. Wagnleitner, eds. 1992. *The European Emigrant Experience in the U.S.A.* Tübingen: Narr.
Holton, G. 1993. "From the Vienna Circle to Harvard Square: The Americanization of a European World Conception." In *Scientific Philosophy: Origins and Developments,* ed. F. Stadler, 47–74. Boston: Kluwer.
Hughes, H. S. 1975. *The Sea Change: The Migration of Social Thought, 1930–1975.* New York: McGraw-Hill.
———. 1983. "Social Theory in a New Context." In *The Muses Flee Hitler: Cultural Transfer and Adaptation, 1930–1945,* ed. J. C. Jackman and C. M. Borden, 111–22. Washington, D.C.: Smithsonian Institution Press.
Jackman, J. C., and C. M. Borden, eds. 1983. *The Muses Flee Hitler: Cultural Transfer and Adaptation, 1930–1945.* Washington, D.C.: Smithsonian Institution Press.
Johnston, W. M. 1972. *The Austrian Mind: An Intellectual and Social History, 1848–1938.* Berkeley and Los Angeles: University of California Press.
Katz, B. 1991. "The Acculturation of Thought: Transformation of the Refugee Scholar in America." *Journal of Modern History* 63: 740–52.
Kaufmann, F. 1936. *Methodenlehre der Sozialwissenschaften.* Vienna: Springer, 1999.
———. 1944. *Methodology of the Social Sciences.* Oxford: Oxford University Press.
Kendler, H. 1989. "The Iowa Tradition." *American Psychologist,* August, 1124–32.
Koehler, E., and J. Wolenski, eds. 1998. *Alfred Tarski and the Vienna Circle: Austro-Polish Connections in Logical Empiricism.* Dordrecht, Boston, and London: Kluwer.
Kroener, H.-P. 1988. "Überlegungen zur Wirkungsgeschichte der deutschsprachigen wissenschaftlichen Emigration." In *Vertriebene Vernunft II: Emigration und Exil österreichischer Wissenschaft,* ed. F. Stadler, 82–92. Vienna and Munich: Jugend und Volk.
Krohn, C.-D. 1987. *Wissenschaft im Exil: Deutsche Sozial- und Wirtschaftswissenschaftler in den USA und die New School for Social Research.* Frankfurt: Campus.
Krohn, C.-D., P. von zur Mühlen, G. Paul, and L. Winkler, eds. 1998. *Handbuch der deutschsprachigen Emigration, 1933–1945.* Darmstadt: Wissenschaftliche Buchgesellschaft.
Losee, J. 1980. *A Historical Introduction to the Philosophy of Science.* Oxford: Oxford University Press.
Mach, E. 1883. *Die Mechanik in ihrer Entwickelung historisch-kritisch dargestellt.* Darmstadt: Wissenschaftliche Buchgesellschaft, 1982.
Marcuse, H. 1964. *The One-Dimensional Man: Studies in the Ideology of Advanced Industrial Society.* Boston: Beacon Press.
Mises, R. 1951. *Positivism: A Study in Human Understanding.* Cambridge, Mass.: Harvard University Press.

———.1990. *Kleines Lehrbuch des Positivismus: Einführung in die empiristische Wissenschaftsauffassung,* ed. F. Stadler. Frankfurt: Suhrkamp.
Neurath, O., R. Carnap, and C. Morris, eds. 1970–71. *Foundations of the Unity of Science: Toward an International Encyclopedia of Unified Science.* 2 vols. Chicago: Chicago University Press.
Oesterreicher im Exil: USA, 1938–1945: Eine Dokumentation. 1995. 2 vols. Ed. Dokumentationsarchiv des oesterreichischen Widerstandes. Introduction, selection, and redaction: P. Eppel. Vienna: Oesterreichischer Bundesverlag.
Pabisch, P., ed. 1988. *From Wilson to Waldheim: Proceedings of a Workshop on Austrian-American Relations, 1917–1987.* Riverside, Calif.: Ariadne Press.
Platt, J., and P. Hoch. 1996. "The Vienna Circle in the USA and Empirical Research Methods." In *Forced Migration and Scientific Change: Émigré German-Speaking Scientists and Scholars after 1933,* ed. M. Ash and A. Soellner, 224–45. Cambridge: Cambridge University Press.
Regis, E. 1987. *Who Got Einstein's Office? Eccentricity and Genius at the Institute for Advanced Study.* Reading, Mass.: Addison-Wesley.
Reichenbach, H. 1951. *The Rise of Scientific Philosophy.* Berkeley and Los Angeles: University of California Press.
Reisch, G. A. 1995. "A History of the International Encyclopedia of Unified Science." Ph.D. diss., University of Chicago.
———. 2003. "Disunity in the *International Encyclopedia of Unified Science.*" In *Logical Empiricism in North America,* ed. G. L. Hardcastle and A. W. Richardson, 197–215. Minneapolis: University of Minnesota Press, 2003.
Rescher, N. 1996. "H2O: Hempel—Helmer—Oppenheim. Eine Episode in der Wissenschaftsphilosophie des 20 Jahrhunderts." *Deutsche Zeitschrift für Philosophie* 5: 779–805.
Richardson, A. W. 1998. *Carnap's Construction of the World: The Aufbau and the Emergence of Logical Empiricism.* Cambridge: Cambridge University Press.
Runes, D. D., ed. 1944. *The Dictionary of Philosophy.* London: Routledge.
Rutkoff, P. M., and W. B. Scott. 1986. *New School: A History of the New School for Social Research.* New York: Free Press.
Salmon, W., and G. Wolters, eds. 1994. *Logic, Language, and the Structure of Scientific Theries: Proceedings of the Carnap-Reichenbach Centennial.* Pittsburgh: University of Pittsburgh Press; Konstanz: Universitätsverlag Konstanz.
Smith, L. D. 1986. *Behaviorism and Logical Positivism: A Reassessment of the Alliance.* Stanford, Calif.: Stanford University Press.
Spaulding, E. W. 1968. *The Quiet Invaders: The Story of the Austrian Impact upon America.* Vienna: Bundesverlag.
Stadler, F., ed. 1987–88. *Vertriebene Vernunft: Emigration und Exil österreichischer Wissenschaft.* 2 vols. Vienna and Munich: Jugend und Volk.
———, ed. 1993. *Scientific Philosophy: Origins and Developments.* Dordrecht, Boston, and London: Kluwer.
———, ed. 1995. *Wissenschaft als Kultur: Österreichs Beitrag zur Moderne.* Vienna and New York: Springer.
———. 1996. "Wissenschaft und oesterreichische Zeitgeschichte." *Oesterreichische Zeitschrift für Geschichtswissenschaften* 1: 1–23.
———, ed. 1997a. *Phänomenologie und Logischer Empirismus: Zentenarium Felix Kaufmann (1895–1949).* Vienna and New York: Springer.
———. 1997b. *Studien zum Wiener Kreis: Ursprung, Entwicklung, und Wirkung des Logischen Empirismus im Kontext.* Frankfurt: Suhrkamp. Translated as *The Vienna Circle:*

Studies in the Origins, Development, and Influence of Logical Empiricism. Vienna: Springer, 2001.

———. 1998a. "Der 'Wiener Kreis.'" In *Handbuch der deutschsprachigen Emigration 1933–1945,* ed. C.-D. Krohn, 813–23. Darmstadt: Wissenschaftliche Buchgesellschaft.

———. 1998b. "Vienna Circle." In *Routledge Encyclopedia of Philosophy,* ed. E. Craig, 606–14. London: Routledge.

———. 2001. *The Vienna Circle: Studies in the Origins, Development, and Influence of Logical Empiricism.* Vienna and New York: Springer.

———. 2002. "The *Wiener Kreis* in Great Britain: Emigration and Interaction in the Philosophy of Science." In *Intellectual Migration and Cultural Transformation: Refugees from National Socialism in the English-Speaking World,* ed. E. Timms and J. Hughes, 155–80. Vienna and New York: Springer.

Stadler, F., and P. Weibel, eds. 1995. *The Cultural Exodus from Austria.* Vienna and New York: Springer.

Strauss, H. A., K. Fischer, C. Hoffman, and A. Söllner, eds. 1991. *Die Emigration der Wissenschaften nach 1933: Disziplingeschichtliche Studien.* Munich, London, New York, and Paris: K. G. Saur.

Suppe, F., ed. 1977. *The Structure of Scientific Theories.* Urbana: University of Illinois Press.

Timms, E., and R. Robertson, eds. 1995. *Austrian Exodus: The Creative Achievements of Refugees from National Socialism.* Edinburgh: Edinburgh University Press.

Uebel, T. E., ed. 1991. *Rediscovering the Forgotten Vienna Circle: Austrian Studies on Otto Neurath and the Vienna Circle.* Dordrecht, Boston, and London: Kluwer.

———. 1992. *Overcoming Logical Positivism from Within: The Emergence of Neurath's Naturalism in the Vienna Circle's Protocol Sentence Debate.* Amsterdam: Rodopi.

Weibel, P., ed. 1998. *Jenseits von Kunst.* Vienna: Passagen Verlag.

Zilian, H. G. 1990. *Klarheit und Methode: Felix Kaufmanns Wissenschaftstheorie.* Amsterdam: Rodopi.

———. 1997. "Felix Kaufmann: Leben und Werk." In *Phänomenologie und Logischer Empirismus: Zentenarium Felix Kaufmann (1895–1949),* ed. F. Stadler, 9–22. Vienna and New York: Springer.

Zilsel, E. 2000. *The Social Origins of Modern Science.* Ed. D. Raven, W. Krohn, and R. S. Cohen. Dordrecht: Kluwer.

10
The Linguistic Doctrine and Conventionality: The Main Argument in "Carnap and Logical Truth"

"Carnap and Logical Truth" (1963, written in 1954) is a compendium of W. V. O. Quine's arguments against Rudolf Carnap. Of the paper's ten sections, the first six are devoted to one main argument, of some intricacy and enormous rhetorical force. That argument is against what Quine calls the linguistic doctrine of logical truth and against a form of conventionalism in logic. From this Quine concludes that "the very distinction between a priori and empirical begins to waver and dissolve" (397). Though some of this argument is derived almost verbatim from earlier work and some had earlier been suggested strongly, most of this argument is new. Not only does the argument form the basis of much of Quine's subsequent writing, but also many of his readers have found these passages, even when taken in isolation, to be utterly convincing against Carnap. Thus, it will be worthwhile to examine and weigh this main argument with great care. Of course, the last four sections of Quine's paper deserve attention as well, but what is new there largely presupposes all or part of the first six sections.

We may well begin, if only for further reference, with an outline of this main argument. Quine opens with some considerations that seem to weigh in favor of the linguistic doctrine of logical truth (§1). He then goes on to argue (§§2–3) that the obviousness of elementary logic undermines whatever force those prior considerations seemed to have and makes the linguistic doctrine empty when applied to elementary logic. Set theory is not obvious, and so the linguistic doctrine has some plausibility there. In §4, by an argument partly quoted from his "Truth by Convention" (1936), Quine argues that elementary logic cannot be gotten from convention and that thus the claim that logic is true by convention adds nothing to the linguistic doctrine (which is empty). Convention is involved in set theory, but Quine stops short of saying outright that it is true by convention. That phrase is explicitly denied to the truths of geometry, and the case of mathematics is also discussed. After some useful distinctions between the legislative and the discursive in §5, Quine urges in §6 that the hypotheses of natural science are conventional in the same way as set theory, and that the eventual confrontation with empirical data, so characteristic of natural science, is likewise present for mathematics, set theory, and even elementary logic. Thus, he concludes, there is no distinction between sentences that confer truth by convention and those that do not, and "the very distinction between a priori and empirical begins to waver and dissolve" (397).

It should be noted that Carnap's evaluation in "W. V. O. Quine on Logical Truth" (1963) of Quine's paper differs somewhat from the one I shall offer. What I am calling the main argument, Carnap calls "detailed informal discussion" (915), and he gives at most a brief comment. The bulk of his reply (not to mention that part that spilled out into "Meaning and Synonymy in Natural Languages" [1956]) is devoted to §§8–10 in which Carnap describes as Quine's "more specific arguments" (915) and later as "Quine's important arguments" (917). The later sections of Quine's paper are indeed important, but the themes there are largely those of his "Two Dogmas of Empiricism" (1951) and can usefully be treated separately.

Moreover, what I am calling the main argument hangs together as a unified whole that has been enormously influential. It would be hard to overestimate the significance of its conclusion or the rhetorical force of Quine's argument. There are, to be sure, a few passages that may seem a bit unclear or puzzling. Perhaps the structure of Quine's argument or one of its premises is occasionally inconspicuous, but the net effect is overwhelming: The premises seem invulnerable; the argument seems cogent; and the conclusion seems devastating. Perhaps perversely, then, I shall argue that this is not so. Quine may offer views, interesting views, that differ from Carnap's but not a decisive case against him. The apparent seamlessness of Quine's argument depends on several factors: a crucial unclarity in one of its central terms, a famous but irrelevant argument against the conventionality of logic, and crucial assumptions toward the end that, in effect, simply beg the question against Carnap. In order to make my argument, I shall need to look more closely at Quine's.

On the Obviousness of Elementary Logic

Quine begins his paper by admitting that it is hard to frame his dissent from Carnap's philosophy in Carnap's own terms. So Quine will discuss instead what he calls "the linguistic doctrine of logical truth" and later make extensive use of the phrase 'truth by convention'. Neither of these are phrases that Carnap used, but perhaps the first is not too misleading. The same can hardly be said for the latter phrase, however, and we will need to examine it more closely when we reach that portion of Quine's text.

In any case, Quine opens by suggesting that one of the attractions of the so-called linguistic doctrine of logical truth is that it provides an answer to the question, "How is logical certainty possible?" This question, he says, is logically prior to and less tendentious than Kant's question of how synthetic a priori judgments are possible. Perhaps the modest query ought to have concerned the a priori rather than certainty. Carnap's philosophy is designed to give an account of the a priori, though the account he gives makes it a relativized and revisable a priori. The account is that the truth of such a priori claims as 'Brutus killed Caesar or did not kill Caesar' does not depend in any way on the killing. Instead, the claim is automatically true, given the language. Quine returns to this issue only in the waning paragraphs of the main argument.

A further consideration on behalf of the linguistic doctrine is provided by reflection on two cases, each of which seems to show that the logical truths are inseparable from the meanings of the logical vocabulary. The first of these concerns alternative logics. Suppose that someone advances a logic apparently at variance with our own. If in addition, the logical particles of each system can be so paraphrased in the other system that their behavior is duplicated, "then we are pretty sure to protest that he was wantonly using the familiar particles 'and' and 'all' (say) where we might unmisleadingly have used such-and-such other familiar phrasing" (Quine 1963, 386). Carnap is, indeed, interested in alternative logics, but not primarily in these cases where some straightforward translation is possible. Quine notes that there may be a failure of translatability but says no more about it. But it is only such nonequivalent alternatives that seem to raise interesting epistemological questions about which system, if any, is correct. It is only such alternatives that challenge our own complacency about the logical claims we accept. It is only these that provide a motive for Carnap's denial that there is a uniquely correct logic, that is, for the principle of tolerance, which is the cornerstone of the linguistic doctrine.

The second case designed to show the "inseparability of the truths of logic from the meanings of the logical vocabulary" (Quine 1963, 387) concerns supposedly prelogical peoples. If it is claimed that such people accept certain simple self-contradictions as true, what must be intended is that what they accept is to be translated as a self-contradiction. But surely, such a circumstance counts overwhelmingly against any such translation. As Quine says, "prelogicality is a trait injected by bad translators" (1963, 387). Whether or not such considerations provide motivation for the linguistic doctrine, it is not a case that Carnap discussed. He might have agreed that the proposed translation was a bad one without thinking that all who might be called prelogical are the victims of bad or uncharitable translation. They might simply fail to accept what we accept as logical truth. What they do accept might have no satisfactory translation, and to insist that they really believe as we do is less charitable than merely provincial. In this, the two cases are much alike. Again, except for the caveat (also expressed by Quine) that these cases may be only tangentially tied to Carnap's own reasons for adopting the linguistic doctrine, there is little here that Carnap would need to disagree with. In the evaluation of Quine's argument, it is not necessary to interject more plausible candidates for Carnap's reasons. Quine has already agreed that there may be such (1963, 385). Instead it will be enough to trace the threads of Quine's own argument.

Quine begins §2 by delimiting the scope of 'logical truth' by listing the logical vocabulary and letting the logical truths be those that involve only logical words essentially. He then distinguishes two parts of logic: elementary logic and set theory. Elementary logic comprises truth function, quantification, and identity theory, and its vocabulary consists of the truth-function signs, quantifiers and their variables, and '='. Set theory involves, in addition, the membership sign, '∈'.

When Quine speaks of elementary logic, he seems to mean a doctrine rather than a domain (that is, rather than an area of inquiry or subject matter), and the doctrine that he means is the classical first-order predicate calculus extensionally construed. He does not mean intuitionist logic, higher-order logics, or even modal logics, though the vocabulary involved could still be counted as quantifiers and truth-function signs. Contrariwise, when Quine speaks of set theory, there seems to be no one particular doctrine that he has in mind.

Moreover, Quine claims that there is an even greater contrast between elementary logic and set theory:

> Every truth of elementary logic is obvious (whatever this really means), or can be made so by some series of individually obvious steps. Set theory, in its present state anyway, is otherwise. I am not alluding here to Gödel's incompleteness principle, but to something right on the surface. Set theory was straining at the leash of intuition ever since Cantor discovered the higher infinities; and with the added impetus of the paradoxes of set theory the leash was snapped. Comparative set theory has now long been the trend. . . . What we do is develop one or another set theory by obvious reasoning, or elementary logic, from unobvious first principles which are set down, whether for good or for the time being, by something very like convention. (1963, 388)

It is very difficult to assess this passage without knowing what the word 'obvious' means. Quine's first use of the word is accompanied by a parenthetical phrase that suggests that it might not mean anything at all, and this reading is confirmed later in the essay (390). Against this, Quine's statement that elementary logic is obvious is confident and unblushing, which it would probably not be if he thought that it meant nothing at all. Indeed, obviousness figures in Quine's later argument so essentially that Quine simply cannot count it as meaningless. Even if we agree that it means something, this still does not tell us what. Apparently, obviousness is the sort of thing that can be gainsaid by the appearance of paradox. Well, since antiquity, the notion of truth has had its share of paradox. Does this mean that Tarski's T-sentences, which Quine has long taken as paradigms of lucidity, lack obviousness? Or if someone once developed a version of elementary logic that, however natural, turned out to be inconsistent, would this mean that elementary logic is not really obvious after all? Or if the first set theory developed had deftly avoided paradox and it had become standard, would it, too, be obvious, even if alternative consistent set theories became known?

I do not think that Quine wants us to answer such questions. Elementary logic is obvious, and set theory is not, and that is the end of the matter. Or is it? Since Quine does not tell us what he means (and offers no argument for his declaration that logic is obvious), we have little to fall back on but common and for the most part philosophically untutored speech. 'Obvious' is used in common speech most characteristically for what is observable, that is, readily, almost unavoidably observable. Thus, when Colonel Mustard stumbles across the body in the otherwise empty but well-lighted hall, his judgment, "My God, it's Miss Peacock" is likely to be right. Nor could he have failed to notice the glistening pool of blood. The body and the blood are obvious. It is possible to extend this notion of obviousness a bit to include what is readily or imperceptibly deduced from what is directly observed.

Philosophers, sensing an opportunity, have for centuries extended the notion still further to cover what is manifest to reason as well as to the eye. Thus, mathematics, geometry, logic, and metaphysics, including theology, were declared obvious, at least when the mind was properly prepared. This usage, too, has escaped into ordinary language. What obviousness in these contexts and obviousness in the observational ones are supposed to share is the absence of any need for further justification. When properly placed, you cannot fail to notice what is obvious. And when you notice, your judgment is bound to be right (or virtually so bound).

What needs no further justification nonetheless requires explanation. If the blood is obvious to Colonel Mustard because he sees it, then his powers of seeing must be scouted, their limits marked, and his success explained. Our explanations along these lines are getting better and better. Similarly, if Socrates is right that geometry is obvious to Meno, then Socrates (or at least Plato) has to explain how Meno gets it right: hence, the doctrine of recollection. This doctrine has not fared well in recent centuries, but modern Platonists have fearlessly endowed us with powers equally fantastic. Details have been hard to come by, as has any real integration with our other scientific beliefs. Unlike for observation, the explanations here are not improving.

Now, Quine may have a limited Platonism in his ontology, but *not* thus in his epistemology. That is, he in no way assimilates what is obvious to what is observable by postulating a special power of direct intuition. In fact, he explicitly rejects a "doctrine of ultimate and inexplicable insight into the obvious traits of reality" (1963, 390) calling it a "pseudodoctrine." My point is not that Quine should somehow be more Platonist. The point is, rather, that the claim that elementary logic is obvious is not free. No matter how vague, no matter how nearly devoid of explanatory content the claim is, to make it without serious qualification, as Quine does, is to shoulder the burden of providing an explanation, or at least issuing a promissory note for one. The difficulties of paying off on that note are so formidable that one might well be moved to revise or even withdraw completely the claim that elementary logic is obvious.

A comparison with Carnap's position is in order at this juncture. His linguistic doctrine was born in the context of worry over alternative logics. Given that there are several logical systems that are each internally consistent, which is correct? Since we seem to lack a power of direct metaphysical insight, what non-question-begging argument could possibly show which is correct? That is, what argument that does not already presuppose some logic or other could show this? Rather than declaring one of the candidates obvious or retreating to skepticism, the heart of the linguistic doctrine is that there is no uniquely correct logic (Carnap 1934, iv–vi) and hence no need or possibility of an argument that classical logic, say, is the true one and no need or possibility of an explanation of our success in adopting it. Some logics are handier than others, but not truer.

Once we have adopted a language and thereby a logic, there appears a kind of ersatz obviousness about that logic. Following Quine in setting the Gödel sentences aside, we can say that each logical sentence we accept either is a primitive sentence, and hence needs no further justification, or follows from the primitive sentences by a series of steps each of which needs no further justification. Thus, *from within,* our logic has a kind of direct or potential obviousness. This same sort of obviousness from within would likewise be available if our language choice had embodied an intuitionist logic, or for that matter any set theory you choose. They are all internally obvious in the same sense. Presumably, this is not what Quine has in mind in calling elementary logic (but not set theory) obvious. This leaves only the external sense, and Carnap simply denies that elementary logic is obvious in that sense.

Save for having learned that Quine cannot count 'obvious' as meaningless, we have made only negative progress in determining what he does mean: He does not mean that we have some direct intuitive power corresponding to observation by which we apprehend the truths of elementary logic, and he does not mean that elementary logic is obvious in Carnap's internal sense. Further clues of a more positive sort appear in §6. There he says, "Set theory, currently so caught up in legislative postulation, may some day gain a norm—even a strain of obviousness, perhaps—and lose all trace of the conventions in its history" (1963, 395). Moreover, we are told, our elementary logic could have been instituted as a conventional deviation from something earlier, though it was not. And enough reference is made to today's dissident logicians to suggest that if they are successful in convincing enough people, elementary logic will become conventional and not obvious. If obviousness is something that can be gained or lost, it is not so much a feature of the truths themselves or of our powers with respect to them as it is an indicator of how widely these claims are believed or how seriously people take the prospect of revising them. This, then, suggests one thing that Quine might mean by 'obvious': A truth is obvious just in case it is universally or at least very widely believed. We might pursue this line further. Quine contrasts the obvious and the conventional. He also holds that convention is a feature that is tied to claims at the moving front of science and that leaves no enduring trace on the claims or beliefs thus introduced. This suggests that Quine might take as obvious any claim that is sufficiently behind the lines that traces of its history are lost or that it is no longer seriously considered as a candidate for revision. All this is speculative, as it must be when Quine does not tell us what he means. But the alternatives just described have more than just a Quinean flavor. At the very end of "Truth by Convention," Quine says, "We may wonder what one adds to the bare statement that the truths of logic and mathematics are a priori, or to the barer still behavioristic statement that they are firmly accepted, when he characterizes them as true by convention in such a sense" (1936, 123–24). Quine's 'firmly accepted' seems so close to what we have just come to as a gloss on what Quine means by 'obvious' that our speculations seem somewhat confirmed. For now, that will have to do.

Is the classical first-order predicate calculus, extensionally interpreted, obvious in this Quinean sense? The answer is that in 1954, when "Carnap and Logical Truth" was written, this logic was probably more widely accepted than it was either twenty years before or twenty years after. In 1934, when Carnap published *Logische Syntax der Sprache,* not only was intuitionism a force to be reckoned with, but Ludwig Wittgenstein's earlier views were sufficiently influential even within Carnap's immediate circle that any quantifiers that could not be replaced with finite conjunctions or disjunctions were rejected by some. It was for this reason that Moritz Schlick construed laws of nature not as universally quantified sentences of a special sort but, rather, as rules of inference telling us what genuine, that is, molecular, sentences could be inferred from which others. By 1974, modal and other intensional logics were increasingly fashionable, and while the first-order predicate calculus was not exactly rejected, it was often thought not to represent ordinary or scientific discourse. If Quine were ever to announce that classical elementary logic was obvious in the sense of being widely and firmly accepted, he chose the most opportune time to do so.

Let us return to the text to see what use Quine makes of the claim that elementary logic is obvious. Quine begins §3 by wondering how to "warp the linguistic doctrine of logical truth around into something like an experimental thesis" (1963, 389). To Carnap, this would be warping it indeed, for it was never intended as an experimental thesis. The linguistic doctrine, or more accurately the principle of tolerance that is its core, was always intended as a proposal for—or if adopted, as an analytic claim in— the metalanguage. In neither case is it an empirical claim. Carnap's position is not open to Quine, of course, for Quine has only two categories, the empirical and the meaningless. So the attempted warping can be thought of as charity of a sort, not that Carnap would want it. The general question of whether the viability of Carnap's position requires the possibility of the empirical determination of meaning relations in natural languages is a vexed one. Quine says yes, and Carnap and many of his followers say no, and besides, if the task were to be undertaken, it would be a job for social scientists, not philosophers. My own view is that the *possibility* of empirical criteria sufficient to determine meaning relations in natural languages is required for the success of Carnap's program because one cannot even make a proposal unless there is at least the possibility of determining whether it has been adopted. This does not require that English, considered as a natural language, have all of the features that Carnap thinks it does or wants it to have. It is sufficient for the business of making proposals that we could in principle tell whether it has those features. I have expanded on these considerations elsewhere and need not do so here (Creath 1991).

The specific warping of the linguistic doctrine that Quine suggests as a first approximation is "Deductively irresoluable disagreement as to a logical truth is evidence of deviation in usage (or meanings) of words" (1963, 389, emphasis omitted). Quine notes that this is "not yet experimentally phrased" (389), and indeed, it sounds more like an attempt to provide empirical or behavioral criteria for an exceptionally weak claim about meanings. The point, one gathers, is that a difference in meaning might be thought to be (part of) an explanation of any deductively irresoluable disagreement as to a logical truth. Quine's response, which we will get to in a moment, is that a much simpler explanation will do, and hence there is no need for philosophically extravagant talk of meanings.

Before making that response, Quine includes the following paragraph:

> The philosopher, like the beginner in algebra, works in danger of finding that his solution-in-progress reduces to '0 = 0'. Such is the threat to the linguistic theory of elementary logical truth. For that theory now seems to imply nothing that is not already implied by the fact that elementary logic is obvious or can be resolved into obvious steps. (1963, 389)

Carnap, in his reply, points out that this "seems not to be a refutation of [the linguistic doctrine] but rather a proof of it" (Carnap 1963, 917). After all, the linguistic doctrine implies itself, and so by Quine's last sentence above the fact that logic is obvious must also imply the linguistic doctrine. Carnap also takes the remark about '0 = 0' to be evidence that Quine thinks the linguistic doctrine is true, and he expresses some vexation that Quine does not say so explicitly. Howard Stein (Stein 1992, 277) has endorsed Carnap's reading of Quine's paragraph, so perhaps it would not be inappropriate to say that a very much more charitable reading is possible. Quine's point in the '0 = 0' remark is not to concede the truth of the linguistic doctrine but only to underscore his contention that it comes to no more than something utterly trivial, that is, the claim that elementary logic is obvious. Quine's last sentence is easily understood when we reflect that his general point in this section is that the linguistic doctrine is explanatorily unnecessary, that whatever observational news the linguistic theory might be thought to explain is explained more simply by the bromide that elementary logic is obvious. The misunderstanding can be avoided by the simple expedient of inserting the words 'at the observational level' after the word 'nothing' in Quine's last sentence. The fault may be Quine's, but it is at most the fault of excess brevity.

There is another passage, no less in need of explanation, that Carnap could point to that seems to show that Quine thinks the linguistic doctrine is true. He says,

> Consider, however, the logical truth 'Everything is self-identical', or '$(x)(x = x)$'. We *can* say that it depends for its truth on traits of the language (specifically on the usage of '='), and not on traits of its subject matter; but we can also say, alternatively, that it depends on an obvious trait, viz., self-identity, of its subject matter, viz., everything. The tendency of our present reflections is that there is no difference. (Quine 1963, 390)

If Quine accepts the obviousness of self-identity, and there is no difference between that and the claims's depending for its truth only on language, then Quine seems to be acquiescing in the latter as well. Again it is necessary to amend the letter of what Quine says if there is to be any hope of preserving some intelligible spirit. Again a remedy is to insert 'at the observational level', this time after the last word, 'difference'. So construed, this passage is no more than a repetition of Quine's general claim that there is no observational or behavioral difference between a claim's being firmly accepted and its being true in virtue of meaning. The passage does not add anything to the rest of Quine's argument and hence need not be evaluated separately.

Quine most definitely does have an argument against the linguistic doctrine that appears immediately before the questionable passage just discussed:

> The considerations which were adduced in §I, to show the naturalness of the linguistic doctrine, are likewise seen to be empty when scrutinized in the present spirit. One was the circumstance that alternative logics are inseparable practically from mere change in usage of logical words. Another was that illogical cultures are indistinguishable from ill-translated ones. But both of these circumstances are adequately accounted for by mere obviousness of logical principles, without help of a linguistic doctrine of logical truth. For, there can be no stronger evidence of a change in usage than the repudiation of what had been obvious, and no stronger evidence of bad translation than that it translates earnest affirmations into obvious falsehoods. (Quine 1963, 389–90)

In assessing this argument, we may set the truth of the final two sentences temporarily aside and concentrate first on the structure of the argument as a whole. The linguistic doctrine is said to be empty, but only in the sense of explaining no more than does the obviousness of logical principles. So it is somewhat surprising when Quine says a few lines later, "I have been using the vaguely psychological word 'obvious' non technically, assigning it no explanatory value" (1963, 390). Surely, there can be no more than stylistic differences between saying that the circumstances are adequately *accounted for* by the obviousness of logical principles and saying that they are *explained* by such obviousness. Apparently either Quine has forgotten what he wrote only two paragraphs earlier, or his claim that he assigns 'obvious' no explanatory value is hyperbole for the idea that the circumstances from §1 that he used it to explain are so unimportant as to be negligible. In effect, this turns 'no explanatory value' into 'virtually no explanatory value'.

At last, we can return to the question, so central to Quine's argument, of whether it is really true that the two circumstances of §1 "are adequately accounted for by the mere obviousness of logical principles, without the help of a linguistic doctrine of logical truth." If elementary logic were obvious in the sense earlier called Platonic, then there would indeed have to have been a change of usage when someone announces logical principles apparently at variance with our own. If there is some power that causes us to judge rightly and if we have no reason to think that person blind or otherwise deficient in this power, then we must be only words apart. Similarly, there could no more be prelogical peoples than there could be a people unable to detect the sourness of a lemon once they had bitten into one. But, of course, Quine would deny that logic is obvious in this sense.

Logic is obvious in Carnap's internal sense, too, but that presupposes the linguistic doctrine. The internal obviousness of (classical) elementary logic does *not* for Carnap guarantee that there will not be logicians or earlier peoples whose logic is very different from our own. But it does guarantee that it would be inappropriate to translate what they earnestly affirm as, say, a simple contradiction. This is because translation is supposed to preserve the consequence relation, which the cases at hand would presumably not. Carnap could admit as well that if elementary logic were obvious in the external sense, which it is not, then the circumstances of §1 could also be explained. This, however, is because if logic were obvious in this sense we would all (including would-be dissident logicians and "prelogical" peoples) be in frameworks such that the same logic would be obvious in the internal sense. And of course this, too, presupposes the linguistic doctrine.

What of the senses of 'obvious' that Quine can accept? Suppose that a belief in the classical first-order predicate calculus with identity (interpreted extensionally) is firmly accepted or very widespread but that no assumption is being made that this is due to a power comparable to observation that everyone has for divining such truths. Why must a dissident logician really be in agreement with us? If that is all that obviousness comes to, then there seems to be no reason at all. Remember that obviousness is something that a doctrine can acquire or lose. What is firmly accepted by oneself or even by all one's immediate colleagues is often firmly rejected by others. What is obvious to us may be quite otherwise to someone else. The world may seem obviously flat or to be obviously orbited by the sun. It would be a grave mistake to rule out an erstwhile Copernicus by semantic sleight of hand. Since obviousness can be gained or lost, why should there not have been a people before that blessed day when classical logic became obvious? Why should there not be a post-(classical)-logical people? The same sort of comments, questions, and objections are raised by all the readings of 'obvious' earlier seen as available to Quine. If classical elementary logic is obvious only in the sense of being sufficiently behind the moving front of science as to have lost all traces of its history or as to be no longer a serious candidate for revision, then this is even more clearly tied to our time, place, and scientific group. It says little about what the translation practices should be concerning dissidents or people outside the group. When 'obvious' is interpreted in ways open to Quine, the claim that logic is obvious has almost as little explanatory power as his more hyperbolic statements would suggest.

The claim that elementary logic is obvious gets its apparent invulnerability from the variety of senses in which it can be meant and the variety of points of view from which it can be defended. It gets its power from not keeping those senses distinct and those different points of view firmly in mind. From Quine's point of view, the claim that elementary logic is obvious can be defended only in a fairly narrow range of senses of the word 'obvious'. Unfortunately for his argument, in these specific senses the claim that elementary logic is obvious does not have the consequences that he claims it to have and that his argument needs it to have. The claim may, indeed, have the needed consequences, but only in senses outside this narrow range. That fact gives the claim its apparent power but can offer no aid or encouragement to Quine's own argument.

A Quinean Reply about Obviousness and Why It Fails

The preceding remarks will be interpreted by some as substantially or even primarily an attack on the notion of obviousness. So interpreted, my remarks may seem to invite the following reply on behalf of Quine: Obvious is obvious; it has no special meaning. The word is taken from ordinary English, where it is used unproblematically and with no special philosophic freight. So understood, (classical elementary) logic is obvious. If one dislikes the notion of obviousness (and Quine can grant that it is pretty feeble), well, the linguistic doctrine is no better. Indeed, any argument against obviousness just shows how empty the linguistic doctrine is.

I think that this reply is indeed how Quine and his defenders would respond to the preceding section. But I also think that it decisively misses the point of my argument. Thus, I shall risk some repetition in order not to be misunderstood.

Though I do not much like Quine's simultaneous use of and refusal to commit himself to the notion of obviousness, I have no quarrel with that notion itself. I do, however, note that it is ambiguous. In one sense a Platonist can say that elementary logic is obvious. But even setting aside the Platonist's account of what makes logic obvious, Quine does not say that logic is obvious in this sense. Obviousness in the Platonist's sense is not a feature that claims even could gain or lose. There is also another sense, in which Carnap can say that classical logic is obvious, but Quine cannot say so in this sense either. Not only does this sense of 'obvious' presuppose the Carnapian picture, which Quine proposes to dispense with, but this sense of classical logic's obviousness does not preclude equal and parallel obviousness for intuitionistic logic. Nor is Quine allowed to say (though in fact he might) that logic is just the sort of thing that needs no justification. Precisely because he wants to emphasize parallels between logic and general theories such as relativity and quantum mechanics, logic will require the same sort of justification that they require. And quantum mechanics is *not* obvious in the sense of requiring no justification. To be sure, Quine works in the coherentist tradition so that the justification will be in terms of the overall contribution that the theory makes to our intellectual economy (which is where simplicity, conservatism, and empirical adequacy come in). So far as one can see, this leaves Quine basically only one sense in which he can say that logic is obvious. This comes to what one might call entrenchment and what I did call being so far behind the lines of the moving front of science that we no longer seriously entertain revisions. In this sense Quine *can* say that elementary logic, extensionally and objectually interpreted, is obvious. Even here, however, one suspects that the claim is at best autobiography, for logic so interpreted is not in the wider community so entrenched that alternatives are not seriously considered.

Having finally arrived at a notion of obviousness that we can attribute to Quine, I have no quarrel with it at all. It is a fine notion. I have no desire to deprive anyone of any notion. What I do want is to challenge cavalier claims about what the obviousness of elementary logic, in Quine's sense, can explain. Quine concedes that the linguistic doctrine can account for the two little stories with which his paper begins. He asserts that the obviousness, in his sense, of elementary logic can account, in particular, for these two stories and, in general, for everything that the linguistic doctrine can. This assertion is repeated but in no way defended by the Quinean reply sketched above. That some theory is deeply entrenched *among ourselves* is no reason at all to think that people in other communities or at other times could not possibly disagree with us. It is, I think, highly doubtful that Quine could ever make good on either the particular or the general promissory notes about what obviousness can explain. And this is why his defenders are likely, when challenged, to deny that he ever meant anything at all by 'obvious'.

Convention in Logic and Set Theory

Beginning with §4 Quine turns to the broad theme of convention in logic. He himself, however, does not think he has changed topics, for he says, "The linguistic doctrine of logical truth is sometimes expressed by saying that logical truths are true by linguistic convention" (1963, 391). It

THE LINGUISTIC DOCTRINE AND CONVENTIONALITY 247

must be emphasized at the outset that 'true by convention' is a phrase that Carnap himself never used. Carnap called his view in the 1920s "critical conventionalism" (Richardson 1998, 159–82), but in that period he would not have counted logic as conventional. From the beginning of his mature philosophy in about 1932 the term 'convention' is rarely used and then very cautiously. Usually Carnap's remarks tend to narrow the domain of what some may have thought to be conventional. He does say that tolerance itself could be termed the principle of the conventionality of language forms (Carnap 1942, 247), that is, the idea that the choice among alternative linguistic frameworks is conventional. Accepting a framework commits one to accepting various statements, and these Carnap would call true in virtue of meaning. This does not imply that there is, in modern terms, some proposition whose truth depends on our choosing, say, a true proposition such that, had we chosen otherwise, that selfsame proposition would have been false. I think that Quine (who does not much like propositions) understands this point very clearly, but some of his modern readers have not. So talking about convention can be misleading, and talking about truth by convention can be very misleading. But tolerance is at the center of Carnap's philosophy, and convention is involved. So let us grant Quine the phrase 'true by convention' but resolve not to be misled by it.

Whether elementary logic is true by convention is the first of two related themes of §4; the historical roots of conventionalism in non-Euclidean geometries is the other. Concerning elementary logic, Quine's central claim is that

> it is impossible in principle, even in an ideal state, to get even the most elementary part of logic exclusively by the explicit application of conventions stated in advance. . . . Briefly the point is that the logical truths, being infinite in number, must be given by general conventions rather than singly; and logic is needed then to begin with, in the metatheory, in order to apply the general conventions to individual cases. (1963, 391–92)

In short, you need logic to get logic. This claim would be false, as would be the passage it abbreviates, were the word 'logic' taken in *each* of it occurrences to mean what Quine usually means, namely, classical first-order predicate logic with identity. What Quine must mean is that you need logic of *some* kind to get logic of *any* kind, though the logic you need and the logic you get may be different. In any case, Quine concludes from this that speaking of logic as conventional throws little light on the linguistic doctrine.

Setting aside his conclusion for the moment, Quine's claim that you need logic to get logic is, when properly understood, perfectly true and convincing—as long as we remember that it concerns only what can be generated explicitly by someone with no logic of any kind, not even in a metalanguage. Carnap can agree, moreover, that without some logical structure in the metalanguage, there could be no possibility of an explicit statement of a convention generating infinitely many sentences at the object level. But a metalanguage without such structure would, for Carnap, be no language at all, and hence Quine's claim would be little more than an instance of the more general truth that without a language you cannot say much of anything.

All of this is fine, but it is also utterly irrelevant to Carnap's claim that logic is conventional. Carnap's claim is not about origins; it concerns, rather, what is available right now. Using the resources we now have, we can describe our current logical structures for the object level and describe alternatives to them as well. These various alternatives are in no sense equivalent. That we consider such logics inequivalent is shown in part, as in Quine's opening story, by our refusal to translate the one into the other. Moreover, according to Carnap, any attempt to argue that one of these is the uniquely correct logic, whether on the basis of many facts or of none, is at best circular because the notions of both argument and fact presuppose one of the logics at issue (Goldfarb and Ricketts 1992, 65). In this sense, which is of course very different from Quine's, Carnap, too, can insist that you need logic to get logic.

Likewise, Carnap's conception of what it is for logic to be conventional is very different from the conception against which Quine seems to be arguing. We might say that for Carnap, our logic is conventional because we could now abandon that logic in favor of some other without having to justify that change save in terms of practical utility. Putting this in Quine's phrasing, we *now* have "deliberate choice unaccompanied by any attempt at justification other than in terms of elegance and convenience" (Quine 1963, 396).

Because Carnap is not interested in origins, the question of what can be generated explicitly is of little concern to him. But *should* he care? If conventions for logic cannot arise, how can logic be conventional now? Quine's argument, however, does not in fact show that conventions for logic cannot begin. At most, it shows that they cannot arise by explicit statement under certain conditions. Could conventions emerge inexplicitly? Not long after "Carnap and Logical Truth" was published, David Lewis (1969) showed that the answer is yes in a sense of 'convention' relevantly similar to Carnap's. Quine considered this possibility more seriously in "Truth by Convention" (1936), the paper from which this argument is drawn. We might adopt the conventions "through behavior," he said, leaving the explicit statement thereof for a later, more sophisticated stage when a suitable logic is in place. As he put it,

> It may be held that the verbal formulation of conventions is no more a prerequisite of the adoption of the conventions than the writing of a grammar is a prerequisite of speech. . . . So conceived the conventions no longer involve us in a vicious regress. . . . It must be conceded that this account accords well with what we actually do. (1936, 123)

These concessions make no appearance in "Carnap and Logical Truth," but in both papers he dismissed the possibility as lacking in explanatory force. In calling logic conventional, however, Carnap surely attempted no explanation of anything, at least not in a causal or other standard scientific sense of 'explanation'. Nor is it obvious that he should have. The demand for explanation in any other sense is far from clear and seems to beg the question against Carnap.

I have argued elsewhere that Quine's first formulation of this argument in "Truth by Convention" was put forward somewhat tentatively, that Quine did not intend it to be decisive against Carnap (Creath 1987). By the time of "Carnap and Logical Truth," however, Quine's position had hardened. The argument had not changed, but his background beliefs had, and hence his tone changed as well. It is the tendency of our present reflections that Quine's argument against the conventionality of elementary logic here is far from decisive and hence adds nothing substantive to his arguments against the linguistic doctrine, arguments that have come to seem rather thin indeed.

At the end of §3, Quine remarked favorably on the linguistic doctrine as applied to set theory, and he does the same now for conventionalism as applied to set theory. This topic, however, is temporarily set aside in favor of an impressively concise reconstruction of complex historical events surrounding the development of non-Euclidean geometries. Here, he thinks, are to be found the roots of conventionalism in the philosophy of mathematics. Quine invites us to view the geometric postulates as never both true and conventional. He does not put this forward as an argument that this is the only plausible reading of the history or that historical events have philosophic consequences (though the suggestion is certainly there for the taking). As I said, it is more of an invitation.

I, in turn, shall not argue that Quine is wrong either in his reading or in the implicit suggestion. This is not because I think he is right but because such an argument would be much too large an undertaking. Quine's discussion depends crucially, both for its content and its plausibility, on a distinction between interpreted and uninterpreted expressions. This seems both harmless and reassuringly familiar. Certainly, some such distinction is in order. Quine is undoubtedly right that expressions utterly uninterpreted are not plausibly called true. Whether what are sometimes called uninterpreted systems—especially either those of David Hilbert or where "implicit definition" is claimed—are really wholly uninterpreted is another matter. In any case, the exact meaning of Quine's distinction here is sufficiently unclear that it would be difficult to specify the details of his historical claims or to assess their ultimate plausibility. It is important that the applicability of the historical discussion to Carnap is doubtful because he develops mathematical geometry not as a set of uninterpreted postulates but, rather, within a system of ordered n-tuples of what are in effect real numbers. We could, of course, try to reconstruct the historical events ourselves and assess the role of convention therein, but this would tell us little about Quine. Thus, I shall bracket Quine's historical discussion, treating it as chiefly enriching his other arguments, which stand or fall by themselves.

Returning, then, to set theory in §5, Quine reiterates the conventionality he sees:

> [In set theory] We find ourselves making deliberate choices and setting them forth unaccompanied by any attempt at justification other than in terms of elegance and convenience. These adoptions, called postulates, and their logical consequences (via elementary logic), are true until further notice.
>
> So here is a case where postulation can plausibly be looked on as constituting truth by convention. (1963, 393–94)

That he calls the postulates true is puzzling, and adding 'until further notice' is doubly so. What shall we say at the time, if any, of that further notice? Shall we say that what was true is no longer so? But the sentences in question are presumably eternal sentences, always true if ever true. Carnap might say such things because on his view, adopting the postulates might endow the expressions with certain meanings, and abandoning the postulates would be to change the meanings. But this line is not available to Quine. In changes of ordinary scientific theory, Quine would say simply that what was thought to be true is later thought to be untrue and that we were wrong in one of the cases. He is very shortly to argue that the cases of set theory and natural science are very much alike. It strains credulity to think that Quine just meant 'thought to be true' but wrote 'true' in a moment of telegraphic concision. Quine would have been too sensitive to the importance of just that difference at just this moment. Nonetheless, it is equally difficult to find any other way to reconcile Quine's various commitments here.

There is one other feature of the above quotation that is worth noting, and that is the shift from the use of 'postulates' to the use of 'postulation', that is, from the sentences thus instituted to the act of instituting them. The change is permanent for the remainder of Quine's discussion. Thus, when he develops a useful distinction between legislative and discursive postulation, it is of the act that he speaks and not of the sentences or enduring beliefs. He also develops a parallel distinction between legislative and discursive definition that need not concern us here, but it, too, pertains to the act. It is only legislative postulation that institutes truth by convention. On this basis he concludes, "So conceived, conventionality is a passing trait, significant at the moving front of science but useless in classifying the sentences behind the lines. It is a trait of events and not of sentences" (1963, 395). Note, however, that the claim that conventionality is a trait only of events is underwritten by no argument specifically about conventionality; rather, it is facilitated by the prior choice of postulation rather than postulates as a topic. The only possibility that Quine entertains of projecting the trait upon the sentences is to speak of a "sentence as forever true by convention if its first adoption as true was a convention" (1963, 395). This he quite rightly rejects as involving the most unrewarding historical conjecture.

It would not be unfair at this point to suggest that Quine's discussion is, for want of a better phrase, guided by a certain picture of how our beliefs are to be justified, or alternatively put, a picture of what a general epistemology would look like. Put extremely schematically, a belief system, for Quine, is best represented by a set of sentences, those that the believer is disposed to assent to. Justification in turn concerns only how (for what causes) new sentences are added to the set or old ones deleted from it and hence concerns only the moving front of science. There are causal relations among the various assent dispositions, and none is ever immune from revision. But there are no genuinely epistemic or justificatory relations among the sentences "behind the lines" of the moving front of science. This picture was to be modified somewhat over the years, especially in the change from the radical holism of "Two Dogmas of Empiricism" (1951) to the more modest holism of *Pursuit of Truth* (1990). Insofar as one can tell, however, the picture that guides "Carnap and Logical Truth" is the unmodified one.

I shall not say here that Quine's picture is wrong, but Carnap's guiding image of the epistemic situation is very different. Many of the differences between them over the linguistic doctrine and the conventionality of logic, set theory, and the like can be traced to such differences of basic approach. Put informally and slurring over important details, Carnap's basic picture holds that logical, mathematical, and other analytic sentences that we accept do not need (epistemic) justification, that is, some sort of argument that we have chosen aright. Indeed, these sentences are presupposed by the very procedures of justification. There is a sense in which not all systems are equally good, differing as they do in simplicity, ease of use, and the like. But such practical considerations do not justify a given system in the sense of providing grounds for thinking that it is the uniquely correct one. Carnap would count this picture itself as fitting into the category of the analytic; more precisely, it is a proposal that if adopted would be analytic. So it, too, needs no justification. Of course, if someone were to take a different picture, we would have to allow that person to do so, requiring of them no justification either. This is the point of the principle of tolerance.

The nonanalytic claims of empirical science, by contrast, do require justification by the observational evidence. The amount of this justification, at least for nonobservational claims, is best represented by the degree of confirmation by the evidence thus far. That confirmation will be less than one and greater than zero in a system of probabilities, and it may be quite small when a hypothesis is first considered. The confirmation of individual claims will continue to fluctuate as new observational evidence continues to flow in. In consequence, some of our beliefs are now well confirmed by the evidence; other are less so. This is not a psychological claim; it is not, for example, a claim about how unwilling we may be to give up one belief or another. Moreover, it has nothing to do with how the beliefs were first acquired or how long ago; it has to do only with how they now stand with respect to certain abstract, logical relations they bear to the available evidence. This difference in confirmation among our beliefs, even if well behind the lines, is important but difficult to represent in Quine's picture, especially in its radical holist version. Even at the level of empirical hypotheses, Carnap wants to distinguish between logical (epistemic) considerations and practical ones such as ease of use. The former are relevant to truth and justification, and the latter are not.

As noted, one can, on Carnap's account, be guided by a general picture of the epistemic situation without having to justify that picture, and so he need not attempt anything of the sort for himself. Quine, by contrast, must treat such "guiding pictures," quantum mechanical claims, and reports of planets around other stars as all on a par. For Quine, there is a fact about all these matters, and the claims are not to be admitted into the body of our belief without the same sort of explicit and detailed justification that physicists provide. That can be indirect, but it needs to be given explicitly. The salience of this need for explicitness rises with the remoteness of the claim from the observational evidence. Presumably, Quine would reckon guiding pictures, including his own, as pretty far from direct experience. Thus, what is needed for Quine's picture is something along the lines of what is offered in the scientific literature for quantum mechanical or astronomical hypotheses.

I have belabored the differences between the underlying approaches of

Carnap and Quine as well as between what each picture commits its holder to in the way of justifying that picture. I have done so in order to underscore the fact that nowhere in "Carnap and Logical Truth" and nowhere in any publication before it does Quine make any attempt to argue for his picture of the epistemic situation. Certainly, the picture is not entailed by Duhemian considerations alone. Quine's guiding image may be attractive, and it is certainly permissible to presume it rather than to argue for it. To do so, however, is to forgo the claim to be making an *argument* against Carnap rather than to be presenting an alternative to him.

Quine opens §6 by remarking that obviousness, earlier attributed to elementary logic, is an aspect that set theory could one day acquire. As we discussed earlier, obviousness, in Quine's sense, is a feature that sentences have just by being sufficiently behind the lines of inquiry. In becoming obvious, however, set theory would cease to be conventional. Thus, from a sufficiently ahistorical point of view, the situations of elementary logic and of set theory are very much on a par. Elementary logic could once have been conventional (though Quine suggests that it was never so), and it could once again become so, should dissident logicians become more influential. Correspondingly, as noted, set theory could become obvious. The two domains are now, in fact, in different positions with respect to the moving front of science, but that is at best a historical contingency. With the caveat that he embraces neither Quine's conception of convention nor the guiding picture that seems to be behind it, Carnap would find nothing in Quine's remarks with which to disagree.

Quine then summarizes what he thinks he has established so far: The linguistic doctrine of elementary logical truth is without substance, as is the claim that such logic is true by convention. The notion of truth by convention attaches only to legislative postulation, and this process is now to be found in set theory.

But Quine goes on:

> And do we not find the same continually in the theoretical hypotheses of natural science itself? What seemed to smack of convention in set theory (§V), at any rate, was "deliberate choice, set forth unaccompanied by any attempt at justification other than in terms of elegance and convenience"; and to what theoretical hypothesis of natural science might not this same character be attributed? For surely the justification of any theoretical hypothesis can, at the time of hypothesis, consist in no more than the elegance or convenience which the hypothesis brings to the containing body of laws and data. (1963, 396)

This last sentence is presented as though it were obvious. We struggled with obviousness earlier on, but this claim does not seem a likely candidate for being so far behind the lines as to be no longer a serious candidate for revision. This passage may express Quine's general epistemic picture, but he can hardly expect Carnap to accept it.

The justification of a theoretical hypothesis, whether at the time of hypothesis or at any other, consists, for Carnap, not in elegance or convenience but in certain logical relations to evidence. If at the time of hypothesis, there is no relevant evidence, then it is true that there are only a priori considerations like those at play in logic and set theory. Informally, one might say that in such a case, the hypothesis and the sentences of logic would be alike in having no justification. But this gives no aid or comfort to the idea that logic and the hypotheses of natural science are really on a par. The logic has no justification because it needs none. For the hypothesis, there is simply no relevant evidence at the moment. One might wonder why, in such circumstances, bother with the hypothesis? But at least the hypothesis could acquire the justification it needs. As relevant evidence comes in, its degree of confirmation will change, and that can continue to fluctuate. Not so with the statement of logic.

Quine would demur:

> The situation may seem to be saved, for ordinary hypotheses in natural science, by there being some indirect but eventual confrontation with empirical data. However, this confrontation can be remote; and, conversely, some such remote confrontation with experience may be claimed even for pure mathematics and elementary logic. . . . For a self-contained theory which we can check with experience includes, in point of fact, not only its various theoretical hypotheses of so-called natural science but also such portions of logic and mathematics as it makes use of. (1963, 397)

Again, Quine can hardly expect Carnap to accept this. Ordinary confirmation can be remote, but whether to say that logic and mathematics are confirmed is precisely the issue. For Carnap, they are already presupposed in confirmation and indeed in the very empirical hypotheses themselves. Logic and mathematics are without empirical content and when conjoined to ordinary hypotheses leave the consequence class thereof unaltered. Thus, logic and mathematics are not confirmed by experience, even remotely. This, in turn, is not an argument against Quine and is not intended as such. But it is a reminder that what Quine says is not obvious in his sense or in any other. Apparently, we need an argument, but he offers none.

It may seem that I am unfair to Quine in demanding an argument from him but not from Carnap. The charge would be unfounded, for I make no such demand. I note only that Quine's position seems to demand it of Quine. Carnap's position, conversely, does not demand it of Carnap, and Carnap's position demands no argument from Quine, either. Naturally,

Carnap would be happy to treat Quine's statement, unaccompanied as it is by attempt at justification, as a legislative postulation.

Of course, Quine knows perfectly well that Carnap would not accept his bald assertion. So Quine may be doing something more subtle. He may be trying to provoke Carnap into a counterargument, into an argument that Quine's picture is mistaken. Had Carnap attempted to make such an argument, it would have been to concede, however tacitly, that there was a fact of the matter. Carnap made no such attempt, nor shall I.

Quine concludes this section (§6)—and indeed, the whole main argument of the paper—with the remark that "the very distinction between a priori and empirical begins to waver and dissolve, at least as a distinction among sentences" (1963, 397). Indeed, in his picture, it has dissolved. But in our examination of the first six sections of Quine's paper, what has also dissolved is the idea that he has presented any *argument* against Carnap. When we looked at Quine's discussion of what he called the linguistic doctrine of logical truth, we saw that it hinged on what Quine might mean by the word 'obvious'. When the senses of that word available to Quine were sorted out, we could see that far from having an argument against the doctrine, Quine had at most provided some weak support for it in the form of the stories about translation with which he began. When we turned to the conventionality of ordinary logic, we found that Quine's famous argument against this is irrelevant to Carnap's conception of conventionality. Finally, in examining the conventionality of set theory and the possibility that it differs from ordinary empirical science therein, we found Quine consistently to be presupposing a guiding picture about the epistemic situation. This picture is clearly different from Carnap's. But Quine's picture was at most expressed without any attempt at the justification it would need. Thus, it at most begs the question against Carnap. Of course, that fact, if it is a fact, hardly shows that Carnap is right or that Quine is wrong in their own views. But it may provide grounds for continuing to explore both guiding pictures and to use both as we try to find our philosophical way.

Note

I would like to thank the National Science Foundation (Research Grant Number SBR-9515398) for research support and Arizona State University for research and sabbatical leaves while this was written.

References

Carnap, R. 1934. *Logische Syntax der Sprache*. Vienna: Springer. Translated as *The Logical Syntax of Language*. Trans. A. Smeaton. London: Routledge and Kegan Paul, 1937.

———. 1942. *Introduction to Semantics*. Cambridge, Mass.: Harvard University Press.
———. 1956. "Meaning and Synonymy in Natural Languages." In *American Philosophers at Work: The Philosophic Scene in the United States,* ed. S. Hook, 58–74. New York: Criterion Books.
———. 1963. "W. V. Quine on Logical Truth." In *The Philosophy of Rudolf Carnap,* ed. P. A. Schilpp, 915–22. La Salle, Ill.: Open Court.
Creath, R. 1987. "The Initial Reception of Carnap's Doctrine of Analyticity." *Nous* 21: 477–99.
———. 1991. "Every Dogma Has Its Day." *Erkenntnis* 35: 347–89.
———. 1994. "Functionalist Theories of Meaning and the Defense of Analyticity." In *Logic, Language, and the Structure of Scientific Theories,* ed. W. Salmon and G. Wolters, 287–304. Pittsburgh: University of Pittsburgh Press; Konstanz: Universittsverlag Konstanz.
Goldfarb, W., and T. Ricketts. 1992. "Carnap and the Philosophy of Mathematics." In *Science and Subjectivity: The Vienna Circle and Twentieth Century Philosophy,* ed. D. Bell and W. Vossenkuhl, 61–78. Berlin: Akademie Verlag.
Lewis, D. 1969. *Convention*. Cambridge, Mass.: Harvard University Press.
Quine, W. V. O. 1936. "Truth by Convention." In *Philosophical Essays for A. N. Whitehead,* ed. O. H. Lee, 90–124. New York: Longmans.
———. 1951. "Two Dogmas of Empiricism." *Philosophical Review* 60: 20–43.
———. 1963. "Carnap and Logical Truth" (1954). In *The Philosophy of Rudolf Carnap,* ed. P. A. Schilpp, 385–406. La Salle, Ill.: Open Court.
———. 1990. *Pursuit of Truth*. Cambridge, Mass.: Harvard University Press.
Richardson, A. 1998. *Carnap's Construction of the World*. Cambridge, Mass.: Cambridge University Press.
Stein, H. 1992. "Was Carnap Entirely Wrong, After All?" *Synthese* 93: 275–95.

11
Languages and Calculi

W. V. O. Quine prefaces "Carnap and Logical Truth" (1963; written in 1954) with a confession of sorts: "My dissent from Carnap's philosophy of logical truth is hard to state and argue in Carnap's terms. This circumstance perhaps counts in favor of Carnap's position" (1963, 385).[1] Quine's hesitancy is well placed. Here and in "Two Dogmas of Empiricism" (1953d), Quine reviews explanations of analyticity in terms of convention, in terms of legislative postulation, and in terms of a linguistic doctrine of logical truth. Rudolf Carnap, in his reply to Quine, rejects these candidate explanations. He recognizes his own view in none of them. In both essays, Quine brings Pierre Duhem's point about theory testing to bear on accounts of analytic truth in terms of empirical meaning. Carnap is unmoved. In §82 of *The Logical Syntax of Language* (1937), he has applied his understanding of logical truth to develop an account of theory testing built around Duhem's point.

Finally, the issue appears to be joined in §4 of "Two Dogmas," where Quine takes up the explanation of analyticity implemented in Carnap's technical work, analyticity as truth guaranteed by the semantic rules of a formal language. Quine says of Carnap's formal work,

> Obviously any number of classes K, M, N, etc. of statements of L_0 can be specified for various purposes or for no purpose; what does it mean to say that K, as against M, N, etc., is the class of the 'analytic' statements of L_0? (1953d, 33)

A couple of pages later, Quine adds,

> Appeal to hypothetical languages of an artificially simple kind could conceivably be useful in clarifying analyticity, if the mental or behavioral or cultural factors relevant to analyticity—whatever they may be—were somehow sketched into the simplified model. (36)

Carnap responds to these points in "Quine on Analyticity" (1990; written in 1952). The manuscript begins,

> It must be emphasized that the concept of analyticity has an exact definition only in the case of a language system, namely a system of semantical rules, not in the case of an ordinary language, because in the latter the words have no clearly defined meaning. (1990, 427)

He goes on to charge Quine with obscurity, "since it is not clear whether [Quine] is asking about the elucidation explicandum 'analytic' or about an explicatum" (430). The latter, the explicatum, is the definition of L-truth given by the semantic rules for a formalism. As Carnap views matters, a definition of L-truth for a formal language explicates—makes precise, replaces for scientific purposes—the informal notion of analyticity in application to that language. As Quine has not charged Carnap's work with any technical inadequacies, it must be the explicandum, the informal notion of analyticity, whose obscurity Quine finds objectionable. Although Carnap finds the informal notion serviceable, he freely concedes its vagueness, and he holds forth no prospect of serious clarification by the use of notions like understanding and meaning, themselves inexact. Indeed, the vagueness of the informal notion motivates Carnap's development of mathematically precise characterizations of L-truth for formal languages.

Quine's point in §4 of "Two Dogmas," then, must appear to Carnap to be unproductive burden shifting. Quine asks for a clarification of analyticity, and Carnap mentions semantic rules. Quine then complains of unmotivated precision, and Carnap talks about truth solely in virtue of meaning. Quine asks for a clarification.

I shall argue that this unsatisfying standoff is, to a considerable extent, the product of Carnap's and Quine's different understandings both of logical notation and of the application of logic to the language of science.[2] This difference is in turn linked to Carnap's and Quine's disagreements over the philosophy of mathematics.

Following Burton Dreben, I hold that the Carnap-Quine analyticity debate is centrally a debate about the philosophy of mathematics.[3] The Carnap-Quine exchanges on analyticity, however, obscure the centrality of mathematics to their debate, for these exchanges focus on broad-gauge analyticity, on meaning postulates, on statements like "All bachelors are unmarried." In *The Logical Syntax of Language,* Carnap develops a philosophy of mathematics that he basically adheres to for the rest of his career: Mathematical truths are analytic, formal truths—notational auxiliaries to the sentences of substantive, empirical science. Nothing Quine says shakes Carnap's confidence in his account of the analyticity of mathematics. This account is not, however, readily extendable to the sentences Carnap dubs "meaning postulates." Carnap thus thinks that he can rebut Quine's critique by explaining what analyticity is for these sentences. Quine's view of the situation complements Carnap's. Quine finds Carnap's *Logical Syntax* account of the analyticity of mathematics devoid of philosophical interest on account of the strength of the metalanguage in which the account is given. But at least when discussing mathematics, we can identify the truths whose status is under discussion. This may deceive us that we understand the alleged special status. By focusing on meaning postulates, Quine hopes to highlight the utter vacuity of Carnap's general conception of analyticity.

In order to get at the difference about logic and logical notation that I believe underlies the analyticity debate, I propose to put broad-gauge analyticity to one side, and with it Carnap's treatment of modality and his application of semantics to inductive logic. Working from within Carnap's viewpoint, I shall discuss his account of analytic truth as formal truth guaranteed by language and the application of this account to mathematics. I first present Carnap's view of the relationship between calculi (formal languages) and hypothetical or actual used languages in the context of his project of *Wissenschaftslogik,* the logic of science. Next, I take up Carnap's conception of analyticity and its application to mathematics, first in *The Logical Syntax of Language* and then in his semantics period. And finally, I explore the differences between Carnap and Quine on logic and logical notation.

Calculi and Used Languages

In an address to the International Congress for the Unity of Science held in Paris in 1935, Carnap describes three stages in the emergence of scientific philosophy. The first, Kant's contribution, is the rejection of speculative metaphysics in favor of theory of knowledge. The second stage is the rejection of the synthetic a priori in favor of a thoroughgoing empiricism. The third and final stage is the current task for scientific philosophy. Carnap says,

> It seems to me that *theory of knowledge* is in its previous form *an unclear mixture of psychological and logical elements*. That holds as well for the work of our circle, not excluding my own earlier work. There thus arises much unclarity and misunderstanding. From these we see how important it is in so-called epistemological discussions to be explicit as to whether logical or psychological questions are meant.
>
> If we assign synthetic, empirical sentences to factual science *[Realwissenschaft],* then the psychological questions of theory of knowledge belong not in philosophy but in factual science. There remains as the genuine task of philosophy the logical analysis of knowledge—of scientific sentences,

theories, and methods—hence, the *logic of science (Wissenschaftslogik)*. (Carnap 1936, 36–37; see also Carnap 1987, 46–47)

Carnap identifies the logic of science with the syntax of the language (or languages) of science. In §2 of *The Logical Syntax of Language,* Carnap characterizes syntax as the study of calculi that

> is concerned with the structure of possible serial orders (of a definite kind) of any elements whatsoever. Pure syntax is concerned with the possible arrangements without reference either to the nature of the things which constitute the various elements or to the questions as to which of the possible arrangements of these elements are anywhere actually realized. . . . Pure syntax . . . is nothing more than combinatorial analysis, or, in other words, the geometry of finite, discrete, serial structures of a particular kind. (1937, §2, 6–7)

The core of a syntactic description of a calculus is a specification of its formulas via formation rules and a description of a consequence relation over these formulas via transformation rules. The content of the 'combinatorial analysis' that syntax delivers is clarified when, thanks to Kurt Gödel's work, we appreciate that syntax can be interpreted in arithmetic so that "[t]he definitions and sentences of syntax arithmetized in this way do not differ fundamentally from the other definitions and sentences of arithmetic" (Carnap 1937, §19, 57).[4] *Wissenschaftslogik* is an application of pure syntax: It is the syntactic description, investigation, and comparison of calculi suitable for use in science.

Just here, Carnap's separation of the psychological from the logical, applied now to language, raises questions. What relevance can the mathematical study of the structure of calculi have to the statements and theories propounded by science? Carnap's answer in §2 of *Logical Syntax* is that languages are instances of calculi: The sentences of a language are concrete instances of the "possible serial orders" whose structure pure syntax studies. But what are languages, as Carnap conceives of them? What makes a language, so conceived, an instance of a particular calculus?

In the early 1930s, influenced by Otto Neurath and perhaps in reaction to some features of Ludwig Wittgenstein's *Tractatus,* Carnap adopts a naturalistic, broadly behaviorist approach to language. In *The Foundations of Logic and Mathematics,* Carnap says:

> A language, as, e.g., English, is a system of activities or, rather, of habits, i.e., dispositions to certain activities, serving mainly for the purposes of communication and of co-ordination of activities among the members of a group. (Carnap 1939, §2, 3; see also Carnap 1942, §1, 3; 1937, §2, 5)

Carnap's treatment of the relationship between calculi and languages in his syntax period writings is very cursory, and things do not change after the shift to semantics. Pure syntax studies calculi in abstraction from any spatiotemporal series that may realize them. Descriptive syntax is concerned with the syntactic description of "empirically given expressions" of used languages.[5] Carnap compares the relationship between pure and descriptive syntax to that between mathematical and physical geometry. We can correlate the expression types of a calculus with orthographic or phonemic utterance types (see Carnap 1937, §25, IIB, 79–80). Such rules are parallel to Hans Reichenbach's coordinating definitions that give physical significance to the vocabulary of unapplied, pure mathematical geometry, for the correlation of expression types to utterance types gives empirical content to the claim that an actually uttered sentence is a formula of the calculus or is a valid formula of the calculus. We can also exploit such a correlation to envision the use of a language whose sentences are the formulas of some particular calculus.

In *Logical Syntax,* Carnap speaks casually of a calculus "being in agreement with the actual historical habits of speech" of a linguistic community (1937, §62, 228). It is fairly clear what this comes to as regards the formation rules of a calculus: Roughly speaking, with respect to a correlation of expression types with utterance types, it turns out that the sentences uttered in the linguistic community are formulas of the calculus.[6] However, Carnap gives no indication as to when the transformation rules of a calculus are in agreement with the speech habits of the speakers of a language.[7] Presumably, agreement between speech habits and transformation rules is a matter of speakers' standing dispositions to accept some sentences and to reject others, together with their dispositions to infer certain kinds of sentences from certain other kinds. Such dispositions will not, it appears, very closely constrain the calculi that agree with them, at least when we are dealing with calculi of interest for *Wissenschaftslogik*. To explain this point, I need to say something about Carnap's transformation rules.

The transformation rules specify a consequence relation over the sentences of the language. There are two kinds of transformation rules and so two kinds of valid sentences. The L(ogical)-rules establish a logical consequence relation. The P(hysical)-rules extend that relation via the selection of sentences as P-axioms. The P-rules thus formalize in the language some body of empirical theory. Carnap maintains that in devising a calculus to serve as a language of science, it is a matter of convention how much of accepted science is built into the calculus by P-rules:

> We may, however, also construct a language with *extra-logical rules of transformation*. . . . Whether in the construction of a language we formulate only

L-rules or include also P-rules, and, if so, to what extent, is not a logico-philosophical problem, but a matter of convention and hence, at most, a question of expedience. If P-rules are stated, we may frequently be placed in the position of having to alter the language; and if we go so far as to adopt all acknowledged sentences as valid, then we must be continuously expanding it. But there are no fundamental objections to this. (1937, §51, 180)

So if one calculus is in accord with speakers' dispositions to affirm what the calculus labels descriptive sentences, other calculi that supplement or diminish the P-valid sentences of the first will accord with those dispositions equally well.

What about L-rules? What does agreement between the L–consequence relationship of a calculus and speech dispositions come to? Carnap insists that in the calculi that are candidates for use in *Wissenschaftslogik,* the L-rules should mark each mathematical formula of the calculus as L-valid or L-contravalid. That is, the transformation rules should reproduce in Carnap's generous syntactic terms a bivalent true/false distinction over the mathematical formulas of the calculus.[8] It is implausible to hold that a classical mathematician who uses Carnap's Language II, whose mathematical part is Type ω arithmetic, is in any sense disposed to affirm or deny each mathematical sentence of this language. And nothing Carnap says suggests he believes mathematicians are so disposed. Speech habits thus do not fix the L-rules for a calculus instantiated by a language.[9]

We see, then, that as regards transformation rules, the agreement between a calculus and speech habits in virtue of which a language can be taken to instantiate a calculus is rather loose. For a speaker's habits to agree with a calculus, Carnap appears to require little more than that the speaker not be disposed to affirm any contravalid sentence nor to deny any valid sentence.[10] So having coordinated the utterance types of a language with the formulas of a calculus, there will be many distinct consequence relations over those formulas that agree with the speech habits of speakers of the language. Speech habits do not even in principle fix a unique calculus instantiated by the language.[11] If I imagine the speech habits of investigators in agreement with a given calculus, I imagine speech habits that also agree with any number of variant calculi.

How, then, are we to think of the relationship between calculi and actual or hypothetical used languages? The *Wissenschaftslogiker* describes a calculus. Via a correlation of formal expressions of the calculus with utterance types, the consequence relation of the calculus and attendant syntactic distinctions are projected onto the statements of individuals whose speech habits, with respect to this correlation, agree with the calculus in the loose fashion just described. The results of this projection are set forth in the sentences of descriptive syntax that, for example, may describe a specific utterance as an utterance of a contradictory formula of the calculus. The *Wissenschaftslogiker* is not, of course, interested in describing calculi that conform to the speech habits of actual groups.[12] When a calculus is envisioned as a language for science, hypothetical, idealized investigators are imagined whose speech habits agree with the calculus so that their used language may be taken as an instance of the calculus. It is this possibility that gives the study of abstract calculi the desired application in *Wissenschaftslogik*.

We need one more notion in order to apply calculi to *Wissenschaftslogik*, that of an observation predicate. Carnap divides descriptive vocabulary into observation predicates and nonobservation predicates. This classification has to do with the use of language. As a first approximation, an observation predicate of a language is one that speakers of the language, under suitable circumstances, agree in applying or denying to demonstrated items. It thus belongs, Carnap says, "to a biological or psychological theory of language as a kind of human behavior, and especially as a kind of reaction to observation" (1936–37, 454).[13] When a calculus is projected onto the language of an actual or hypothetical investigator, some of the descriptive predicates of the calculus are correlated with observation predicates of the language. Equivalently, in thinking of a calculus as a language for empirical science, we label certain predicates as observation predicates, specifying the observationally ascertainable conditions under which each applies to observationally discriminable items (or locations).

Given the identification of a used language as an instance of a particular calculus, we can apply the vocabulary of logical syntax to describe both the statements an investigator accepts as true and changes in those statements. For example, we can represent a change in an investigator's speech habits as the rejection of a hypothesis on the basis of contradictions between observation sentences logically implied by a theory that includes the hypothesis and observation reports that appear in the investigator's protocol. We can go on to evaluate whether this change restores consistency to the investigator's theory. In this way, the logical syntax of a calculus is imposed like a grid on an investigator's used language, on her speech habits. With this grid in place, we represent an investigator's acceptance and rejection of sentences as the epistemic evaluation of hypotheses. Without this grid, we simply have changes in speech dispositions. In a revealing passage in *The Foundations of Logic and Mathematics* (1939) about the relation of used languages to semantic systems, the successors of the calculi of *Logical Syntax,* Carnap says,

The facts [about linguistic behavior] do not determine whether the use of a certain expression is right or wrong but only how often it occurs and how often it leads to the effect intended, and the like. A question of right or wrong must always refer to a system of rules. Strictly speaking, the rules which we shall lay down are not rules of the factually given language B; they rather constitute a language system corresponding to B which we will call the *semantical system B-S*. (Carnap 1939, §4, 6–7; see also Carnap 1942, §5, 14; 1990, 432, final paragraph)

In all this, we should not assimilate Carnap's conception of descriptive syntax and semantics to contemporary projects in linguistics and psycholinguistics, an assimilation Carnap's rhetoric in places abets. Carnap's pragmatics is the interdisciplinary science of language as a behavioral and biological phenomenon. The pragmatics for a used language will include, I assume, a grammar demarcating the sentences of the language. It will also include descriptions of the application conditions for the observation predicates of the language. To this extent then, pragmatics includes some statements that Carnap counts as parts of descriptive syntax and semantics. However, Carnapian pragmatics, at least as I am construing it, does not include those statements made available by the coordination of a language with a calculus/semantic system of sufficient power to be of interest for *Wissenschaftslogik*.[14] It will not include statements like "Such and so utterance is an utterance of an L-valid formula of system S." Carnap—behaviorist that he is[15]—does not think of such statements as playing any role in the explanation of linguistic behavior. For this reason, I am reluctant to take the loose relation between calculi and used languages to be just another instance of underdetermination of theory to evidence. The projection of calculi onto actual or hypothetical used languages is a stipulation solely for the purpose of *Wissenschaftslogik,* of recasting epistemic evaluations as the syntactic (later semantic) descriptions of sentences—and so of investigators' linguistic behavior—that this projection makes available.[16]

We can now understand better the final purification of philosophy that *Wissenschaftslogik* is to provide. The coordination of calculi and languages yields an understanding of the linguistic activity of scientists as the formulation and empirical testing of theories. In *Wissenschaftslogik,* calculi are devised, investigated, and compared. Via the potential projection of calculi onto languages, old epistemological distinctions are explicated by syntactic surrogates. And for Carnap, in this setting, explication is replacement. The resulting descriptions of scientific language then combine descriptions of the use of language, especially the affirmation and denial of sentences as a response to observation, with the logical overlay provided by syntax.[17]

Analyticity and Mathematics

A central task for *Wissenschaftslogik* is to explicate in logical terms the role of observation in the evaluation of scientific theories. By the early 1930s, Carnap has accepted that the vocabulary of science is not explicitly definable in terms of observation predicates, logic, and mathematics. Testability will, however, remain a matter of logical links between sentences of the calculus and sentences capable of being used to report observations. To give an account of testability, Carnap needs to distinguish within a calculus those sentences that, so to speak, forge these logical links from the sentences whose testability they secure.[18] He needs to elucidate the distinction between the sentences of factual science *(Realwissenschaft)* and the auxiliary sentences of formal science *(Formalwissenschaft)*. In "Formalwissenschaft und Realwissenschaft" (1935), Carnap describes the distinction to be explicated in these terms:

> These auxiliary statements *[Hilfsätze]* have no factual content or, to speak in the material idiom *[inhaltlich gesprochen]*, they do not express any matter of fact, actual or non-actual. Rather they are, as it were, mere calculational devices *[bloße Rechenausdrücke]*, but they are so constructed that they can be subjected to the same rules as the genuine (synthetic) statements. In this way, they are an easily applicable device *[handhabendes Hilfsmittel]* for operations with synthetic statements. (1935, 34; translation, Carnap 1953, 126)[19]

As the use of the term *"Formalwissenschaft"* suggests, Carnap wants the pure mathematical sentences of a calculus to be among notational auxiliaries for operating with the sentences of substantive science. In giving them this status, Carnap will thus explain how pure mathematics, while not itself testable, plays an essential role in observationally testable theories. In this way, Carnap will vindicate an idea from the *Tractatus:*

> In life it is never a mathematical proposition which we need, but we use mathematical propositions *only* in order to infer from propositions which do not belong to mathematics to others which equally do not belong to mathematics. (Wittgenstein 1922, 6.211)[20]

Synthetic sentences are, one might say, the genuine sentences about reality, the ones that represent facts (see Carnap 1937, §14, 41). Carnap, although using such rhetoric, is well aware that such characterizations of the sentences of factual, empirical science are themselves objectionable pseudo-object sentences.[21] He does not himself seriously rest the distinction between synthetic sentences and formal auxiliaries on any "in virtue of" notions, such as "truth in virtue of facts." To do so would sully the purity of *Wissenschaftslogik* with an injection of language-transcendent ontology. Instead, Carnap seeks to explicate and replace the pseudo-object characterization of the difference between factual sentences and notational auxiliaries with a metaphysically innocent syntactic substitute.

In *Logical Syntax,* a calculus includes a consequence relation given by transformation rules. The transformation rules may include P-rules in addition to L-rules. Both kinds of rules are purely syntactic. In Part IV of *Logical Syntax,* entitled "General Syntax," Carnap looks for a way to separate L-notions from P-notions for an arbitrary language based on the consequence relation for the language. This will give him the desired distinction between the sentences of formal science and those of factual science. To this end, Carnap develops a language-general definition of logical versus descriptive vocabulary. With this definition in place, the L-valid sentences are defined as those valid sentences that contain only primitive logical expressions or those valid sentences with descriptive signs that remain valid under substitution of signs for their primitive descriptive signs (see Carnap 1937, §§50–52).

Carnap motivates his syntactic definition of the logical-descriptive distinction with the following remark:

> If a material interpretation is given for a language S, then the symbols, expressions, and sentences of S may be divided into logical and descriptive, i.e. those which have a purely logical, or mathematical, meaning and those which designate something extra-logical—such as empirical objects, properties, and so forth. This classification is not only inexact but also non-formal, and thus is not applicable in syntax. But if we reflect that all the connections between logico-mathematical terms are independent of extra-linguistic factors, such as, for instance, empirical observations, and that they must be solely and completely determined by the transformation rules of the language, we find the formally expressible distinguishing peculiarity of logical symbols to consist in the fact that each sentence constructed solely from them is determinate. (1937, §50, 177)

Carnap begins here with a pseudo-object characterization of the difference between logical and descriptive. He rejects this characterization with one couched in less metaphysical, more epistemic terms: Observation and other 'extra-linguistic factors' have nothing to do with the use of logico-mathematical terms; the use of these terms is, after a fashion, completely fixed by the language. Carnap now seeks to make this vague idea precise in syntactic terms. A sentence of a calculus is determinate if it is either valid or contravalid. Carnap now defines the primitive logical expressions along the following lines: The primitive logical vocabulary is the largest vocabulary of uncompounded, undefined expressions such that there are sentences constructed solely from that vocabulary and any such sentence is determinate. The transformation rules of a calculus thus completely fix the 'meaning'—that is, the significance, the potential use—of the logico-mathematical vocabulary in that they partition the sentences containing just that vocabulary into valid and contravalid. In contrast, whereas the validity of some sentences containing descriptive vocabulary is fixed by the transformation rules, other descriptive sentences are indeterminate.[22]

I have talked about the transformation rules of a calculus *fixing* the validity or contravalidity of various formulas. How does Carnap understand this idea of fixing in *Logical Syntax*? In *Logical Syntax,* Carnap does not think of definability in semantic terms. In particular, there is no general notion of the extension of a predicate, that is, of the class of items of which a predicate is true. For the transformation rules in the syntax language to fix the validity of an object calculus formula is for the validity of that formula to be a consequence in the syntax language of the syntax language definition of validity for the object calculus. So the transformation rules fix the validity of formula s in that, where σ is a canonical name (formal numeral for the Gödel number) of s, the syntax language sentence

$$\sigma \text{ is valid}$$

is itself valid, indeed L-valid, in the syntax language.[23] For validity/contravalidity to be bivalent over the mathematical formulas of the object calculus, the consequence relation in the syntax language must be strong indeed. To build mathematics into the object calculus, the mathematics of the object calculus and then some must be present in the syntax language.

Carnap does not claim that the mathematical sentences of a calculus owe their validity, that is, their truth, to linguistic stipulation. Such a claim would be vitiated by vicious circularity on account of the mathematics used in framing the stipulation.[24] Carnap places no such explanatory burden on his syntactic constructions; and in his reply to Quine in "Quine on Logical Truth" (1936c), he wisely forswears the rhetoric of linguistic fiat and convention (1963c, 916.) I see Carnap, in transposing epistemology into *Wissenschaftslogik,* to be rejecting equally Platonist and conventionalist accounts of mathematics. The definition of L-validity stands on its own, making precise a way in which the mathematical sentences of a language are formal auxiliaries to the substantive sentences.

Carnap's philosophy of mathematics is, I hold, unaffected by the shift to semantics, although the basic continuity in his position is obscured by his use of an apparatus of propositions and intensions in his semantics metalanguage.[25] I take his basic and stable view to be articulated in *The Foundations of Logic and Mathematics.* There, the core of the semantics for a calculus takes the form of a Tarski-style truth definition. To provide a truth definition for a calculus with descriptive vocabulary, we need to add this descriptive vocabulary (or translations of it) to the old syntax language for the calculus. We then use these translations to specify the designations of the object calculus's descriptive vocabulary. This use of descriptive vocabulary in the metalanguage is the most important modification of Carnap's syntax-era views occasioned by his embrace of Tarski. No change, however, is made in the transformation rules implicit in the old syntax language. In particular, these remain restricted to L-rules. Semantics, then, does not rely on any nonlogical truths containing the descriptive vocabulary in the metalanguage.

How does a Tarski-style truth definition for a descriptive calculus distinguish the analytic formulas of a calculus from the factual, synthetic ones? Carnap maintains that the L-truths of an object calculus are those sentences whose truth is in the semantics metalanguage a consequence of the truth definition for the calculus. So where s is an object calculus sentence and σ is a standard name for s in the semantics language, s is L-true in the object calculus if

$$\sigma \text{ is true}$$

is a consequence in the semantics language of the truth definition. A Tarskian truth definition will deductively yield a T-sentence for each object calculus formula,

$$\sigma \text{ is true iff } S^*,$$

where "S^*" is replaced by a translation of s in the metalanguage. "σ is true" will then be a consequence of the truth definition, when its translation S^* is L-true in the semantics language. Thus, the sample Tarski truth definition I use as an illustration in my classes implies

⌜der Schnee ist weiß ∨ ~der Schnee ist weiß⌝ is true.

Carnap's semantics language inherits the strong consequence relation of his syntax language. Hence, a truth definition in the semantics language will yield the truth and falsity of the logico-mathematical formulas of the object calculus.

This characterization of object calculus L-truth is, however, not a definition of L-truth for the object calculus in the semantics metalanguage. For the characterization *mentions* metalinguistic L-consequence. Carnap thinks of this characterization as expressing an adequacy condition for a proper semantics language definition of L-truth for an object calculus S. As such, he takes it to stand to a definition of L-truth as Tarski's Convention T stands to a truth definition (see Carnap 1942, §16, 84). How then are we to formulate a definition of L-truth in the semantics language?

Given a list of the primitive logical vocabulary of the calculus, we could adapt Carnap's procedure in general syntax, with truth playing the role of validity. The logical truths of a calculus S are those truths that essentially contain only logical vocabulary, that is, those truths that remain true under substitution of expressions for primitive descriptive expressions. This is, of course, exactly the way Quine defines elementary logical truth in section 2 of "Carnap and Logical Truth," "without," as Quine puts it, "any thought of any epistemological doctrine" (Quine 1963, 109). Carnap ends up doing something similar in his sketch of the form of semantic rules in the Schilpp volume. There, Carnap appends to a truth theory for a language a model-theoretic definition of L-true as true in all models.[26] The characterization of the models for the language holds fixed the construal of the logical expressions while allowing the designation of the descriptive constants to vary. In this way, using the set theory in the semantics language, Carnap can give in the semantics metalanguage itself a definition of L-truth that meets his adequacy criterion.

This approach to the definition of L-truth for a language depends on a demarcation of the primitive logical expressions of a calculus. Carnap's old general syntax characterization of the logical vocabulary does not carry over to semantics. For the general syntax characterization exploits the fact that the transformation rules for a calculus with descriptive predicates will leave some sentences indeterminate, neither valid nor contravalid. In semantics, a definition of truth replaces the syntax-era definition of validity. Because the truth predicate defined in Carnap's favored truth definitions is bivalent, there are no analogs to the syntax-era indeterminate sentences after the shift to semantics. In *Introduction to Semantics,* Carnap presents the general semantic characterization of the logical-descriptive dichotomy as an open question—one that Carnap never answers (see Carnap 1942, §13, 58–59). In the meantime, Carnap is happy simply to stipulate the primitive logical vocabulary as a part of the description of a semantic system.

Carnap versus Quine

Let us now turn to the tangle Carnap and Quine get into over Carnap's explanation of analyticity, of L-validity, in terms of semantic rules. We saw how in §4 of "Two Dogmas," Quine asks Carnap to explain analyticity by reference to "the mental or behavioral or cultural factors relevant to analyticity." Carnap is perplexed by this challenge. I have noted how Carnap thinks that there will be many calculi equally in accord with an individual's or group's speech dispositions. Indeed, he concedes that from the perspective of psychology, the difference between the formal auxiliaries of a language and synthetic sentences is one of degree (see Carnap 1935, 31; translation Carnap 1953, 123). Nothing changes here with the shift to semantics. Carnap does not expect any sharpening of analyticity from pragmatics, from within the empirical study of linguistic behavior. His attitude is clearly expressed in *The Foundations of Logic and Mathematics*. There, to introduce the notion of L-truth, Carnap contrasts the English sentences "Australia is large" and "Australia is large, or Australia is not large." He says that to ascertain the truth value of either sentence, we have to know the language. He then observes that in the case of the first sentence, a person, in addition to knowing English, has to ascertain some facts about Australia, whereas as regards the second sentence, "just by understanding [it] we become aware that it must be right." This, Carnap tells us, is the difference between factual and logical truth. He continues, "These unprecise explanations can easily be transformed into precise definitions by replacing the former reference to understanding by a reference to semantical rules." (1939, 12). Even as Carnap tries to meet Quine's challenge to clarify in pragmatic terms the informal notion of truth solely in virtue of meaning, he maintains that the application of semantic concepts like analyticity in *Wissenschaftslogik* does not depend on such clarification (see Carnap 1963c, 919, last paragraph of §B).

What, then, of Quine's objection to the explanation of L-validity in terms of semantic rules? Quine dismisses the specification of L-truth that Carnap appends to a truth theory for a calculus. Carnap appears to be pinning a label on an arbitrary subclass of truths of the calculus. Quine sees no point to this exercise in the absence of some further explanation of the label. As we have seen, Carnap's definition of L-validity for a calculus is not the selection of an arbitrary subclass of truths: The L-valid formulas of a calculus are to be those whose truth is an L-consequence of the truth definition in the semantics language.

Quine does not find this response on Carnap's part at all persuasive. Carnap's adequacy criterion mentions L-validity in the semantics language to explain L-validity in an object calculus. The criterion is particularly uninformative as regards full-width analyticity, when meaning postulates for descriptive predicates are counted among the L-valid sentences. But let us continue to put meaning postulates to one side in order to focus on logic and mathematics. Here, Quine's objection is a version of the point he makes in §7 of "Carnap and Logical Truth" against Carnap's characterization of L-validity in *Logical Syntax*. I have noted how the notion of L-consequence invoked in Carnap's adequacy criterion is a strong notion of consequence that secures the L-validity in the semantics language of the logic and mathematics expressible there. The adequacy criterion then comes to this: A sentence in an object calculus is L-valid just in case the semantics language sentence that says that it is true is a quantificational consequence of the truth definition together with the totality of logico-mathematical truths of the semantics language. Quine observes that in the same way, we can mark out any class of truths in the object calculus that can be identified in terms of vocabulary, and he concludes that this conception of analyticity is "uninteresting" because "[n]o special trait of logic and mathematics has been singled out after all" (Quine 1963, §7, page 125 in 1976b).

I have emphasized that Carnap is fully aware of the technical situation. Because he places no explanatory burden on his definition of L-validity, the charge that his characterization is 'uninteresting', that is, fails to establish any 'epistemic' asymmetry, will not move him. Quine's accusations will appear to Carnap to misunderstand the transformation of epistemology into *Wissenschaftslogik*.

Any calculus/semantic system that is a candidate to be the language for existing natural science must link theories to observation sentences. Because Carnap characterizes testability in terms of these links, on his view of the application of pure logic in *Wissenschaftslogik,* it makes no sense to speak of the links as confirmed or disconfirmed. This point is essential to the understanding that *Wissenschaftslogik* delivers of empirical testability. In expedient languages, these links, the L-consequence relation, generate L-true and L-false sentences in the language.[27] Consequently, these L-determinate sentences as well are neither confirmed nor disconfirmed by the predictive success or failure of theories stated in the language. I noted that Carnap lost a language-general characterization of the distinction between logical and descriptive sentences with his shift from syntax to semantics. He is not overly concerned with this loss, because he thinks that when we are constructing a semantic system as a candidate for the language of science, it will be intuitively clear what vocabulary to treat as logical and what vocabulary as descriptive (see Carnap 1942, §13, 58). What is the intuition here? What guides Carnap in building this distinction into various semantic systems in the ways that he does?

This question can be sharpened. Carnap is convinced from the outset that there is an important epistemic distinction between mathematics and factual science, a distinction that *Wissenschaftslogik* should explicate. However, Quine, pressing Duhem's point, urges that mathematical truths play the same role in the deduction of observation sentences as the more abstract physical laws do. If we restrict the principles of logic, the notational auxiliary, to elementary quantificational logic, this is true. Why does Carnap continue to insist that mathematics belongs to formal science? And why does Carnap stop at mathematics? Why not swell the class of L-valid truths further to include basic, abstract physics? Although Carnap does not address this question, I think an answer is implicit in his work.

If we have to draw a line in a semantic system between factual sentences and formal auxiliaries, it would be silly (pointless, inexpedient) to draw it between truth-functional and quantificational logic. Carnap thinks it is similarly silly, similarly inexpedient to separate elementary quantificational logic from mathematics. Carnap wants to characterize formal science in notational terms, that is, in terms of its vocabulary. Having identified certain undefined vocabulary as auxiliary (logical) as opposed to descriptive, we introduce a notion of form and define the formal truths as any truth that essentially contains only members of the selected vocabulary. So any truth containing only the auxiliary vocabulary is a formal truth. It seems expedient to maximize the scope of formal science in a calculus/semantic system that is a candidate for a language for science. In this way, we diminish the scope for the logically arbitrary choices in theory construction that Duhem's point forces on empirical science. Hence, besides the vocabulary of elementary quantificational logic, we should by all means include the vocabulary for pure mathematics as a part of the vocabulary of formal science. Why not go further? Why not add the vocabulary of basic physics to the vocabulary fixing formal science?

In the languages for unified science that Carnap favors, the theoretical predicates of physics designate physical magnitudes that assign values of these magnitudes to locations in space-time, which are designated by means of numerical coordinates.[28] The general laws of physics state constraints on these assignments of values, but they do not themselves determine these values. We can discover these values only by adding to the general laws the results of measurement, of observation. This is a feature of scientific method that Carnap wants to represent—wants to elucidate—within *Wissenschaftslogik*. Carnap accordingly takes as descriptive singular statements that assign a value to a physical magnitude at a location. Indeed, these are his paradigms of nonobservational descriptive statements. For this reason, the vocabulary of abstract physics should not be included in the auxiliary, logical vocabulary. Mathematics contrasts with physics here. In applications of mathematics in science, singular equations of pure mathematics are established on the basis of general mathematical principles. There is, then, no similar bar to relegating the mathematical vocabulary to formal science.

The Principle of Tolerance encapsulates Carnap's approach here. There is no right or wrong in the selection of a calculus/semantic system to adopt as a language for science, for any talk of right and wrong and any properly precise epistemic description presuppose the adoption of a calculus with its L-consequence relation. Rather, this selection is a practical decision concerning a course of action, the establishment of standards of right and wrong. As a practical decision, it is guided by considerations of expedience. And for Carnap, expedience dictates the inclusion of mathematics in formal science.

Quine rejects all this, and his rejection bespeaks a very different conception both of logic and of the use of logic in scientific philosophy than the one we have found in Carnap. The application of logic for Quine involves no actual or hypothetical projection of calculi onto the linguistic habits of investigators, no stipulative identification of a calculus as a language for science. As Quine views matters, in the course of doing science, we introduce new terminology and other notational innovations in the interest of clarity, perspicuity, and simplicity. The development and use of logical notation—truth-functional connectives and the quantifier-variable notation for generality—is such a linguistic change, continuous with the use of letters to replace pronouns in colloquial mathematics and with the use there of parentheses to indicate the order of application of mathematical operations.

Apart perhaps from some worries about quantum mechanics, Quine believes the notational framework of quantifiers, truth functions, and predicates suffices for the formulation of the statements and demonstrative reasoning across the sciences. Any alleged losses in expressive power occasioned by the austerity of elementary quantificational logic are counterbalanced by the uncontroversial familiarity and lucidity of the framework. This last point is central to Quine's approach. The selection of this canonical notation brings a sort of conceptual clarification and simplicity to science generally. Science, Quine tells us, has the task "of specifying how 'reality' really is" (Quine 1976a, 232).[29] When we scientifically minded philosophers reflect on the answers that our science proposes to general ontological questions, we find ourselves stymied by the irregularities, context dependencies, and especially the possibilities for nominalization in colloquial language. We can give clear sense to general ontological questions, Quine urges, by envisioning the regimentation of our theories into canonical notation. The existential generalizations implied by a regimented theory tell us what kinds of things the theory says there are. Quine argues that when we reflect on the predicates that figure in science, especially the best-established sciences, we recognize that we can construe our quantifiers to range over nothing but physical objects and classes.[30] Moreover, when we see how one body of theory can be interpreted within another body, regimentation into quantificational notation opens the prospect of simplification. The identification of ordered pairs with certain classes is Quine's paradigm here (see Quine 1960, §53).

The clarification and simplification effected by the adoption of Quine's recommended canonical notation is, in his eyes, a very powerful consideration in its favor. Here, there is a close parallel to Carnap's view of the benefits to moving in science from the use of colloquial language to something approximating an explicitly described formal language. But there is an important divergence as well (see Quine 1963, §8). As Quine views matters, the advantages of clarity and simplicity attaching to his choice of a canonical notation are not different in kind from the advantages that commend any fairly abstract scientific theory:

> The same motives that impel scientists to seek ever simpler and clearer theories adequate to the subject matter of their special sciences are motives for simplification and clarification of the broader framework shared by all the sciences. Here the objective is called philosophical, because of the breadth of the framework concerned; but the motivation is the same. The quest of a simplest clearest overall pattern of canonical notation is not to be distinguished from a quest of ultimate categories, a limning of the most general traits of reality. (Quine 1960, 161)

In so viewing matters, Quine is oblivious to Carnap's distinction between pure logic—the elaboration of calculi or semantic systems—and the application of pure logic in *Wissenschaftslogik* via the coordination of the systems of pure logic with actual or hypothetical used languages. Consequently, Quine does not accept the terms in which Carnap seeks to explicate empirical testability.

We have seen that for Carnap, judgments of correct and incorrect, true and false, are legitimate only when made in the context of the selection of a calculus/semantic system as a language for science. Only this selection imposes definite standards of right or wrong on amorphous linguistic behavior. In this way, Carnap admits a logical pluralism within *Wissenschaftslogik* and accommodates an antipsychologistic *Wissenschaftslogik* with a naturalistic account of language as a biological, behavioral phenomenon.

In contrast, Quine's understanding of truth is disquotational:

> Where it makes sense to apply 'true' is to a sentence couched in the terms of a given theory and seen from within the theory, complete with its posited reality. Here there is no occasion to invoke even so much as the imaginary codification of scientific method. To say that the statement 'Brutus killed Caesar' is true, or that 'The atomic weight of sodium is 23' is true is in effect simply to say that Brutus killed Caesar or that the atomic weight of sodium is 23. (Quine 1960, 24)

For Quine, the statements of science, as they are, are true or false. However, if we wish to discuss logical relationships among them, we do well to regiment them into canonical notation. For the chief benefit of the development of logical notation is the clarity it promotes in the science of logic itself. Indeed, this benefit is the basis of the other advantages already mentioned. The regularity and simplicity of a language built up from predicates, (object) variables, truth-functional connectives, and quantifiers yields a tractable notion of logical form, and with it a notion of logical truth that admits a complete proof procedure. Semantics figures here in the use of truth and satisfaction predicates in the definition of logical truth and in the statement of logical laws that generalize over the forms of statements. Whereas Carnap takes talk of designation in pure semantics to set forth an arbitrary relation between meaningless expression types and (typically) extralinguistic items, Quine's satisfaction predicate, like his truth predicate, is disquotational. Both truth and satisfaction predicates are explained by reference to the Tarski paradigms, and they are thus applicable only to predicates in the used language (see Quine 1953b; see also Quine 1960, 273). Here, it is important to observe that Quine's fundamental characterization of logical truth is framed not model theoretically but in terms of lexical substitution. The equivalence of this characterization with the usual model-theoretic one for suitably rich languages gives set theory its uncontroversial application in investigations of elementary logical truth (see Quine 1986, 53–56; 1982, 211–12). On this view of truth, logic, and logical notation, there is neither a basis nor a need to distinguish factual truths from notational auxiliaries.

Quine, then, has no place for a special principle of tolerance about logic; there is merely fallibilist humility that should accompany any statement, including statements of logical laws (see Quine 1960, 25, the final sentence). Our language, our speech habits, encompasses the use of logical notation for the regimentation of some statements into others for various purposes. It encompasses semantic ascent as well. And logic, in its different way, encompasses language; for via regimentation and semantic ascent we can discuss logical relationships among arbitrary sentences. We have here a reciprocal containment of logic in language and language in logic parallel to the reciprocal containment of epistemology in natural science and natural science in epistemology that Quine portrays in "Epistemology Naturalized."[31] This view of reciprocal containment lies at the core of Quine's rejection of Carnap's *Wissenschaftslogik,* of Carnap's version of scientific philosophy. The contrast here between the two thinkers is vividly illustrated in their correspondence about Quine's paper "Carnap and Logical Truth." Carnap, in preparing his reply to this paper, queried Quine,

> Now there is a point where I should like some clarification of what you mean so that my reply can be more specific. . . . The question is which of your discussions are meant to refer to (a) natural languages, and which to

(b) codified languages, language systems based on explicitly formulated rules.... The distinction is of great importance for my discussion, because from my point of view the problems of analyticity in the two cases are quite different in their character. (Carnap to Quine, July 7, 1954, in Quine and Carnap 1990, 435)

Quine replied,

It is indifferent to my purpose whether the notation be traditional or artificial, so long as the artificiality is not made to exceed the scope of "language" ordinarily so-called, and beg the analyticity question itself.... The languages I am talking about comprise natural languages and any (used, or interpreted) artificial notations you like, e.g. that of my *Mathematical Logic* plus extralogical predicates. They are not uninterpreted notations. Each predicate has its unique extension, and correspondingly for the logical signs. (Quine to Carnap, August 9, 1954, in Quine and Carnap 1990, 437–38)

Notes

I have benefited from formative conversations on the topic of this paper with Richard Creath, Warren Goldfarb, Alan Richardson, and especially Michael Friedman. I am also indebted to André Carus, Burton Dreben, Gary Ebbs, and Peter Hylton for criticisms of earlier drafts of this paper.

1. Quine's preface to "Carnap and Logical Truth" is not included in the republication of this essay in Quine 1976b.
2. I presented Quine's demand for a criterion of analyticity as a challenge internal to Carnap's philosophy in Ricketts 1982. I now think I was mistaken to have done so.
3. Dreben forcefully made this point in a symposium on analyticity at the Eastern Division meeting of the American Philosophical Association in 1994.
4. Here, it should be observed that Carnap includes in arithmetic the full, open-ended resources of set theory. His 'combinatorial' analysis of calculi is thus in no way mathematically restricted. See Carnap 1937, §§ 34i and 60d.
5. See Carnap 1937, §2, 7. In §1, Carnap opines that it is not, as a practical matter, feasible to describe the syntax of natural languages on account of their complexity and irregularity.
6. Carnap says nothing about grammar construction, that is, nothing about the demarcation of a class of utterance types as the sentences of some group's language. He assumes this task to have been done prior to the correlation of the expression types of a calculus and the utterance types of a language. If pressed here, I assume Carnap would be happy to avail himself of the account Quine sketches in §§2–3 of Quine 1953c.
7. Things do not change significantly with the shift to semantics. Until Quine challenges him on analyticity, Carnap's remarks about the empirical basis for descriptive semantics are more or less restricted to the basis for the assignment of extensions to observation predicates.
8. Carnap has an accurate and deep understanding of Gödel's incompleteness theorems. He is fully aware that this desideratum precludes identifying L-validity with derivability in any formal system. In effect, Carnap's syntax language must contain set-theoretic resources for

a truth definition for the purely mathematical part of the object calculus. For a discussion of the motivations for Carnap's requirement of bivalence for the logico-mathematical sentences of a calculus, see Goldfarb and Ricketts 1992.

9. Indeed, it might be urged that speech dispositions at best pin down a formal system for the language, Carnap's L-derivability, not his much stronger notion of L-validity. For this distinction, see Carnap 1937, §§10, 14, 34a, 47, and 48.

10. Beyond this, perhaps we might require that the speakers' dispositions over the logico-mathematical sentences fix a formal system that is sound with respect to the notion of L-validity. Furthermore, if we suppose that the P-rules take the form of an effective demarcation of nonlogical axioms, then we might also require that the speakers be disposed to affirm these P-axioms.

11. I wrongly claimed that Carnap thought otherwise in Ricketts 1994, 191.

12. Via informal regimentation, the statements of an actually used language may be identified with the formulas of a calculus. In this way, a *Wissenschaftslogiker* may describe a calculus and then stipulate that her used language is, via regimentation, to be taken as an instance of the calculus.

13. On the following page, Carnap presents his rough-and-ready explanation of observation predicates, remarking that this distinction is not a sharp one. See also Carnap 1932, 182.

14. For Carnap's conception of pragmatics as the empirical science of language, see the discussion in Carnap 1939, §§2 and 3, especially the last paragraph of §3. Carnap does not himself distinguish those statements of descriptive syntax and semantics that are a part of pragmatics from those that are not. Indeed, in Carnap 1942, §5, 13, Carnap suggests that all of descriptive semantics and descriptive syntax belong to pragmatics. I think that Carnap's treatment of pragmatics and the relation between semantic systems and languages is more detailed and more nuanced in Carnap 1939 than in Carnap 1942.

15. However, Carnap's treatment of belief and kindred notions as nondispositional, theoretical notions in Carnap 1955 represents a shift toward a more cognitive, less behavioral view of psychology.

16. My discussion of the relationship of pure and descriptive syntax and semantics in this paragraph and this section has been stimulated by Ebbs 1997, §§57–59. Ebbs does not make my distinction between pragmatics proper and the descriptive syntax and semantics that result from the coordination of calculi with used languages in agreement with them. Accompanying this distinction, I sense a difference between my interpretation and Ebbs's of Carnap's distinction between pure and descriptive syntax/semantics.

17. André Carus has remarked to me that Carnap himself drops the term *"Wissenschaftslogik"* after the shift to semantics. In *The Logical Syntax of Language*, Carnap boldly identifies philosophy with the logical syntax of the language of science. However, even before the shift to semantics, he acknowledges that the notion of an observation predicate figures importantly in the logic of science, and this notion is not one from pure syntax. For an early discussion of the interplay here of purely logical and empirical notions in the logical of science, see Carnap 1932, 177–83. More importantly, the identification in *Logical Syntax* of philosophy with the syntax of the language of science is tied to Carnap's analysis of ontological claims as pseudo-object sentences. However, the shift to semantics forces Carnap to give up the notion of a pseudo-object sentence; see Ricketts 1996, 246–47. Carnap discusses how he views the logic of science after the shift to semantics in *Introduction to Semantics* (1942, 250). I believe that the continuities in Carnap's conception of his enterprise, especially as regards the topic of this paper, are overwhelming. For this reason, I use the term *"Wissenschaftslogik"* to describe Carnap's enterprise from logical syntax onward.

18. It is for this reason, I believe, that Carnap says that the distinction between logical and factual sentences is indispensable in any account of scientific method; see Carnap 1963c, 922; 1963a, 932.

19. I am grateful to Michael Friedman for calling my attention to this important paper.

20. The way Carnap develops this theme, however, distinguishes his philosophy of mathematics from other positivist approaches. For a discussion of the differences, see Goldfarb 1996.

21. For further discussion, see Ricketts 1994, 180–81.

22. For a discussion of various anomalies and difficulties with the general syntax definition of L-validity, see Creath 1996.

23. As Carnap's syntax languages lack descriptive vocabulary, consequence and L-consequence coincide in them.

24. Gödel objects to Carnap along these lines in "Is Mathematics Syntax of Language?" (1995). In effect, Quine urges a vicious circularity here in the final section of "Truth by Convention" (1936). For further discussion of this charge, see Goldfarb and Ricketts 1992.

25. A defense of this claim would require an extended discussion of Carnap's view of and use of intensional notions in semantics.

26. See Carnap 1963b, 900–903. See also the discussion of definitions of L-truth in Carnap 1942, § 16, especially the second strategy on 86–87.

27. In "Formalwissenschaft und Realwissenschaft" (1935, 34; translation 1953, 126–27), Carnap considers the possibility of transforming familiar calculi into calculi with the same P-valid sentences, but no L-valid sentences, that is, into calculi in which L-determinate sentences are excluded by complicated, noneffective formation rules and logical connections among P-sentences are forged directly by inference rules.

28. See Richardson 1994, 78ff. for a discussion of this point and its significance.

29. "The Scope and Language of Science" (1976a) is coeval with "Carnap and Logical Truth."

30. For an early expression of this theme, see Quine 1976a, 242–44.

31. See Quine 1969, 83. For an evocative portrayal of the theme of reciprocal containment in Quine's philosophy, see Dreben 1994.

References

Carnap, R. 1932. "Erwiderung auf die vorstehenden Aufsätze von E. Zilsel und K. Duncker." *Erkenntnis* 3: 177–83.

———. 1935. "Formalwissenschaft und Realwissenschaft." *Erkenntnis* 5: 30–37.

———. 1936. "Von Erkenntnistheorie zur Wissenschaftslogik." In *Acts du Congrès internationale de philosophie scientifique, Sorbonne, Paris, 1935, 1. Philosophie Scientifique, Sorbonne, Paris, 1935*. Vol. 1, *Philosophy scientifique et empirisme logique*, 36–41. Paris: Hermann and Cie.

———. 1936–37. "Testability and Meaning." *Philosophy of Science* 3: 419–71; 4: 1–40.

———. 1937. *The Logical Syntax of Language*. Trans. A. Smeaton. London: Routledge and Kegan Paul. Original publication, 1934.

———. 1939. *The Foundations of Logic and Mathematics*. Chicago: University of Chicago Press.

———. 1942. *Introduction to Semantics*. Cambridge, Mass.: Harvard University Press.

———. 1953. "Formal Science and Factual Science." In *Readings in the Philosophy of Science*, ed. H. Feigl and M. Brodbeck, 123–28. New York: Appleton, Century, Crofts.

———. 1955. "On Some Concepts of Pragmatics." *Philosophical Studies* 6: 89–91.
———. 1956. *Meaning and Necessity*. 2nd ed. Chicago: University of Chicago Press.
———. 1963a. "E. W. Beth on Constructed Language Systems." In *The Philosophy of Rudolf Carnap*, ed. P. A. Schilpp, 927–33. La Salle, Ill.: Open Court, 1963.
———. 1963b. "My Conception of Semantics." In *The Philosophy of Rudolf Carnap*, ed. P. A. Schilpp, 900–905. La Salle, Ill.: Open Court, 1963.
———. 1963c. "Quine on Logical Truth." In *The Philosophy of Rudolf Carnap*, ed. P. A. Schilpp, 915–22. La Salle, Ill.: Open Court, 1963.
———. 1987. "The Task of the Logic of Science" (1934). In *Unified Science: The Vienna Circle Monograph Series Originally Edited by Otto Neurath*, ed. B. McGuinness, 46–66. Dordrecht: Reidel.
———. 1990. "Quine on Analyticity" (1952). In *Dear Carnap, Dear Van: The Quine-Carnap Correspondence and Related Work*, ed. and with an introduction by R. Creath, 427–32. Berkeley and Los Angeles: University of California Press, 1990.
Creath, R. 1996. "Languages without Logic." In *Origins of Logical Empiricism*, ed. R. N. Giere and A. Richardson, 251–65. Minneapolis: University of Minnesota Press, 1996.
Dreben, B. 1994. "In Mediis Rebus." *Inquiry* 37: 441–48.
Ebbs, G. 1997. *Rule-Following and Realism*. Cambridge, Mass.: Harvard University Press.
Giere, R. N., and A. W. Richardson. 1996. *Origins of Logical Empiricism*. Minnesota Studies in the Philosophy of Science, vol. 16. Minneapolis: University of Minnesota Press.
Gödel, K. 1995. "Is Mathematics Syntax of Language?" In *Collected Works*, ed. S. Feferman, J. Dawson Jr., W. Goldfarb, C. Parsons, and R. Solovay, vol. 3, 324–62. Oxford: Oxford University Press.
Goldfarb, W. 1996. "The Philosophy of Mathematics in Early Positivism." In *Origins of Logical Empiricism*, ed. R. N. Giere and A. Richardson, 213–30. Minneapolis: University of Minnesota Press, 1996.
Goldfarb, W., and T. Ricketts. 1992. "Carnap and the Philosophy of Mathematics." In *Wissenschaft und Subjektivität*, ed. D. Bell and W. Vossenkuhl, 61–78. Berlin: Academie Verlag.
Quine, W. V. O. 1936. "Truth by Convention." In *Philosophical Essays for A. N. Whitehead*, ed. O. H. Lee, 90–124. New York: Longman. Reprinted in *Ways of Paradox*, 107–32. Cambridge, Mass.: Harvard University Press, 1976.
———. 1953a. *From a Logical Point of View*. Cambridge, Mass.: Harvard University Press.
———. 1953b. "Notes on the Theory of Reference." In *From a Logical Point of View*, 138–39. Cambridge, Mass.: Harvard University Press, 1953.
———. 1953c. "The Problem of Meaning in Linguistics." In *From a Logical Point of View*, 47–64. Cambridge, Mass.: Harvard University Press, 1953.
———. 1953d. "Two Dogmas of Empiricism." In *From a Logical Point of View*, 20–47. Cambridge, Mass.: Harvard University Press, 1953.
———. 1960. *Word and Object*. Cambridge: MIT Press.
———. 1963. "Carnap and Logical Truth" (1954). Reprinted in *The Philosophy of Rudolf Carnap*, ed. P. A. Schilpp, 385–406. La Salle, Ill.: Open Court, 1963; and in *Ways of Paradox*, 107–32. Cambridge, Mass.: Harvard University Press, 1976.
———. 1969. "Epistemology Naturalized." In *Ontological Relativity and Other Essays*, 69–90. New York: Columbia University Press.
———. 1976a. "The Scope and Language of Science" (1954). In *Ways of Paradox*, 228–45. Cambridge, Mass.: Harvard University Press, 1976.
———. 1976b. *Ways of Paradox*. Cambridge, Mass.: Harvard University Press.
———. 1982. *Methods of Logic*. 4th ed. Cambridge, Mass.: Harvard University Press.

———. 1986. *Philosophy of Logic*. 2nd ed. Cambridge, Mass.: Harvard University Press.
Quine, W. V. O., and R. Carnap. 1990. *Dear Carnap, Dear Van: The Quine-Carnap Correspondence and Related Work*. Ed. and with an introduction by R. Creath. Berkeley and Los Angeles: University of California Press.
Richardson, A. 1994. "Carnap's Principle of Tolerance." *Aristotelian Society Supplementary Volume* 68: 67–82.
Ricketts, T. 1982. "Rationality, Translation, and Epistemology Naturalized." *Journal of Philosophy* 79: 117–36.
———. 1994. "Carnap's Principle of Tolerance, Empiricism, and Conventionalism." In *Reading Putnam*, ed. P. Clark and B. Hale, 176–200. Oxford: Blackwell.
———. 1996. "Carnap: From Logical Syntax to Semantics." In *Origins of Logical Empiricism*, ed. R. N. Giere and A. Richardson, 231–50. Minneapolis: University of Minnesota Press, 1996.
Schilpp, P. A., ed. 1963. *The Philosophy of Rudolf Carnap*. The Library of Living Philosophers, vol. 11. La Salle, Ill.: Open Court.
Wittgenstein, L. 1922. *Tractatus Logico-Philosophicus*. London: Routledge and Kegan Paul.

Contributors

Richard Creath is professor of philosophy at Arizona State University. He has written numerous papers on Rudolf Carnap, W. V. O. Quine, and philosophy of science and has edited *Dear Carnap, Dear Van: The Quine-Carnap Correspondence and Related Work* as well as coedited (with Jane Maienschein) *Biology and Epistemology*. He is the general editor of *The Collected Works of Rudolf Carnap*.

Michael Friedman is Frederick P. Rehmus Family Professor of Humanities at Stanford University. His publications include *Foundations of Space-Time Theories, Kant and the Exact Sciences, Reconsidering Logical Positivism, A Parting of the Ways: Carnap, Cassirer, and Heidegger*, and *Dynamics of Reason*.

Rudolf Haller, professor of philosophy, lives in Graz, Austria. His main research areas are analytic philosophy, aesthetics, epistemology, and Austrian philosophy. He is the author of more than 350 publications, including books on Alexius Meinong, Ernst Mach, Otto Neurath, and Ludwig Wittgenstein, and he has also been the editor of several series and journals, such as *Grazer Philosophische Studien, International Journal of Analytic Philosophy,* and *Studien zur Österreichischen Philosophie*.

Gary L. Hardcastle is assistant professor of philosophy at Bloomsburg University and has also taught at the University of Wisconsin–Stevens Point, the University of San Diego, and Virginia Polytechnic Institute and State University. He is the author of articles on philosophy of science, epistemology, and history of psychology.

Don Howard is professor of philosophy and director of the Program in History and Philosophy of Science at the University of Notre Dame. Among his recent publications is *Einstein: The Formative Years, 1879–1909,* coedited with John Stachel. His research areas include the history and philosophy of physics and the history of the philosophy of science.

Diederick Raven teaches history and theory of anthropology at the University of Utrecht. He is coeditor of *Cognitive Relativism and Social Science* and coeditor, with Wolfgang Krohn and Robert S. Cohen, of *The Social Origins of Modern Science* (which contains all of Zilsel's wartime essays) and is in the closing stages of his new book, the *Cultural Roots of Science.*

George Reisch received his Ph.D. in philosophy and history of science from the University of Chicago and is an independent scholar. His research is on the history of logical empiricism and the Unity of Science movement. He has taught at Northwestern University and the Illinois Institute of Technology.

Alan W. Richardson is associate professor of philosophy at the University of British Columbia. He is the author of numerous essays and the monograph *Carnap's Construction of the World: The "Aufbau" and the Emergence of Logical Empiricism.* He is coeditor of *Origins of Logical Empiricism* (Minnesota, 1996) and *Cambridge Companion to Logical Empiricism* (forthcoming).

Thomas Ricketts is professor of philosophy at Northwestern University. His areas of research are history of analytic philosophy and philosophy of language.

Friedrich K. Stadler is associate professor at the University of Vienna, where he is head of the Department of Contemporary History and a member of the Center for Interdisciplinary Research. He is also the founder and director of the Institute Vienna Circle/Institut Wiener Kreis. His primary fields of research are history and philosophy of science, science studies (particularly exile and emigration of German-speaking scientists and intellectuals), central-European and Austrian intellectual history, and theory and philosophy of history. Among his many publications in English are *The Vienna Circle, Scientific Philosophy: Origins and Philosophy, The Cultural Exodus from Austria* (coedited with P. Weibel), *Encyclopedia and Utopia: The Life and Work of Otto Neurath* (coedited with E. Nemeth), *History of Philosophy of Science: New Trends and Perspectives* (coedited with M. Heidelberger), and *The Vienna Circle and Logical Empiricism* (editor).

Thomas E. Uebel teaches philosophy at the University of Manchester, England. His research interests include history of philosophy of science as well as systematic epistemology and philosophy of social science; he has served as member of the Steering Committee of the International Society for the History of Philosophy of Science from 1996. His books include *Overcoming Logical Positivism from Within, Otto Neurath: Philosophy between Science and Politics* (with N. Cartwright, J. Cat, and L. Fleck), and *Vernunftkritik und Wissenschaft*.

Index

Compiled by Alex Korolev

abduction, 82n.34
absolutism, 97, 201
Adler, Friedrich, 29–30, 44, 81n.23
Analysis (journal), 95, 98, 116
analytic/synthetic distinction, 5–10, 20n.14, 115–16, 209; Carnap and Quine on, 6–10, 16–17, 257–58; Feigl on, 115
analyticity, xiii, 7–9, 115, 176, 257–59, 265, 269–71, 276, 276nn.2–3; behavioral criterion for, 7–8, 10
Anschluss, the, 129–31, 218, 224
anticommunism, 28, 34, 71–72, 199, 210, 211n.2
antifoundationalism, 40
antinaturalism: in philosophy of social science, 135
anti-Semitism, 33, 82n.35, 130–31, 143, 217
a priori: Dewey on, 14; Lewis's "pragmatic a priori," 13–14; Morris on, 5, 13; Quine's rejection of, 7, 234–35, 240; relativized, 13–16, 236; synthetic, 12, 154, 236, 259
Aquinas, Thomas, xxii, 71
Aristotle, xxii, 71, 77
arithmetic, 260, 262, 276n.4
atomism, epistemological, 45, 80n.19
Aufbau (construction), 95–96, 104–8, 110, 110n.2, 182, 200, 225
Ayer, A. J., 158

Bacon, Francis, 50, 59
Bauer, Otto, 44, 80n.17, 144
Bauhaus, the, 107
behaviorism, 219, 264; neobehaviorism, 222
Bergmann, Gustav, 4, 52, 66, 218, 221

Biringuccio, Vannoccio, 137–38
Birkhoff, Garrett, 171–75
Black, Max, 226
Blumberg, Albert E., 81n.27, 122–24, 126n.1
bolshevism, 33–34
Boltzmann, Ludwig, 79n.6, 151
Boring, Edwin, 172–73, 176, 178, 191n.10, 192n.13
Bridgman, Percy Williams, xix, 48, 122–23, 171–74, 177, 179, 181, 186, 189n.1, 190n.1, 190n.4, 191n.10, 191n.12
Brunswik, Egon, 222
Buddhism, 202

calculus (formal language), xiii, 259–74, 276n.6, 277n.8, 277n.12, 277n.16, 278n.27; classical first-order predicate, 237, 240–41, 244, 247
Cantor, Georg, 237
Carnap, Rudolf, viii–ix, xii–xiii, xvii–xviii, xx, 2–19, 19n.6, 20n.17, 20n.20, 21n.24, 22, 26, 31, 33–39, 41, 45–47, 52–54, 56, 61–62, 72–73, 78, 80n.12, 80n.14, 81n.24, 95–110, 110n.2, 111nn.4–5, 111n.7, 111n.9, 111nn.11–12, 112n.12, 119, 122–25, 149–50, 153–60, 164, 170, 172–73, 175–78, 180, 182–85, 189, 189n.1, 191nn.9–10, 192n.17, 197–211, 211n.1, 212nn.4–5, 212n.7, 212nn.10–13, 217–20, 223, 225–28, 234–37, 239–42, 244–45, 247–55, 257–76, 276n.2, 276nn.4–8, 277nn.8–9, 277n.11, 277nn.13–17, 278n.18, 278n.20, 278nn.23–27; on mathematics, 175;

285

286   Index

on philosophy of science, 227–28; principle of tolerance. *See also* analytic/synthetic distinction: Carnap and Quine on; principle: of the conventionality of language forms (Carnap); principle: of tolerance (Carnap)
Carnap and Logical Truth (Quine), 234, 240, 248–49, 251, 253, 257, 269, 270, 275, 276n.1, 278n.29
causality, xx, xxiii, 73, 78n.1, 121–22, 124, 174, 228, 249, 251
certainty, logical, 8, 236
Chicago Logic of Science Discussion Group. *See* University of Chicago's Logic of Science Discussion Group
Church, Alonzo, 189n.1
Cohen, Morris Raphael, xiv, 48
coherence, theory of truth, 98–100, 246
communism, 26, 56, 74
completeness, 206
concept, formation of, 179
confirmation, viii, xxiv, 3, 8–9, 13–14, 17–18, 64, 101–3, 106–7, 109, 115–16, 164, 175, 252, 254, 271
conservatism, 31, 55, 57, 246
context of discovery/justification, 3, 5, 54–55, 65, 82n.34
convention, 8, 42, 54, 80n.18, 97, 150, 203, 234–35, 237, 240, 246–51, 253, 257, 262, 267. *See also* truth: by convention
conventionalism, xiii, 33, 42, 47, 76, 78n.1, 80n.19, 154, 234–35, 247, 249–51, 255, 267
Convention T, 268
Cooley, John, 173, 176
Copernicus, 61, 244
correspondence, theory of truth, 99
corrigibility. *See* incorrigibility
Creath, Richard, xiii, 18, 19, 183, 249, 278n.22

D'Acconti, Alessandra, 150, 165n.3
Dahms, Hans-Joachim, 30, 78n.2, 82n.32, 146n.21, 216, 218–19, 224
Darwin, Charles, 81n.26
"debabelization" of science, xii, 170, 172, 175, 177, 180–83, 187–89, 191n.7, 192n.15
decision theory, 64, 221

definability, 267
definition: explicit, 105; implicit, 250; legislative/discursive, 240, 251, 253, 255, 257; stipulative, 267
democracy, 25–26, 31–33, 56–57, 74, 83n.42, 203
Derrida, Jacques, 27
de Santillana, Giorgio, 171–72, 186, 189–90n.1
Descartes, René, viii, xxii, 4, 136, 137
Dessau Bauhaus. *See* Bauhaus, the
Dewey, John, xiv, 1–3, 11, 14, 16–18, 19n.2, 25, 27–28, 45–56, 58–62, 64–66, 69, 72–74, 77, 78n.3, 79nn.3–4, 81n.26, 82n.29, 82n.31, 82n.34, 83n.42, 83n.46, 163–64, 203, 220; rejection of traditional metaphysics and epistemology of, 16. *See also* naturalism: Dewey's biological-evolutionary
dialectical materialism, 29
Dilthey, Wilhelm, 135
discovery/justification, context of. *See* context of discovery/justification
discursive/legislative distinction, 234, 251
doctrine of recollection, the, 238
Dreben, Burton, 258, 276n.3, 278n.31
Driesch, Hans, 152
dualism, 4, 50
Duhem, Pierre, 29, 39–42, 44–45, 80n.21, 81n.22, 116, 119, 151, 153, 253, 271–72; on theory testing, 257. *See also* theory testing
Düring, Eugen, 29
Durkheim, Emile, 138

Eddington, Arthur, 33
Einstein, Albert, 29–30, 32–33, 37, 80n.20, 82n.35, 118–20, 122, 151, 153, 160, 170, 181
Engels, Friedrich, 29, 45, 68
Enlightenment, 25, 27, 49, 60, 161, 217
epistemology, 8, 15–16, 18, 54, 101–2, 122, 182, 227, 239, 251, 259, 267, 269, 271, 275; historical, xvi
Epistemology Naturalized (Quine), 275
Erkenntnis (journal), 13, 15, 33–34, 80n.12, 99, 101, 104, 124, 190n.1, 198, 212n.14, 221
exile studies, 221–22

existentialism, 27
experience, 16–17, 40–42, 50–52, 54, 61–62, 95–98, 118–19, 125, 252, 254
Experience and Prediction (Reichenbach), 16, 54, 190n.3
experimentalism, 64
explanation, viii–ix, xiii, xxi, xxiii–xxiv, 13, 43, 106, 208–10, 220, 228, 238–39, 241–43, 245, 249, 257
explication, 104–9, 111n.12

fallibilism, 275
falsificationism, xxviiin.10, 40–41
fascism, 26, 55–56, 74, 217
Feigl, Herbert, xi–xii, xiv, xvii–xiii, 4, 52, 73, 79n.6, 81n.27, 82n.32, 95, 111n.11, 115–29, 150, 164, 165n.3, 171–72, 176, 189–90n.1, 191n.10, 208, 218–20, 223, 228; on mind-body problem, 124–26
feminism, 27
Feyerabend, Paul, 115, 220, 222
formalism, logical, 157, 258
formation rules. *See* rules: formation and transformation
foundationalism: Carnap's, 6, 12, 95, 183; epistemic, 8–9, 209
Foundations of Logic and Mathematics (Carnap), 212n.10, 260, 263, 267, 270
Foundations of the Social Sciences (Neurath), 201, 208, 212n.5
Foundations of the Unity of Science (Neurath, Carnap, and Morris), 201, 208, 212n.5
Frank, Philipp, xii, xvii–xxi, xxviinn.6–8, 4, 27, 34, 44–45, 47, 53, 56, 64–66, 70–72, 78n.1, 80n.14, 81n.22, 82n.33, 82nn.39–40, 83nn.42–43, 118, 123–24, 149–65, 165n.4, 165n.6, 165nn.10–11, 170, 171–74, 176, 185–89, 189n.1, 190n.1, 190n.4, 191n.7, 191n.10, 192nn.16–17, 200, 204, 212n.13, 217–19, 223, 229n.2; on the role of social factors in the theory choice, 61–63
Frege, Gottlob, 159
Frenkel-Brunswik, Else, 222
Friedman, Michael, xi, xxvin.1, 19, 79n.6, 166n.13, 184, 191n.8, 212n.4, 278n.19
fundamentalism, religious, 56

Galilean transformations, 154
Galilei, Galileo, 137
Galison, Peter, xvii, xxi, xxiii–xxiv, 19, 111n.11, 165n.9, 181, 185, 187–88, 192n.16, 197, 203, 210–11, 212n.14, 213n.14
game theory, 174–75, 221
General Theory of Knowledge (Schlick), 118
general theory of relativity. *See* relativity, theory of: general
geometry, 33, 111n.9, 154, 234, 238, 249–50, 260–61; non-Euclidean, 81n.23, 247, 249
Giere, Ronald, xv, 2–5, 18, 19, 19n.1, 20n.7, 55, 73, 79n.6, 109, 111n.11, 191n.8, 217, 224
Gieryn, Thomas, xv–xvi
Gödel, Kurt, 4, 95, 126n.6, 218, 221, 237, 260, 278n.24; sentences, 239. *See also* incompleteness theorems
Goldstein, Kurt, 170, 172, 189n.1, 192n.13
Gomperz, Heinrich, 227
Goodman, Nelson, 108, 173, 176
grammar, 249, 264, 276n.6
Grünbaum, Adolf, 78, 220

Hahn, Hans, 31–32, 34, 38, 44, 53, 65, 79n.6, 95, 118, 149–53, 155–56, 159, 165nn.3–4, 217, 219, 225
Haller, Rudolf, xi, xxvi, 115–28, 150, 183, 216, 218, 222
Hanson, Norwood Russell, 209
Hardcastle, Gary L., xii, 19, 171, 191n.7, 191n.12, 192n.13, 192n.15, 219
Harding, Sandra, 27
Heidegger, Martin, 25
Helmer, Olaf, 109, 125, 229n.1
Helmholtz, Hermann von, viii–ix, 30
Hempel, Carl G., xi–xii, xiv, xvii–xviii, 3, 4, 52, 73, 94–95, 98–110, 110n.1, 110n.3, 111n.5, 111n.10, 125, 157–59, 164, 165n.7, 166n.13, 189n.1, 208, 220, 227, 229n.1
Henderson, L. J., 171–72, 174, 189n.1
Hilbert, David, 151, 153, 250
Hintikka, Jaakko, 115
historical materialism, 139

Index

historical-sociological naturalism (Neurath), 99–100, 102–5, 108
holism, 8, 17, 39–41, 45, 54–55, 77, 78n.1, 80n.19, 80n.21, 98, 109, 116, 217, 251–52
Holton, Gerald, xix, 71, 82n.33, 165n.9, 170, 185–86, 190n.1, 190n.4, 191n.6, 191n.8, 192n.16, 219
Howard, Don, viii–xi, xvii, 19, 45, 80nn.18–19, 80n.20, 165n.5, 166n.14, 199
Hughes, H. Stuart, 216, 223
humanism, 27, 137
Hume, David, xxii, 16, 50, 121, 225
Husserl, Edmund, 111n.12
hypothetico-deductive method, 119

idealism, xxiii, 4, 97
identity, 116, 125–26, 237, 242
implicit definition, 250
incompleteness theorems, 237, 276. *See also* completeness
incorrigibility, 40–41
individualism, epistemological, 51
induction, 55, 64, 117, 119, 121, 207
inductive logic. *See* logic: inductive
Institute for the Unity of Science. *See* International Institute for the Unity of Science
instrumentalism, 45, 48, 76, 165, 201
International Encyclopedia of Unified Science, xii–xiii, xxiii–xxiv, xxviin.9, 4, 10, 15, 46–47, 53, 56, 125, 135, 160, 163, 177–78, 184, 192n.14, 197–202, 204–11, 212nn.11–12, 212n.14, 213n.14, 219, 229n.2
International Institute for the Unity of Science, xii, 53, 56, 61, 63, 175, 184, 186, 188–89, 190–91n.6, 192nn.17–18, 199–200, 219–20
internationalism, 11, 203, 205, 211, 218–19, 226, 228, 229n.2
Inter-Science Discussion Group, xii, 175, 184–87, 189, 190n.6. *See also* International Institute for the Unity of Science
intersubjectivity, 48, 52, 95, 125
Introduction to Semantics (Carnap), 176, 201, 269, 277n.17
intuition, 237, 239

intuitionism, 240, 246
intuitionistic logic, 237, 239
irrationality, 27
ISDG. *See* Inter-Science Discussion Group
IUS. *See* International Institute for the Unity of Science

James, William, 203, 219
Journal of Philosophy, 124, 221
Journal of Unified Science, 190n.1, 198, 202, 208, 212n.14, 221
Juhos, Belá, 79n.6, 218
justification/discovery, context of. *See* context of discovery/justification

Kant, Immanuel, viii–ix, xxii, 81n.23, 111n.12, 119, 151, 153, 206, 236, 259
Katz, Barry, 223
Kaufmann, Felix, 79n.6, 217–18, 220, 224
Kegley, Charles W., 70–71
Köhler, Wolfgang, 95, 120, 229n.1
Koyré, Alexandre, 82n.40, 137
Kuhn, Thomas, S., vii, xi, xix–xx, xxii, xxiv–xxv, xxviin.9, 1, 20n.13, 81n.24, 82n.39, 98, 105–6, 109, 171, 209–10, 222, 229n.2; model of scientific change of, 209, 229n.2

Langer, Susan, 123, 189n.1
language, private/public, 125
Language II (Carnap), 262
laws: of logic, 14, 201, 275; of nature, 121, 124, 182, 184, 189, 240, 253, 271–72; probabilistic, 105, 121; in sociohistorical sciences, 142
legislative postulation. *See* definition: legislative/discursive
legislative/discursive distinction. *See* definition: legislative/discursive
Lenin, Vladimir, 29, 45, 83n.43
Lewin, Kurt, 95, 189n.1, 229n.1
Lewis, C. I., xiv, 13–14, 17, 19n.2, 55–56, 59, 123, 172, 174
Lewis, David, 248
liberalism, 25–27, 30–31, 34, 44, 49, 52, 56–57, 61, 74, 83n.42, 108, 217
linguistic doctrine of logical truth, the, xiii, 234–37, 239, 241–47, 249, 251, 253, 255, 257

linguistic framework, 9, 17–18, 81n.24, 247
Locke, John, 16
logic: alternative, 236–37, 239, 243–44, 247; elementary, 234–35, 237–47, 249–50, 253; foundations of, 176; inductive, 111n.12, 112n.12, 259; intuitionistic, 237, 239; modal, 108, 182, 237, 241; obviousness of elementary, xiii, 234–35, 237–46, 255. *See also* prelogicality
logic of science. *See Wissenschaftslogik*
Logic of Science Discussion Group. *See* University of Chicago's Logic of Science Discussion Group
Logical Foundations of the Unity of Science (Carnap), 182
L(ogical)-rules, 261–62, 266, 268
Logical Syntax of Language (Carnap), 12, 225–26, 240, 257–61, 263, 266–67, 270, 277n.17
Logische Aufbau der Welt (Carnap), 210
Logische Syntax der Sprache (Carnap). *See Logical Syntax of Language* (Carnap)
L-truth, 268–71
L-validity, 262, 264, 266–67, 269–72, 276n.8, 277nn.9–10, 278n.22, 278n.27

Mach, Ernst, 26–27, 29–30, 44–45, 79n.6, 79n.10, 81n.23, 118, 119, 151–52, 154–56, 219, 226
Marburg School of neo-Kantianism. *See* neo-Kantianism (Marburg)
Margenau, Henry, 66, 189n.1
Marx, Karl, 26, 38, 80n.17, 81n.23, 139
Marxism, 25–27, 29–33, 35–39, 41, 44–45, 48, 70, 74, 80n.17, 83n.42, 107–8, 210, 217
materialism: dialectical, 29; historical, 139; philosophical, xxiii, 30, 38, 45, 81n.23
mathematics: relation to empirical science of, 175
McCarthyism. *See* anticommunism
Mead, George Herbert, 10–11, 46–47, 203
meaning: behavioral criterion for, 126, 241, 243; for Carnap, 7; empirical, 123
mechanism, 45, 81n.23, 154

Menger, Karl, 79n.6, 95, 123, 217–18, 222
Meno, 238
metalanguage, 7, 102, 176, 241, 248, 259, 267–69
metalogic, 15, 97
Mill, John Stuart, 49–50
Miller, Dickinson, 122
mind-body problem, xvi, 116–17, 120, 124–25
Minnesota Center for Philosophy of Science, 191n.6, 220
Minnesota Studies in the Philosophy of Science, 220
Mises, Richard. *See* von Mises, Richard
modal logic. *See* logic: modal
Monist, The (journal), 219
Morgenstern, Oskar, 174, 221
Morris, Charles W., xi–xii, xiv–xv, xvii–xviii, xxviin.6, 2, 5–6, 10–19, 20n.18, 20n.20–21, 21n.24, 46–48, 53, 56, 61–62, 72, 81nn.27–28, 82n.30, 108, 125, 157, 176–78, 189n.1, 192n.14, 192n.16, 197–200, 202–8, 210–11, 211n.1, 212nn.8–14, 213n.14, 219, 223, 226; as advocate of scientific philosophy, 12, 18; analysis of religion of, 204–5; behaviorist turn of, 204, 207; philosophy of, compared to Carnap's, 11–19. *See also* semiotic: Morris's
multiplicity, 41–43, 184

Nagel, Ernest, xiv–xv, xix, 48–49, 65–66, 71, 73–74, 81n.27, 82nn.37–38, 108, 189n.1, 208, 220, 223; on Dewey's views on science and value, 58–61
naturalism, 3–4; Dewey's biological-evolutionary, xi, 49–51, 60–61; Kuhnian historical and sociocultural, 98; methodological, 12–13, 21n.24; Neurath's epistemological, xi, 43, 45, 81n.24, 81n.26, 82n.38, 149, 201, 209; posteriorist, 7; Quine's, 7, 16
Nazism (National Socialism), 34, 80n.12, 130, 152, 171, 185, 217, 225
neobehaviorism, 222
neo-Kantianism (Marburg), ix, 28, 30, 42, 76, 79n.8, 110
neopositivism, 76, 216
neopragmatism, 76, 228
neo-Thomism, 26, 56, 71

Index

Neurath, Otto, xi–xiv, xvii–xviii, xxii, xxiv, xxviin.1, xxviiin.6, 11, 16, 19, 20n.18, 26–48, 53–56, 61–62, 64–66, 74, 77, 78n.1, 79n.3, 79nn.6–7, 80n.12, 80n.14, 80nn.16–17, 80nn.19–20, 81n.21, 81n.23, 81n.26, 82n.29, 82n.31, 83n.46, 95–110, 110n.2, 111n.5, 116, 118, 120, 124–25, 130–32, 145n.1, 146n.23, 149–53, 155–63, 165n.4, 165n.7, 177–78, 180, 183–84, 188–89, 189n.1, 192n.15, 192n.17, 197–209, 211, 211n.1, 212nn.4–5, 212n.7, 212nn.11–13, 213n.14, 217–19, 224–25, 260; on logical empiricism, 4; on semiotic, 206
New Realism, 4
Newtonian mechanics. *See* physics: Newtonian

objectivism, 31, 38, 43–45, 48, 51, 61, 74, 78n.3, 79n.3, 80n.17, 81n.23, 107, 180, 182
obscurantism, 36
observation, 16, 33, 39–41, 45, 54, 62, 81n.29, 101, 104, 106, 162, 238–40, 242–44, 252, 263, 266, 271; predicate, 263–65, 276n.7, 277n.13, 277n.17
observational-theoretical distinction, 209
obviousness of elementary logic. *See* logic: obviousness of elementary
ontology, viii, xxiii, 76, 201, 239, 266, 273, 277n.17
operationism, xx, 14, 48, 122–23, 171, 191–92nn.12–13; in psychology and the social sciences, 171, 192n.13; Stevens's, 177–83
Oppenheim, Paul, 106, 109, 229n.1
ostension, 95

Parsons, Talcott, 172, 186, 189n.1
Peirce, Charles Sanders, viii, 53, 82n.34, 203, 210
Perry, Ralph Barton, 4, 123
personality: Sheldon's theory of, 204–5
phenomenalism, 40–41, 45, 54, 220
phenomenology, xvi, xxviin.3, 27, 220–21
Philosophical Studies (journal), 220
Philosophy of Science (journal), xi, 12, 57, 60, 64, 66, 70, 73, 179, 199, 221, 224
Philosophy of Science Association (PSA), xxviin.1, 63, 66, 72, 83n.42, 200

physicalism, 16, 36, 38, 44–46, 48, 95–97, 99, 159, 180, 209, 225
P(hysical)-rules, 261–62, 266, 277n.10
physics: foundations of, 181; Newtonian, 118–19. *See also* quantum mechanics
Plank, Max, 30, 32, 120
Plato, xxii, 77, 238
platonism, 201, 238–39, 234, 245, 267
pluralism, 163, 229n.2; logico-linguistic, 108
Poincaré, Henri, viii, 40, 42, 116, 119, 151, 153, 156
Popper, Karl, viii–ix, xxviin.10–xxviiin.10, 40–41, 97, 222
postmodernism, 27
pragmatic: meaning of, for Lewis, Dewey, Morris, Carnap, 17; meaning of, for Quine, 17
pragmatism: relations to logical empiricism, 1–6; as scientific philosophy, 2, 5
predicate logic. *See* calculus (formal language): classical first-order predicate
prelogicality, 236, 244
principle: of the conventionality of language forms (Carnap), 247; of tolerance (Carnap), 11, 97, 111n.4, 236, 241, 247, 252, 272, 275
probability, 3, 94, 108, 109, 111n.12, 117, 119–23, 173–74, 176, 252
protocol-sentence debate, xi, 39, 41–42, 54, 95–101, 108, 110n.2, 158, 227
protocol sentences, 36, 39–42, 44, 180, 200
provisoes, 94, 104–6, 110
pseudo-object statements, 265–66, 277n.17
pseudoproblems, 10, 239
psychoanalysis, xvi, 186, 222
psychologism, 101–2, 112n.12, 180–81, 183
psychology: behavioral, 187, 192n.15; ecological, 222; foundations of, 222
Putnam, Hilary, xxviin.2, 20n.12
P-validity, 278n.27

quantum mechanics, 121, 124, 156, 173, 246, 252, 273
Quine, W. V. O., ix–x, xiii, xxiv, 1–3, 5–10, 14–19, 19n.6, 20n.13, 48, 77, 81n.21, 81n.27, 81n.29–82n.29, 108–9,

115–16, 123, 170–74, 176–77, 185, 189–90n.1, 190n.4, 191n.10, 192n.15, 209–10, 223, 226, 234–55, 257–59, 267, 269–71, 273–76, 276nn.1–2, 276nn.6–7, 278n.24, 278nn.30–31. *See also* analytic/synthetic distinction: Carnap and Quine on; a priori: Quine's rejection of; linguistic doctrine of logical truth
Quine on Analyticity (Carnap), 257
Quine on Logical Truth (Carnap), 235, 267

Ramsey, Frank, 219–20
rationalism, xiv, 11, 48, 51, 78n.3, 198
rational reconstruction, 54, 98, 107, 149, 209
Raven, Diederick, xii, 142
realism, xiv, xxiv, 4, 38, 45, 48, 53, 76, 78n.3, 120, 165, 201
reality, 40–41, 52, 99, 119, 164, 184, 201, 239, 265, 273–74
Red Vienna, 41, 217
reductionism, viii, 180–83, 187; archetypal, 181; noneliminativist, 181
reduction sentences, 104, 182
reflexivity, 39
Reichenbach, Hans, viii, xii, xiv, xvii–xviii, xx, 3–5, 16, 19, 31–35, 46, 52–56, 58, 61, 65–66, 73, 78, 79n.6, 79n.9, 79n.11, 80n.12, 80n.19, 82n.34, 82n.41, 83n.42, 94–95, 119, 124, 146n.15, 150, 154, 157, 189n.1, 190n.3, 212n.13, 218, 220, 223, 229n.1, 261
Reisch, George, xii–xiii, xv, xvii, xxi–xxiv, 19, 20n.11, 20n.13, 20n.18, 72, 78, 81n.24, 81nn.27–28, 83n.42, 83n.47, 125, 178, 184, 191n.11, 192n.14, 206–7, 209, 211nn.2–3, 212n.6, 219
relativism, 41–42, 48, 64, 78n.1, 229n.2
relativity, theory of, xix, xxviii.1, xxviii.7, 79n.10, 120, 246; general, 33, 42, 181; special, 118, 121, 154, 181
Rescher, Nicholas, 220, 229n.1
Rey, Abel, 154, 156
Richards, Ivor A., 173, 176, 186
Richardson, Alan, ix–xi, xiv–xv, xxi, xxviiin.11, 2, 20n.13, 31, 78, 79n.6, 81n.25, 182, 191n.8, 212n.4, 217, 225, 247, 278n.28
Richter, Rudolf. *See* Zilsel, Edgar

Rickert, Heinrich, 135, 159
Ricketts, Thomas, xiii, 19, 248, 276n.2, 277n.8, 277n.11, 277n.17, 278n.21, 278n.24
Rudner, Richard, 64, 66, 70–72
rules: correspondence, 54; formation and transformation, 103, 260, 265–67, 269, 278n.27; L(ogical)-rules, 261–62, 266, 268; P(hysical)-rules, 261–62, 266, 277n.10
Russell, Bertrand, 37, 56, 120, 123, 154, 171–72, 174–76, 219

Santillana, Giorgio. *See* de Santillana, Giorgio
Sarton, George, xix, 132, 136, 145n.5, 171–72, 174, 190n.1, 190n.5
Sartre, Jean-Paul, 27
Schlick, Moritz, xi, xiv, xxviin.1, 31–34, 39–42, 47, 52, 54, 72, 79n.6, 80nn.18–19, 95, 98–102, 108, 110nn.2–3, 111n.6, 111n.8, 116–20, 122–24, 126nn.2–6, 8, 139–40, 149–50, 153–56, 159–60, 165n.3, 165n.7, 180, 190n.6, 217–19, 240
Schuetz, Alfred, 220
Schumpeter, J. A., 171–73, 186, 191n.10
science of science, 177, 180, 192nn.14–15, 207, 219, 228
Science of Science Discussion Group, Harvard, xvii, 170, 172–78, 180, 182, 185–89, 190n.1, 190nn.4–5, 191n.7, 191n.10
scientific philosophy: Carnap's philosophy as, 9, 21n.24
scientific revolution, 137, 179
Sciven, Michael, 220
Sellars, Roy Wood, 48–49
Sellars, Wilfrid, 220
semantic: concepts, according to Carnap, 7–8, 182; foundations of, 176; language, 268–71; rules, 257–58
semiotic: Morris's, 10–11, 47, 192n.14, 205–7; Neurath's critique of, 206
set theory, 120, 234–35, 237–40, 246, 249–51, 253–54, 269, 276n.4; paradoxes of, 237–38, 255, 275
Shapley, Harlow, 171, 186, 190n.2
Sheffer, Henry, 123, 170

Sheldon, William, 204–5
simplicity, 42, 61–62, 119, 121, 162, 246, 273–75; criterion of, 120
skepticism, 239
Smith, Cyril, 137
socialism, 25–26, 29–35, 37, 39, 44, 52, 61, 64, 79n.8, 83n.42, 107, 144, 203, 211
sociology of science, viii, xx, 26, 62, 81n.25, 161–62, 192n.14, 221–22, 228
Socrates, 238
solipsism, 52, 95, 179
special theory of relativity, 118, 121, 154, 181
SSDG. *See* Science of Science Discussion Group, Harvard
Stadler, Friedrich, xiii, 30–33, 38, 79nn.6–7, 80n.16, 82n.35, 129, 139, 143, 184, 216–18, 221–22, 224, 228, 229n.2; three-staged scheme of logical empiricism of, 129–30
Stein, Howard, 242
Stevens, S. S., 171–85, 187–89, 190n.1, 190n.5, 191n.7, 191nn.9–10, 191n.13, 192nn.13–15
stipulation. *See* definition: stipulative
Strauss, Herbert, 130, 145n.2
Strong, C. A., 122, 126n.7
Structure of Scientific Revolutions (Kuhn), vii, xxiv, xxviin.9, 81n.24, 109–10, 209
Suppe, Frederick, 5, 20n.12, 209, 217
Suppes, Patrick, 67, 73
syntax: Carnap on the relationship of descriptive and pure, 261; descriptive/pure, 102, 260–61, 264, 277nn.16–17
syntax language, 267–68, 276n.8
Synthese (journal), 199, 205, 221
synthetic a priori, 12, 154, 236, 259

Tarski, Alfred, 3, 100–101, 103–5; 111n.7, 170–71, 173, 176, 190n.1, 226, 238, 268, 275. *See also* truth: Tarski's semantic account of truth
testability, 102–3, 105, 265, 271, 274
theory choice, 39–41, 43–44, 47–48, 50, 55, 61–63, 78n.1, 79n.3, 83n.47, 106, 120, 161–63
theory-ladenness, 45, 162
theory testing, 17, 116, 257
thermodynamics, 172–74

thing-language (Carnap), 182–83
Tolman, E. C., 223
Tractatus Logico-Philosophicus (Wittgenstein), 96, 99, 260, 265
translatability, 209, 236, 243–44, 255
truth: coherence theory of, 98–100, 246; by convention, 234–35, 246–47, 250–51; correspondence theory of, 99; detached, 50; direct realist conception of, 125; empirical requirement for, 119; Tarski's semantic account of, 3, 5, 100–101, 104, 111n.7, 268. *See also* linguistic doctrine of logical truth, the
Truth by Convention (Quine), 234, 240, 248–49, 278n.24
T-sentences, 238
Two Dogmas of Empiricism (Quine), xiii, xxiv, 2, 8, 17, 235, 251, 257–58, 269

Uebel, Thomas, ix, xii, xvii, 19, 78, 79nn.6–7, 81n.25, 110, 110n.1, 165n.1, 165n.4, 165n.7, 166n.12, 184, 200, 209, 212n.4, 217, 227
underdetermination, xx, 39, 41–45, 47–48, 54–55, 61–62, 78n.1, 79n.3, 80nn.20–21, 81n.21, 162–63, 264
unified science, xii, xvii, xviii, 4, 10–11, 18–19, 33, 36, 96–97, 99, 131, 135, 153, 157, 159–63, 170, 175, 177–85, 187–89, 191n.7, 198–200, 202–7, 209–11, 219–20, 225–26, 228, 259, 272
uniqueness of elementary logic, 78n.1, 100, 163, 236, 239
unity of science. *See* unified science
University of Chicago's Logic of Science Discussion Group, 177–78, 192n.16
utopianism, 29, 210

value theory, 205–7
verification, 8–9, 17, 25, 40–42, 46–47, 51, 54, 103, 119–20, 180
Vienna Circle, x–xiv, xvii, xix, xxviin.1, 21n.27, 31, 33–35, 41–42, 45–47, 53–54, 74, 78nn.2–3, 79n.6, 81n.27, 82n.33, 83n.42, 83n.47, 94–97, 99, 102, 106–8, 110n.3, 111n.11, 115–17, 123–24, 126, 129, 135, 139, 142, 149–60, 163–65, 165nn.3–4, 165n.6, 165n.10, 166n.12, 175–77, 183, 190n.1, 190n.6, 199, 203,

209, 216–22, 224–28, 229nn.1–2; manifesto of, 37–38, 155, 225
vitalism: Hans Driesch's, 152
von Mises, Richard, xx, 44, 123, 171–72, 174–76, 186, 189n.1, 190n.3, 218, 222, 226
von Neumann, John, 221

Waismann, Friedrich, 47, 79n.6, 95, 116, 120, 123–24, 149, 165n.3, 190n.1, 217–18
Weber, Max, 135
Weibel, Peter, 217–18, 222

Whitehead, Alfred North, 122–23, 177
Windelband, Wilhelm, 135
Wissenschaftslogik, 7, 12, 97–103, 111n.4, 216, 218, 221, 225–26, 228–29, 259–67, 270–72, 274–75, 277n.17
Wissenschaftstheorie, 216, 225, 229
Wittgenstein, Ludwig, 20n.13, 96, 99, 103, 123, 158, 219, 240, 260, 265

Zilsel, Edgar, xii, xvii–xviii, xxviin.6, 129–47; on Marxism, 139–40, 142–44; on the social origins of science, 135, 137, 142, 152, 217–18, 222